D0952413

What a reviewer says about
AMATEUR ASTRONOMER'S HANDBOOK
4th edition

Ever since this book was first published in 1955,
it has been the amateur's reference classic on all
matters relating to astronomical instrumentation
and practice, apart from actual observing
techniques. Optically, the book covers practically everything the serious amateur is likely to
need from image characteristics through design
of systems and their aberrations to the latest
catadioptrics. Mechanically, anything relating to
telescopes, their mountings, drives, and all kinds
of associated equipment, are delt with—along
with discussions on photography, vision, and
time reckoning. With such a comprehensive
coverage, extensive revision was due after a
quarter of a century. James Muirden, well-
known writer in astronomy, has very commend-
ably accomplished this mammoth task and has
brought every aspect into line with current
practice. There are few errors, and though the
price reflects its value to the practicing amateur,
I am not just recommending this book—it's an
indispensable reference which completely
outdates all earlier editions.

New Scientist
July 3, 1980

amateur astronomer's handbook

by the same author

★

OBSERVATIONAL ASTRONOMY FOR AMATEURS

INTRODUCING ASTRONOMY

amateur astronomer's handbook

Fourth Edition

by
J. B. Sidgwick

Completely revised
by

James Muirden

Enslow Publishers
Hillside, New Jersey 07205

American edition published by Enslow Publishers, 1980

Revisions for fourth edition © copyright Pelham Books Ltd, 1979

Copyright © J.B. Sidgwick, 1955

Library of Congress Cataloging in Publication Data;

Bibliography: p.
Includes index.
1. Astronomy. I. Muirden, James. II. Title.

QB44.S558 1980   522   80-20596

ISBN 0-89490-049-8

Printed in the United States of America

10 9 8 7 6 5 4 3 2 1

CONTENTS

CONTENTS

TABLE OF ABBREVIATIONS

The following commonly occurring abbreviations and symbols are used in the sense quoted unless the contrary is specified or is obvious from the context.

Å   Ångström unit $= 10^{-8}$ cm

A.E.   *Astronomical* (or *American*) *Ephemeris*

AT   Apparent Time

A.U.   Astronomical Unit

α Right Ascension

B   Latitude

B.   (followed by number): reference to Bibliography, section 41
Bibliographical abbreviations: see section 41

B.A.A.   British Astronomical Association

c/i   colour index

CM   central meridian

D   aperture of objective in cm

Dec   Declination

d/v   direct vision

δ Declination, or diameter of the pupillary aperture of the eye

ET   Ephemeris Time

E/T   Equation of Time

F   focal length of objective

f   focal length of ocular, or 'following'

f/   focal ratio (F/D). $f/10$ is said to be a larger focal ratio and a smaller relative aperture (D/F) than, e.g. $f/5$

𝔣   equivalent focal length

GMAT   Greenwich Mean Astronomical Time

H   hour angle

HP   high power (of oculars)

HR   hourly rate

I   intensity (of radiation)

TABLE OF ABBREVIATIONS

I.A.U.	International Astronomical Union
JD	Julian Date
L	longitude
LP	low power (of oculars)
LPV	long-period variable
λ	wavelength
M	magnification, or integrated magnitude, or absolute magnitude
M'	minimum useful magnification
M''	maximum useful magnification
M_r	minimum magnification for full resolution
m	metre
m	apparent stellar magnitude
m_b	bolometric magnitude
m_p	photographic magnitude
m_v	visual or photovisual magnitude
MSL	mean sea level
MT	Mean Time (prefixed by L=Local, or G=Greenwich)
μ	micron$=10^{-4}$ cm
N	refractive index
NCP	North Celestial Pole
nm	nanometre$=10^{-7}$ cm
NPS	North Polar Sequence
ν	constringence
O.A.A.	*Observational Astronomy for Amateurs*
OG	object glass
ω	deviation
p	preceding
p.a.	position angle
π	stellar parallax, or transmission factor of an optical train
ϕ	latitude
R	theoretical resolution threshold
R'	empirical resolution threshold
r	angular radius of a given diffraction ring or interspace, or angular separation of components of a double star
r'	linear radius of a given diffraction ring or interspace
RA	Right Ascension
ST	Sidereal Time (prefixed by L=Local, or G=Greenwich)
t	time, or turbulence factor
UT	Universal Time
ZD	Zenith Distance
ZHR	zenithal hourly rate

FOREWORD

In 1950 the membership of the British Astronomical Association (the only national association of amateur astronomers in the country) passed the 2000 mark, and at the time of writing it is still growing. A large proportion of these amateurs are active, in the sense that they possess telescopes of some sort and engage in more than desultory observation; of these, a hard core of 'specialists' take a regular part in the organised programmes of the various Observing Sections.

Why the Association's membership, which maintained a fairly constant level at about the 1000 mark during the first forty years of this century, should suddenly have doubled itself within the last ten years I do not know. Presumably the war had something to do with it. But considering the obvious disadvantages of astronomy as a hobby—chief of which are the expense and size of its equipment, and the discomforts and inconveniences of its pursuit—it is a remarkable fact that over 2000 men and women are sufficiently engrossed by it to take up membership of an association of their fellows.

In so far as there is such a thing as a typical amateur astronomer, his evolution might run somewhat as follows. The initial spark of interest may be struck in one of a thousand ways—a book or a magazine article, a radio talk, a snippet of news from one of the great observatories, a personal contact, an obvious astronomical event such as eclipse or a naked-eye comet. Interest once aroused, it will be both fed and increased by a wide range of reading. The first milestone is reached when the amateur realises that the armchair is no place to study astronomy, and he determines to buy himself some sort of telescope, no matter how cheap and unprofessional, in order to see things for himself. From this moment he will be lost to the domestic comforts of the fireside, and is likely to become an increasing trial to his wife. Nine times out of ten his first instrument will be a small refractor, supplied with perhaps three or four oculars and mounted on a tripod altazimuth. Then will follow a period—which can afford to be at least as long as a year, and will probably be longer—during which he 'sight-sees' over the expanse of the night sky. This may at the time appear to him to be primarily a pleasurable and exciting exploration, more or less without direction, but he is in fact laying the foundation of that first-hand familiarity with the night sky which is one of the characteristics of the amateur astronomer, and also one of the most valuable components of his mental equipment. Sooner or later two realisations will be borne in upon him: that he has exhausted the potentialities of his small telescope, and that his now fairly wide though shallow experience of observing the Sun, Moon, planets, double stars and variables, comets, meteors, nebulae, and the rest, is crystallising into an interest directed upon one specific branch of the subject.

This marks the second milestone in the career of this supposedly typical amateur astronomer. In a field as vast as that opened up by the telescope, specialisation is unavoidable if the observer's energies are not to be hopelessly and fruitlessly dissipated. Not only is the amount of time that anyone can devote to telescopic work necessarily limited, but the further pursuit of each branch of observation beyond the 'sight-seeing' level requires specialised techniques, the training of the eye to excel in some particular function, and to a large extent specialised instruments. Great care needs to be devoted to the choice of the best equipment—for, financially, it often means putting all one's eggs in one basket—as well as of the most suitable field of specialisation: one who lives in a great city, for example, would be unwise to devote his time to comet-seeking or the study of the Zodiacal Light, or one who for reasons of finance or available space cannot install a telescope of more than 15-cm aperture, to specialise in planetary observation.

And so the amateur, now in possession of a thorough working knowledge of the night sky and of his instrument, settles down to the regular and intensive observation of a single object or class of related objects. Already he is on the road to becoming an expert in his own chosen field. The leading amateur astronomers are, indeed, men of international repute: one has only to think of such names as Denning, Ellison, Goodacre, Hargreaves, Maunder, Phillips, Steavenson, Stanley Williams—so many might be mentioned that any selection, however random, must be invidious.

Some examples of amateur achievement may not be out of place here: they will put flesh on the bones of the biographical skeleton described above, and at the same time provide both a spur and an encouragement. W. F. Denning, an accountant by profession, became interested in astronomy at the age of twenty-one, when he happened by chance to witness a spectacular Taurid fireball. During the subsequent sixty-two years till his death in 1931 he built for himself a reputation as one of the world's leading authorities on meteor radiants: he was also one of the foremost planetary observers of his time, his work on the rotations of the surface features of Jupiter being of fundamental importance. He discovered five comets and two novae. His classic work on radiants culminated in his *General Catalogue of the Radiant Points of Meteoric Showers*, published in the *Monthly Notices* of the Royal Astronomical Society in 1899, which is still a standard reference work on radiants. He was the first observer of a radiant that moved from night to night, in accordance with theory; and he was the first worker to publish data relating to stationary radiants in a form that permitted profitable discussion of these strange contradictions between theory and observation. The accuracy and reliability of his observations were outstanding—yet he never used more elaborate equipment than a 12·6-in altazimuth reflector.

Burnham, one of the greatest observers of double stars, was an official reporter in the U.S. District Court, Chicago, until eleven years before his death in 1921. His first catalogue—of 81 doubles discovered by himself with a 6-in refractor during the previous three years—was published in 1870, only nine years after he had taken up observing (with a modest 3-in). Two supplementary lists rapidly raised the number of his published discoveries to 182, and his great *General Catalogue of Double Stars* appeared thirty-five years later, in 1906. In all, he discovered 1340 new binaries, besides making thousands of individual measures.

J. Franklin-Adams (1843-1912) became so engrossed in his astronomical hobby that at the age of forty-seven he forsook his business career —he was an insurance broker at Lloyd's—in order to have more time to devote to it. His name, associated with his photographic chart of the whole sky down to magnitude 14·5, is assured of immortality.

Going back somewhat further—to the early years of last century— Pons and Schröter demand mention. Schröter was by profession a lawyer, but his twenty-eight years' study of the lunar surface earned him the title of Father of Selenography. Pons, who at one time was concierge at the Marseilles Observatory, independently discovered no fewer than 37 comets during the twenty-six years from 1801 to 1827—a record of achievement that has never been broken.

Today the greater part of the observational data on which rests our knowledge of, among others, long-period variables, meteors, and the motions of Jupiter's surface markings is provided by amateur observers. The reverse side of the picture is, of course, that certain branches of work are peculiarly unsuited for amateurs, being the natural stamping ground of the professional: generally speaking, they are those that require large apertures, long focal lengths, great precision, or all three, and as examples may be cited the measurement of radial velocities, the photometric study of the extragalactic nebulae, and fundamental positional work.

At some stage during his progress from armchair enthusiast to observational specialist, our typical amateur may be presumed to have joined the B.A.A. and, after his interest has focused on a particular branch of work, to have joined the appropriate Observing Section of the Association. Here he will be brought into contact with other observers in the same field, and a channel will be provided for the publication of his observations. Observations whose records are kept locked away in a drawer might just as well never have been made, so far as their value to science is concerned. They must either be published or periodically sent into the Director of the Section for correlation and combination with the observations of other members and eventual inclusion in the interim *Reports* and *Memoirs* which are published by each of the Sections from time to time. The B.A.A. not only facilitates contact with other observers engaged on the same work, but by means of its geographical list allows any observer to get in touch with other members in his own locality, irrespective of their work and specialised interest. Furthermore it provides an organisation for arranging cooperative observation. This is not only the *sine qua non* of such work as the determination of the real paths of meteors—where parallactic observations from separated stations are required—but in general provides a safeguard against the needless duplication of observations and against gaps in the observational record due to bad weather or bad seeing, while to some extent smoothing out the accidental errors which are ineradicable from the observations of a single observer. The value of cooperation in the pooling of observations has perhaps never been more clearly and convincingly demonstrated than in the work of the Variable Star Section of the B.A.A.

There is, however, more to astronomical research than the mere piling up of observational data, however essential this basic supply of material for discussion may be. An accumulation of facts, as Poincaré remarked, is no more a science than a pile of bricks is a house. It is possible that some observers in this country have got into something of a groove, grinding on and on with the production of observational material without bothering too much about its integration or significance. There is always room for individual purposive research designed to reach a definite objective and solve, or at least throw some light on, a specific problem, no matter how limited it may be. Even a casual flip through such periodicals as the *Journal* of the B.A.A. or *Popular Astronomy* for the past twenty years will bring to light examples of the phenomenal success and of the importance of the results that can be achieved when amateurs do break away from established routine.

The willingness to develop new techniques and to explore and exploit new instrumental developments is perhaps more characteristic of the Americans than of ourselves. It would be interesting, for example, to know how many Schmidt cameras are in amateur hands on the two sides of the Atlantic. The 4-in refractor and the 12-in Newtonian by no means exhaust the instrumental possibilities for the amateur, and a greater degree of daring in the use of available instrumental resources would certainly be an invigorating influence in amateur astronomy. There is no need to go to the other extreme of instrument-fixation exemplified by so many American telescope makers, who become hypnotised by their products as ends in themselves, and while turning out first-class equipment cannot find time away from their laps and their lathes to direct their own telescopes towards the sky. There have been plenty of instrumental developments during the last twenty years which could be successfully exploited by amateurs, but which have, in fact, been to a surprising extent neglected by them. Examples spring to mind: the greater use of photoelectric methods in photometry, other types of reflector than the usual Newtonian, the new and varied offerings of electronics—automatic guiding, the automatic registration of occultations or transits by photocell and chronograph, etc—long-wave solar research (in which amateurs are cooperating usefully in Australia), new optical systems such as the Schmidt and Maksutov, the monochromator, or comparatively rare instruments of enormous scope such as the spectrohelioscope. The astronomical sightseer turning specialist would be very well advised to consider the adventurous outlay of his capital rather than the mere duplication of instruments already in prolific existence.

It is many years since the last edition of Webb's *Celestial Objects for Common Telescopes* appeared, and it is now badly out of date, though its list of objects is still useful to the amateur who is exploring the sky with his first telescope. The inter-war period saw the publication of Hutchinson's *Splendour of the Heavens*, edited by T. E. R. Phillips and W. H. Steavenson, which contained a great deal of information valuable to the amateur astronomer, though it was primarily a descriptive work. The most useful companion to practical observation that has appeared in fairly recent years is undoubtedly the *Astronomisches Handbuch*, edited by Henseling (1921); this, however, has never been translated, and in any case is nowadays virtually unobtainable. There is, however, a gap on the astronomical shelf at just that point where the amateur looks for information and guidance, and it is hoped that the present *Handbook* will go at least some of the way towards filling it.

It may be useful to emphasise that this is neither a formal textbook designed for consecutive reading, nor, on the other hand, a narrowly practical manual of instructions. It is a reference handbook, and a number of its topics constitute the background to telescopic observation rather than the immediate concern of the observer while actually at work: they provide the 'why' behind the 'how'. It is not difficult, I think, to justify the rather detailed treatment given, for example, to telescopic function or the aberrations. The man who understands the operation of his instrument or machine—whether it be a telescope or a motor-car—will use it more efficiently, because more intelligently, than the man who trusts to rule-of-thumb methods. The solution of many of the purely practical problems which confront the amateur observer leads direct to a more thoroughgoing account of the optics and mechanics of his instrument than can be found in most of the books ostensibly written with him especially in mind. That an astronomer is an amateur by no means implies that he wishes, or needs, only to read lists of severely practical instructions on the use of his instrument, unrelated to the so-called 'theoretical' background from which they are derived.

It will be found, therefore, that the various topics figuring in the Table of Contents are treated with different degrees of thoroughness according to their theoretical importance to the telescopist, their immediate value to the practising observer, their intrinsic complexity, and whether or not they are dealt with elsewhere in readily accessible books or journals. Thus no attempt has been made to present more than an introductory bird's-eye view of lens-and-mirror systems, supplemented by a fairly extensive bibliography; the relation between atmospheric condition and telescopic 'seeing', on the other hand, is discussed in more detail, with a correspondingly restricted bibliography. The aim, generally speaking, has been to take up the treatment of a subject at a more elementary level than that of the average 'general reader' book. The Bibliography forms an integral part of the whole, being complementary and supplementary to the text. The extensiveness of the Bibliography may derive some justification from Joad's definition of the educated man as one whose head is not crammed solid with a mass of isolated facts, but who knows exactly where to put his hand on any particular piece of information that he may require.

Inevitably, many topics have been skimpily treated, and many more have been omitted altogether. No mention will be found of telescope making, for example: an adequate treatment of this technical subject would have swollen an already over-long book to prohibitive proportions and, furthermore, a number of excellent books dealing with the subject are easily obtainable; references to these will be found in the Bibliography.

The theodolite and sextant, again, are not described, since their uses are primarily non-astronomical and they are in any case of comparatively little interest to the amateur. The meridian transit telescope is excluded, because it is an instrument that few amateurs possess—or need to possess, since its functions can be performed in other ways. Some instruments which, like the coronagraph, are of little practical significance to the amateur, but are of outstanding theoretical interest, are glanced at briefly in passing.

A device, gadget, instrument, or accessory is often described as being 'easily homemade'. The amateur who can handle ordinary carpenter's tools and is generally dexterous has a very definite advantage over one who is unpractical and kack-handed. There is nevertheless a great deal in the way of ordinary maintenance, modifications, and the construction of accessories of various kinds that anyone can do with simple tools and a little ingenuity and determination. If, of course, the amateur has access to a well-equipped machine-shop, or even to a small lathe, he is to be envied. Some quite astounding examples of ingenuity and improvisation in the building of, for instance, equatorial mountings from such materials as old car parts or water-pipe fittings are to be found described in the pages of the amateur journals.

Amateur Astronomer's Handbook deals solely with the theoretical and instrumental background to observation, together with various matters with practical implications. The actual techniques employed in the observation of the various types of celestial objects are described in *Observational Astronomy for Amateurs*, which may be regarded as a companion volume to the present work. The two books deal with complementary aspects of the amateur astronomer's work, while being individually self-sufficient and mutually independent; cross references between the two volumes* have therefore been reduced to a minimum.

It would be pleasant to think that amateurs of all classes will find something of interest or value in this *Handbook*, but the reader for whom it was specifically written is that amateur astronomer who has been described as approaching his second milestone—the observer who, while gaining practical experience with a small instrument, is beginning to consider in what branch of work he will ultimately specialise.

By way of envoi I cannot do better than quote the words with which a professional astronomer (R. O. Redman), writing in 1945, concluded an article on the post-war prospects of astronomy: 'To attempt at the present juncture a discussion of post-war professional astronomy is to

---

Observational Astronomy for Amateurs is hereafter referred to as *O.A.A.*

risk a charge of futility; to try to deal with post-war amateur astronomy would probably ensure not only the accusation but also the verdict of guilty. It is extraordinarily difficult to say anything useful on the subject at present, and hence a far from unimportant branch of astronomy has been omitted in this article. The enthusiasm and ability of amateurs, to which we have owed so much in the past, may yet prove the principal means of salvation for astronomy in a difficult future.'

It is with pleasure and gratitude that I acknowledge the assistance I have received from various sources: Mr R. H. Tucker, of the Abinger Magnetic Station, for information regarding the Greenwich Time Service; Mr M. J. Moroney for invaluable suggestions regarding section 29; Dr R. L. Waterfield for allowing me to examine Will Hay's plate-measuring machine; the staff of the Royal Astronomical Society's Library. And I must express my particular gratitude to Dr W. H. Steavenson, who undertook to read the complete MS, and to whom I am indebted for innumerable suggestions, corrections, and criticisms. It is hardly necessary to add that the book's errors and shortcomings are my own responsibility.

J. B. SIDGWICK
London,
June 1954.

EDITOR'S PREFACE TO THE SECOND EDITION

Only a few changes have been made, in this edition, and these have been necessary mainly because of the introduction of Ephemeris Time in 1960.

G. FIELDER, *Editor*

EDITOR'S PREFACE TO THE THIRD EDITION

This book is basically not a book about astronomy, but a compendium of things an astronomer needs to know. As such, its contents are, for the most part, essentially physical and mathematical in nature, and their truth is sufficiently fundamental that they will not change with time. In consequence, the changes in this third edition are confined to the chapter on photography, which, dependent as it is upon the properties of the available materials, which are themselves subject to continuous evolution and improvement, will remain subject to review for many years to come.

I am indebted to Mr R. S. Scagell for much valuable advice and assistance.

R. C. GAMBLE, *Editor*

EDITOR'S PREFACE TO THE FOURTH EDITION

The quarter-century which has elapsed since the first edition of this work has rendered extensive revision necessary. However, the overall scheme has not been altered, since I have aimed at preserving the original flavour of a book that has become a classic in its field.

I wish to thank the following, who have advised me over the revision of various topics, or helped in other ways: S. J. Anderson; R. W. Argyle, H. E. Dall; D. G. Hinds; Dr J. L. Houghton; E. J. Hysom; G. J. Kirby; J. H. Mathers; Dr. R. Scaddan; R. S. Scagell; R. H. Taber (Ilford Ltd.); and N. Walker.

JAMES MUIRDEN, *Editor*

TELESCOPIC FUNCTION: LIGHT GRASP

1.1 Light grasp and image brightness

Light grasp is a term, often used rather vaguely, that indicates the quantity of light from a given source that is concentrated by an objective into the image at its primary focus. It is a function of the objective's aperture, D, and is independent of focal length, magnification, or any other factor. More precisely it may be defined as the relative intensity of the light from a point source passed to the retina by the telescope's optical train and by the naked eye. Provided that the diameter of the eye's pupillary aperture, δ, is not smaller than the exit pupil of the telescope, then the instrument's light grasp is equal to

$$\left(\frac{D}{\delta}\right)^2 . \pi$$

where π is the transmission factor of the system. A fair value, in the case of an astronomical telescope, is 60 to 65 per cent, the remaining 35 to 40 per cent of the incident light being lost to the final image through absorption in lenses, reflections at lens surfaces, and imperfect reflection at mirror surfaces (section 6.8).

A convenient criterion of the light grasp of an instrument is the magnitude of the faintest star that it will show to averted vision. Most textbooks contain a table of the minimum magnitudes revealed by various apertures. There is a variety of reasons why such tables tend to be misleading, among them:

(a) The observation, or theoretical assumption, which forms the basis of the extrapolation may be inaccurate or inapplicable to observing conditions at another time.

(b) Atmospheric factors, such as turbulence (which discriminates in favour of small instruments, with, in extreme cases, such anomalous results as the clear visibility of a star with half the aperture of a telescope in which it is practically invisible),* and thin mist or high cloud, which may affect the instrument's performance while still being too diffuse to be detected with the naked eye.

(c) Most objectives, and probably all observers, depart in some degree from the 'normal'.

(d) The magnification employed will (for a reason discussed below, which does not invalidate what was said in the first paragraph) affect the visibility of the faintest stars.

(e) The use of direct or of averted vision.

(f) The use of a reflector or a refractor, the former being inferior, aperture for aperture, even though the silver or aluminium film is in good condition with smaller apertures.

(g) The presence of another and brighter object in the field. (It is assumed that such precautions as the thorough dark adaptation of the eye and the choice of a moonless night are taken.)

Under the most favourable conditions the normal human eye is just capable of glimpsing stars of mag $6 \cdot 5$,[†] whence it can be deduced that the magnitude of the faintest star visible with a telescope of aperture D is[‡]

$$m = 6 \cdot 5 - 5 \log \delta + 5 \log D$$

(If a star in the telescopic field has an apparent brightness other than that of bare visibility, its real magnitude can be derived from this expression by substituting its equivalent naked-eye magnitude—i.e. that of a star which, observed with the naked eye, has the same apparent brightness— for the first term on the right-hand side.)

Now substituting the accepted value of $\delta = 7 \cdot 5$ mm, we have that

$$m = 7 \cdot 1 + 5 \log D$$

which agrees closely with the result arrived at by assuming that a 25-mm objective will show stars to the 9th magnitude. In the following Table (whose data are represented graphically in Figure 1), column (a) tabulates the magnitude thresholds derived from this latter assumption.

*The 2·5-m Mt Wilson instrument, with a theoretical threshold of about mag 18·8, loses nearly 3 mags under bad seeing conditions.

†The value that may be expected with a normal eye whose pupillary aperture is expanded to 7·5 mm. Artificial stars have been seen in the laboratory, against a completely dark background, at mag 8·5. The 2-mag difference is accounted for by the reduced contrast between the natural star and its background of the night sky.

‡If mag 6·5 is taken as the naked-eye threshold, then $2 \cdot 512^{6 \cdot 5} = k.\delta^2$, whence $\log k = 2 \cdot 6 - 2 \log \delta$. But the telescopic brightness of a point source $= k.D^2$. The expression for m is derived by combining these last two expressions.

| D (cm) | Minimum magnitude (uncorrected) | | Refractor | | Reflector Min mag | Pleiades (various) |
| | | | Absorption | Min mag | | |
	(a)	(b)		(c)	(d)	(e)
5·0	10·5	11·5	—	11·5	11·3	11·2
7·5	11·4	12·4	—	12·4	12·2	12·1
10·0	12·0	13·0	−0·1	12·9	12·8	12·7
12·5	12·5	13·5	−0·1	13·4	13·3	13·2
15·0	12·9	13·9	−0·1	13·8	13·7	13·6
17·5	13·2	14·2	−0·1	14·1	14·0	13·9
20·0	13·5	14·5	−0·1	14·4	14·3	14·2
22·5	13·8	14·8	−0·1	14·7	14·6	14·5
25·0	14·0	15·0	−0·2	14·8	14·8	14·7
30·0	14·4	15·4	−0·2	15·2	15·2	15·1
40·0	15·0	16·0	−0·3	15·7	15·8	15·7
50·0	15·5	16·5	−0·4	16·1	16·3	16·2

However, this curve does not agree well with the results of actual observation, yielding results which are consistently too bright by at least 1 mag compared with the vision of normal, if not acute, observers. In fact, a 25-mm OG has a threshold nearer to 10·0 than 9·0, giving the series of magnitude limits tabulated in column (b).

Column (c) is an attempt at predicting the visual limiting magnitude for uncoated refractors, allowing for light wastage as the aperture increases. This is due less to light-loss by absorption (which, from section 6.3, should amount to less than 0·1 mag even in a 50-cm OG, if made of modern material) than to the leakage of light into the secondary spectrum where it contributes nothing to the image—where, in fact, it does the reverse by reducing contrast. Dennis Taylor (B. 3.22, 23) and Linfoot have shown that such leakage may, in some large refractors, amount to about 50 per cent of the light leaving the objective. Column (d) tabulates the figures in (b) reduced by 0·2 mag to allow for normal light losses in a two-mirror reflecting telescope, while column (e) gives the limiting magnitudes obtained from the analysis of observations of faint stars in the Pleiades, below. (See also sections 6, 23.)

Objects most suitable for testing the light grasp of a telescope are the comparison stars of variables, for which charts exist and whose magnitudes have been established with accuracy.

1.2 Effect of visual acuity and magnification

Such tables of magnitude limits may have little relevance when applied to particular individuals, however well they represent the average result from a large number of observers. Examples of exceptionally faint telescopic

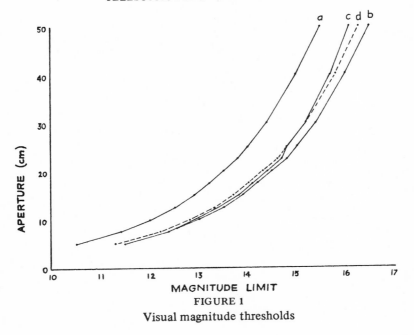

FIGURE 1

Visual magnitude thresholds

limiting magnitudes include: Nova Cyg (1876) at minimum, when of mag 14·6-14·9, with a 15-cm refractor (Steavenson); mag 14·3 with a 12·7-cm refractor, and 14·4 with a 15·2-cm Maksutov-Cassegrain (Nightingale); the outer satellites of Uranus (mags 14·0 and 14·2) with a 10·9-cm refractor (Ward). Ward's observations, if accepted, represent a freak class of vision; the other records, together with those of Peltier, accord approximately with a formula of $m = 8·7 + 5 \log D$.

The whole subject of limiting magnitude is beset by the presence of numerous variable factors; apart from the obvious influence of sky conditions, condition of telescope, visual acuity, magnitude scale used, etc., account must be taken of whether fleeting glimpses or continuous visibility are to be considered as 'detection', while the configuration of other stars in the field can greatly assist the location of one particular test object, while handicapping the search for another. The variable-star observer anxious to make a magnitude estimate, rather than merely to confirm the presence or absence of the variable, will require definite visibility to allow a comparison to be made, and his practical threshold may be a magnitude or so higher than the 'glimpse' limit. For a discussion on the subject of limiting magnitudes with different instruments, see B. 3.27

4

B. 3.18 discusses a test carried out in 1973 on the visibility of faint stars in the Pleiades. The limiting magnitude obtained from the results of a number of observers, using both reflectors and refractors of between 75 mm and 25 cm in aperture, was best represented by the formula

$$m = 7\cdot7 + 5 \log D,$$

but with a standard deviation of 0·4 mag. Taking this formula as representing the performance of typical observers under normal conditions, the following Table (column (*e*)) would seem to suggest average practical acuity with the instruments likely to be found in amateur hands:

D (cm)	5·0	7·5	10·0	12·5	15·0	17·5	20·0	22·5	25·0
m	11·2	12·1	12·7	13·2	13·6	13·9	14·2	14·5	14·7

D (cm)	30·0	40·0	50·0
m	15·1	15·7	16·2

An analysis of limiting magnitudes achieved by six experienced variable-star observers gave an average value of $m = 7\cdot9 + 5 \log D$, but with wide variation (B. 3.4).

I. S. Bowen's formula, $m = X + 2\cdot5 \log D + 2\cdot5 \log M$, is an attempt at correlating the limiting magnitude with the magnification used, so as to bring into consideration the important sky-darkening effect of more powerful oculars. A suitable value of X must be chosen by experience; with $X = 5\cdot5$, for example, we have:

D (cm)	Magnification					
	×20	×50	×100	×200	×300	×400
10	11·3	12·2	13·0	13·8	—	—
15	11·7	12·7	13·4	14·2	14·6	—
20	—	13·0	13·8	14·5	14·9	15·3
25	—	13·2	14·0	14·7	15·2	15·5
30	—	13·4	14·2	14·9	15·4	15·7

as representing the different magnitude limits to be expected, when using different magnifications, assuming that the seeing is sufficiently good to give a virtually point-like stellar image at all powers. It is because such conditions rarely exist that very large telescopes may be seriously handicapped when it comes to glimpsing faint stars, the high magnification required to darken the sky only contributing to increased diffuseness of the stellar image. An extreme example of this is the 1·5-m reflector under the relatively bright sky of Mount Wilson, whose limiting visual magnitude is seldom fainter than 16·5.

It should be emphasised once and for all, however, that no theoretical treatment of the subject of limiting magnitude can be applied rigorously to all telescopes and observers, since sky conditions, efficiency of instrument, and acuity of observer, can all vary over very wide limits.

1.3 Brightness of point source seen with the naked eye

The apparent brightness of most 'point' objects is in fact determined by the law governing the brightness of extended sources, since it is only if the retinal image is small enough to fall upon a single cone that it can be treated as a true point image.

If this condition is fulfilled, however, its apparent brightness is directly proportional to the area of the eye's pupillary aperture (i.e. to δ^2).

1.4 Apparent brightness of extended images

The apparent intensity per unit area of an extended source viewed with the naked eye being proportional to δ^2, the amount of light passed by the objective to the image in its focal plane will be greater than that received on the retina of the naked eye by a factor of $\dfrac{D^2}{\delta^2}.\pi$ (where π is the telescope's transmission factor). When the telescope is used with a magnification M, the illuminated retinal area is increased by a factor M^2. Hence the apparent brightness of the image is increased by a factor

$$\frac{D^2}{M^2\delta^2} \cdot \pi$$

The brightness of the telescopic image is therefore proportional to $\dfrac{D^2}{M^2}.\pi$, and the increase of the telescopic over the naked-eye brightness is:

$$\frac{\text{brightness of telescopic image}}{\text{brightness of naked-eye image}} = \frac{\pi D^2}{\delta^2 M^2} \quad \cdots \cdots \cdots (a)$$

We shall see later (section 3.3) that the minimum magnification (M') that can fully employ the available light grasp of an objective is equal to the aperture divided by that of the eye, or $M'=\dfrac{D}{\delta}$. From this, and from what has gone before, it follows that the brightness of the telescopic image is π times that of the naked-eye image. This means that even under optimum magnification, and neglecting the transmission factor, the telescopic image can only equal, and never exceed, the naked-eye image in brightness. In fact the transmission factor is always less than unity, and

6

hence the apparent brightness of an extended object is always reduced by viewing it telescopically.* Note, in passing, that the relative brightness of the telescopic and naked-eye images of an extended object is independent of the focal length of the objective.

But while the telescopic image of an extended object is always fainter,† area for area, than the naked-eye image, what the telescope can accomplish is the enlargement of the image from an inappreciable to an appreciable size. Also, the sky appears less bright through a telescope than to the naked eye. This fact has practical applications in the observation both of objects with a superabundance of light (e.g. the Moon), and of those whose images lose by lack of illumination if magnification is pressed too far (e.g. the planets).

The extent to which the brightness of extended images is reduced by magnification is brought out in the following Table, which sets out the results of substituting various values for M in equation (b) below.

Magnification (D expressed in cm)	Relative telescopic/ naked-eye brightness (theoretical)	Relative telescopic/ naked-eye brightness (corrected for 60% transmission factor)
$1\cdot2D=M'\ddagger$	1·0	0·6
$3\cdot5D$	0·11	0·07
$5D=M_r\ddagger$	0·06	0·035
$7\cdot5D$	0·026	0·015
$10D$	0·014	0·0086
$12\cdot5D$	0·009	0·0055
$15D$	0·0064	0·0036
$20D$	0·0036	0·0022
$25D=M''\ddagger$	0·0023	0·0014

*Alternatively, from expression (a), since $\dfrac{D}{\delta}=M$

$$\frac{\text{telescopic brightness}}{\text{naked-eye brightness}}=\frac{\pi(M')^2}{M^2} \quad \cdots\cdots\cdots\cdots (b)$$

which reveals the same fact—that $\dfrac{\text{telescopic brightness}}{\text{naked-eye brightness}}$ cannot be larger than unity

(even neglecting to take π into account), since if M, the magnification used, is less than M', the minimum useful magnification, then the exit pupil will be too large to be wholly accepted by the eye. This topic is taken up again in section 3.3.

†This is not true of the Galilean telescope, however; for under all circumstances the diameter of the exit pencil is larger than that of the pupillary aperture, and except for the transmission factor, unit areas of the telescopic and naked-eye images are of equal brightness. (See section 9.2.)

‡These particular values of M are explained in sections 3.3, 3.4.

The reverse side of the picture is exhibited in the following Table, which tabulates the magnifications corresponding to the selected values of $\dfrac{\text{telescopic image brightness}}{\text{naked-eye image brightness}}$ given in the left-hand column, for a 25-mm, 75-mm, 15-cm and 30-cm telescope.

	$D = 25$ mm	75 mm	15 cm	30 cm
1·0	3·0	9	18	36
0·1	9·5	28·5	57	114
0·01	30	90	180	360
0·001	95	285	570	1140
0·0001	300	900	1800	3600

It can readily be seen that a large aperture gives a brighter extended image than a smaller one using the same magnification, and that the large instrument can use a higher magnification than the smaller in the production of an equally bright image.

1.5 Brightness of a point image formed by a lens

The apparent brightness of a point source, unlike that of an area, is increased by passing its radiation through a lens, or mirror and lens, system. (Practical application of this difference: telescopic vision of stars by daylight, when sky brightness is diluted and their own increased to a level at which the contrast is sufficient for them to be seen—for whereas the brightness of a point image varies solely with D, that of the extended image, as we have just seen, varies inversely as M, and can therefore be reduced by increasing M.)

Compared with the naked-eye brightness of a point source, that of its telescopic image is increased by $\dfrac{D^2}{\delta^2}$, since in each case the amount of light received by the retina is a function of the area of the collecting lens. But $\dfrac{D}{\delta}$, as will be seen in section 3, is a measure of the minimum magnification that can be usefully employed. Hence, for point sources,

$$\frac{\text{brightness of telescopic image}}{\text{brightness of naked-eye image}} = (M')^2 \quad \ldots \ldots \ldots \; (c)$$

Thus the brightness of the telescopic image is greater than that of the naked-eye image by a factor equal to the square of the minimum useful magnification—which, in turn, depends solely upon aperture.

8

Magnification may nevertheless make a stellar image appear to be brighter by darkening the background, when the increased contrast between the star and the sky is wrongly interpreted as a brightening of the stellar image.

1.6 Point image brightness and magnification

The relative brightness of the naked-eye and telescopic images of a point object being determined by the relative areas of the objective and the pupillary aperture of the eye, magnification theoretically has no effect upon the observed brightness of a star, nor on the value of the magnitude threshold—provided always that it is not less than the minimum useful magnification, when the aperture is in effect reduced. In practice, however, if M is considerably greater than the resolving magnification M_r* or if the atmospheric turbulence is large, a new factor enters the picture: the expansion of the spurious disc past the size at which it can be considered as a point, to a sensible disc whose brightness is then determined by expression (b), not (c), and is illustrated in the Table on p. 7. The effect of this is to reduce its brightness in proportion to M, and thus to modify the theoretical figures relating aperture and minimum visible magnitude, when the atmosphere is poor or excessive magnifications are used.

The value of M at which this occurs cannot be precisely defined, since the expansion of the spurious disc with increasing magnification is gradual, and at no stage does it suddenly cease behaving like a point and start behaving as an area. Within this upper magnification range, however, the brightness of the image is in inverse proportion to M^2, and it is for this reason that stars near the telescope's light-grasp threshold may be rendered altogether invisible by gross over-magnification.

1.7 Photographic intensity of the image

The photographic and visual intensities of an extended image are not to be confused. We have seen that the visual, or subjective, brightness of an extended image is controlled by the magnification and the aperture (section 1.4). The photographic intensity of such an image in an instrument of given aperture, on the other hand, is a function of its focal ratio: for the area over which the light is spread (proportional to F^2) is as relevant a factor as the amount of light (proportional to D^2). Hence the photographic intensity of the image of an extended object is proportional

---

*The lowest magnification capable of showing all the resolved detail in the primary image—see section 3.3.

9

to $\left(\dfrac{D}{F}\right)^2$, and may be increased by reducing the focal length, or increasing the aperture, or making both changes simultaneously.

It is for this reason that objectives of small focal ratio (large relative aperture) are faster than those of large: since the proportionality is with the square of the relative aperture, an instrument working at $f/8$ is four times as fast as one working at $f/16$. And since the image intensity is independent of the aperture as such, a 25-mm lens working at $f/2$ is faster than the 5-m mirror of the Mt Palomar reflector working at $f/3\cdot3$.

The photographic speed of an objective where point images are concerned is, like the visual brightness (section 1.5), a function of aperture only. Thus a 15-cm lens at $f/5$ is no faster than a 15-cm lens at $f/15$.

1.8 Relative brightness and the scale of stellar magnitudes

The basis on which the scale of stellar magnitudes is constructed is the definition of magnitude such that a star of magnitude m is 2·512* (or $\sqrt[5]{100}$) times as bright as a star of magnitude $m+1$.

Since $\log_{10} 2\cdot512 = 0\cdot4$, we may say of two stars—the brightness and magnitude of the brighter of which are b and m, and of the fainter b' and m'—that

$$\frac{b}{b'} = 10^{0\cdot4(m'-m)}$$

or, taking logs,

$$\log \frac{b}{b'} = 0\cdot4(m'-m) \quad \ldots\ldots\ldots\ldots\ldots\ldots (d)$$

This relationship may be expressed in a variety of forms, some of which it may be useful to have:

$$\frac{b}{b'} = 2\cdot512^{\Delta m} \quad \ldots\ldots\ldots\ldots\ldots\ldots (e)$$

$$\log \frac{b}{b'} = \log 2\cdot512 . \Delta m$$

$$= 0\cdot4 . \Delta m$$

$$\Delta m = \frac{\log b - \log b'}{0\cdot4}$$

$$\Delta m = 2\cdot5 \log \frac{b}{b'} \quad \ldots\ldots\ldots\ldots\ldots (f)$$

If M is the integrated magnitude of two stars, of magnitudes m and m', then

$$2\cdot512^{-M} = 2\cdot512^{-m} + 2\cdot512^{-m'}$$

*More accurately, 2·5119.

Also $\qquad M = m - x$ .... (g)

where $\qquad x = \dfrac{\log\left(1 + \dfrac{1}{2 \cdot 512^{(m'-m)}}\right)}{0 \cdot 4}$

The following Table relates $m' - m$ to b/b' (equation (d)) and to x (equation (g)). Hence it can be used to derive (i) the relative brightness of two stars of known magnitudes, (ii) the magnitude difference of two stars of known brightness difference, (iii) the individual magnitudes of two stars whose integrated magnitude and relative brightness are known (iv) the integrated magnitude of two stars whose individual magnitudes are known.

$m'-m$	b/b'	x	$m'-m$	b/b'	x
0·00	1·00	0·75	1·75	5·01	0·20
0·10	1·10	0·70	1·80	5·25	0·19
0·20	1·20	0·66	1·83	5·40	0·18
0·30	1·32	0·61			
0·35	1·38	0·59	1·85	5·50	0·18
			1·87	5·60	0·18
0·40	1·45	0·57	1·90	5·75	0·18
0·45	1·51	0·55	1·93	5·92	0·17
0·50	1·58	0·53	1·95	6·03	0·17
0·55	1·66	0·51			
0·60	1·74	0·49	2·00	6·31	0·16
			2·10	6·69	0·15
0·65	1·82	0·48	2·20	7·59	0·13
0·70	1·91	0·46	2·30	8·32	0·12
0·75	2·00	0·44	2·40	9·12	0·11
0·80	2·09	0·42			
0·85	2·19	0·41	2·50	10·00	0·10
			2·60	11·00	0·09
0·90	2·29	0·39	2·70	12·00	0·09
0·95	2·40	0·38	2·80	13·00	0·08
1·00	2·51	0·36	2·87	14·00	0·08
1·05	2·63	0·35			
1·10	2·75	0·33	2·90	14·50	0·08
			2·95	15·00	0·08
1·15	2·88	0·32	3·00	15·85	0·07
1·20	3·02	0·31	3·10	17·50	0·06
1·25	3·16	0·30	3·25	20·00	0·05
1·30	3·31	0·29			
1·35	3·47	0·28	3·50	25·00	0·04
			3·70	30·00	0·04
1·40	3·63	0·27	4·00	39·80	0·03
1·45	3·80	0·25	4·25	50·00	0·02
1·50	3·98	0·24	4·45	60·00	0·02
1·55	4·17	0·23			
1·60	4·37	0·22	4·60	69·00	0·02
			4·75	79·00	0·02
1·65	4·57	0·21	4·80	91·00	0·01
1·70	4·79	0·21	5·00	100·00	0·01

Example (*i*)

How much brighter is a star of the 1st magnitude than one of the 10th?

$$m'-m=9$$

From the Table, if $m'-m=5$, then $b/b'=100$,

if $m'-m=4$, then $b/b'=39 \cdot 8$.

Hence if $m'-m=9$, $b/b'=100 \times 39 \cdot 8$

$$=4000 \text{ nearly.}$$

Example (*ii*)

If star A is fifteen times as bright as star B, what is their magnitude difference?

From the Table, if $b/b'=15$, then $m'-m=2 \cdot 95$.

Hence A is nearly 3 magnitudes brighter than B.

Example (*iii*)

If the combined magnitude of a close stellar pair is $4 \cdot 8$, and the estimated magnitude difference of the components is $0 \cdot 5$, what are the magnitudes of the two stars?

$$m'-m=0 \cdot 5$$
$$\therefore \quad x=0 \cdot 53$$
$$\therefore \quad m=4 \cdot 8+0 \cdot 53$$
$$=5 \cdot 33$$
$$\therefore \quad m'=5 \cdot 33+0 \cdot 5$$
$$=5 \cdot 83$$

Hence the magnitudes of the two stars are $5 \cdot 3$ and $5 \cdot 8$.

Example (*iv*)

What is the integrated magnitude of a close double, whose individual magnitudes are $2 \cdot 7$ and $3 \cdot 0$?

$$m'-m=0 \cdot 3$$
$$\therefore \quad x=0 \cdot 61$$

and
$$M=2 \cdot 7-0 \cdot 61$$
$$\simeq 2 \cdot 1$$

Hence the integrated magnitude of the pair is $2 \cdot 1$.

1.9 Colour index

The colour of a star will affect its magnitude as recorded by detectors of different colour response; when observed through a series of filters of similar density, it will appear brightest when the filter's colour transmission characteristics correspond most closely to the star's emission curve. A comparison of its magnitude when recorded through different filters gives information about the emission intensity at different wavelengths.

The well-known UBV convention (now considerably extended) is based on the use of standard filters whose transmission peaks at the following wavelengths:

U   354·0 nm (ultraviolet)
B   433·0 nm (blue-violet)
V   575·0 nm (yellow: 'visual')
R   634·0 nm (red)
I   804·0 nm (near infra-red)

The range of standard filters extends to 20μ, but the measures in B, V and R are of greatest interest to the visual observer. In general, the index B-V is used to form a judgment of a star's tint. If B-V<0, the star is bluish; if >0, the star is yellow or red. Some extreme values for naked-eye stars include: ζ Pup (B-V=−0·3 mag: very blue); Antares (B-V=+1·8 mag; very red). For further information on colour classification, see the introduction to B. 25.11.

TELESCOPIC FUNCTION: RESOLUTION

2.1 Introduction

Diffraction is a phenomenon which, though similar to interference in its general effects, occurs within a single wave system. In a wavefront every point may be regarded as a secondary radiator, itself generating a wave system. Within a pencil, however, these secondary vibrations are eliminated by destructive interference. Only at the edge of the pencil are they not eliminated. For this reason the intensity at the edge of a pencil does not fall sharply to nil, but rhythmically.

It will be seen from equation (d), p. 15, that the angular size of the spurious image of a point source, the basis of all diffraction effects observed with a given telescope, is a function of that telescope's aperture. It should be emphasised that throughout the following discussion the *used* aperture is referred to. If the full aperture is not used, the diffraction pattern will be determined by that fraction of it which is responsible for forming the image. For example, a slit in the focal plane produces the same diffraction effects as a conjugate slit in front of the objective; again, a camera shutter for solar photography, consisting of a slit in an opaque screen that is allowed to fall across the lens, will produce diffraction effects appropriate to its width, irrespective of the aperture of the lens.

Since it is only at the boundary of the wavefront that interference is asymmetrical and diffraction is not eliminated, its effects are most noticeable in those pencils which, so to speak, consist mostly of boundary: in the images of point sources, of boundaries (as in the edges of shadows), and of minute detail generally. The observer should be aware of the fact that diffraction is capable of profoundly modifying the details of the image in shape, size, intensity, and visibility.

2.2 Structure of the star image

The laws of geometrical optics would lead one to expect that a perfectly corrected objective would produce a point image of a point object. This is not so: diffraction within the converging pencil produces an image of finite dimensions and characteristic structure. This diffraction pattern consists (assuming a circular objective) of a central bright disc, termed the spurious disc or Airy disc, surrounded by alternate concentric rings of light and darkness—the diffraction rings and interspaces, respectively—the latter resulting from destructive interference.

The linear radii of the maxima and minima within the image at the focus are given by

$$r' = k \cdot \frac{\lambda F}{D} \quad \dots \dots \dots \dots \dots \dots \quad (a)$$

where k is a coefficient whose value is different for each ring and interspace; for the first interspace, $k = 1 \cdot 220$; for the first ring, $k = 1 \cdot 620$. Hence the theoretical linear radius of the first interspace is given by

$$r'_1 = \frac{1 \cdot 22 \lambda F}{D} \quad \dots \dots \dots \dots \dots \dots \quad (b)$$

In angular measure this (the angle subtended by the image at the centre of the objective) becomes

$$\left. \begin{aligned} r_1 &= \frac{1 \cdot 22 \lambda}{D} \text{ radians} \\ r_1 &= \frac{1 \cdot 22 \lambda}{D} \cdot 206{,}265 \text{ '' arc} \end{aligned} \right\} \quad \dots \dots \dots \dots \dots \quad (c)$$

or

And finally, taking a value for λ corresponding to the yellow region of the spectrum to which the eye is most sensitive (550 nm), the above expression becomes

$$r_1 = \frac{13''84}{D} \quad \dots \dots \dots \dots \dots \dots \quad (d)$$

D being expressed in centimetres.

Using the same value of λ, we have from equation (a) above,

$$r'_1 = 6 \cdot 71 . 10^{-4} \cdot \frac{F}{D} \quad \dots \dots \dots \dots \dots \quad (e)$$

for the linear radius (mm) of the first minimum.

Hence the angular radius, r, is determined solely by D, to which it is inversely proportional; r', the linear radius, on the other hand is a function of F/D, the focal ratio. (See Table, p. 16.)

Theoretically, about 85 per cent of the light forming the whole image is concentrated in the spurious disc. The rings are therefore comparatively inconspicuous—even the first is only 1·7 per cent as bright as the disc

Focal ratio (F/D)	Theoretical radius of 1st interspace (microns)
1	0·67
2	1·3
3	2·0
4	2·7
5	3·4
6	4·0
8	5·4
10	6·7
12	8·1
15	10·1
20	13·4

itself—and may be disregarded in a preliminary discussion of resolving power; they are altogether absent from faint images (see Figure 2). Since λ appears in expression (c), above, it follows that the rings are coloured, their inner edges being blue and their outer red, but this is entirely imperceptible in normal astronomical observation.

The intensity of the spurious disc is theoretically greatest at its centre, falling off rapidly to zero in the first interspace, there being no clearly

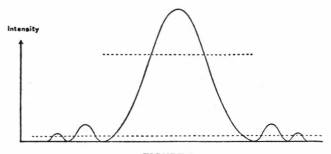

FIGURE 2

Light intensity across the diffraction image

defined 'edge' to the disc. Although f/ratio has no effect upon the angular size of the diffraction pattern it does affect its appearance, and the theoretical appearance just described is in fact only to be seen with f/ratios considerably greater than are usually to be found in astronomical telescopes. At normally encountered f/ratios (say, f/5 to f/20) no intensity gradient across the dic is perceptible, the border between the disc and the first minimum appears nearly sharp, and the rings are brighter than theory would indicate, the first ring being not much fainter than the disc itself.

With good seeing and a 1st-mag star, as many as 5 or 6 rings may be seen with an aperture of about 15 cm.

Theory can define the position of the zone of zero intensity in each interspace—the locus of a point maintaining a constant distance r from the centre of the spurious disc—but cannot define the point at which the intensity of the disc falls below the threshold of visibility: the visible extent of the disc, like the number of rings visible, varies for a given instrument with the brightness of the source, although the discs are in fact the same size, irrespective of brightness. For this reason r_1 cannot be taken as the radius of the spurious disc; a closer approximation would be $\frac{1}{2}r_1$ (Figure 3 (a)).

A

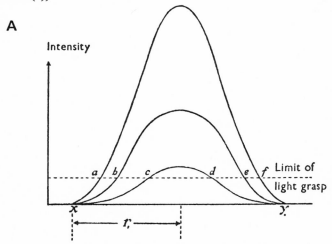

Visible diameters of the three discs are
$$af > be > cd$$
Diameter of the first interspace is xy in each case

FIGURE 3(a)
Effect of light intensity on the diffraction image

B

FIGURE 3(b)
Overlapping diffraction discs

The complete profile of the diffraction pattern of a point source is shown in Figure 2. It can be seen that, with a given instrument, the visibility and number of the rings depends upon the brightness of the star; alternatively, that for a given star it depends upon the light grasp, and therefore aperture, of the instrument employed. Also, more of the disc of a bright star than of a faint star will be visible; this, it should be realised, does not affect the value of r_1, which can only be increased by reducing D.

The appearance of the diffraction pattern as described above assumes that the image is accurately in focus—i.e. that the focus of the ocular coincides precisely with the focal plane of the objective—and that aberrations, if present, are imperceptible. If the focal planes of the ocular and objective are separated, the appearance of the image suffers a progressive modification. The first ring (and to a lesser extent the others) brightens at the expense of the disc, while the first interspace becomes faintly luminous. As the separation of the focal planes increases, a stage is reached when all the light has been drained from the disc into the rings; the image then consists of a set of exaggerated rings enclosing a dark central disc. With further separation, a new development begins. The rings start to expand, and a bright point appears at the centre of the image; this point expands into a bright ring. Again a bright point appears at the centre of the pattern, and in turn expands into a ring. Thus successive new rings are thrown off from the centre of the image. In the case of a reflector a stage is reached at which the throwing out of new rings from the centre is succeeded by the appearance of a ring of characteristic appearance (broad, with an irregular edge) which then remains, further rings being projected from its outer edge. These extra-focal rings, being extremely sensitive to aberrations in the objective, are usefully employed in the testing of OGs and mirrors.

The structure of the in-focus diffraction pattern produced by a Newtonian is modified by the central obstruction in the way of the incident light caused by the diagonal. The diameters of the disc and rings are slightly reduced, while the rings are relatively thicker and noticeably brightened at the expense of the disc. The interspaces being unaffected, the apparent brightness of the rings is increased still further by contrast. The relationship between the relative diameter of the central obstruction of an optical system, and the resulting percentage of the total transmitted light that is imaged in the Airy disc (assuming a perfectly corrected system), is as follows:

Obstruction	0	0·05	0·10	0·15	0·20	0·25	0·30	0·40	0·50
Intensity of Airy disc (%)	83·8	83·2	81·8	79·5	76·4	73·2	68·2	58·4	47·9

The effect of a central obstruction on the diameter of the Airy disc is relatively small; at $D/2$, the diameter of the first interspace is 0·818 of its value with full aperture, and in the limiting case of practically total obstruction, it is still about 0·63 of the full value. The brightening of the rings matches the dimming of the Airy disc; with an obstruction of $D/2$, the brightness of the first ring is increased by a factor of 4, while that of the disc is reduced by a factor of 2.

The modifications of the image caused by diffraction at the supports of the flat may conveniently be mentioned here. In the case of a bright image, a four-armed spider produces a four-pointed cross centred upon the diffraction disc. The length of the arms of this cross is inversely proportional, and their brightness directly proportional, to the thickness of the supports. The diffraction cross causes slight loss of contrast in an extended image.

The effect of this relative thickening and brightening of the rings upon the definition of the reflector's image is to some extent controversial: Pickering, for example, considered it to be one of the causes of the reflector's slightly inferior definition, while Hargreaves denied that it has any appreciable effect upon planetary definition whatever. It is probable that this difference of opinion springs from a consideration of different cases, for Dall, in a careful investigation (B. 7.23),* has clearly shown that if the obstruction is relatively small (for an instrument of given f/ratio) little or no deterioration of planetary definition is encountered; if it exceeds about $D/5$ an impairment of definition becomes increasingly evident, especially through the loss of contrast near the limit of vision.

The visibility of double stars of equal, or nearly equal, magnitude is certainly facilitated by the type of diffraction pattern produced by the reflector,† though this is not accompanied by improved accuracy in their measurement—an observational field in which the refractor is certainly superior. Close unequal doubles, on the other hand, tend to be more easily resolved with a refractor than with a reflector of equal aperture, owing to the exaggeration of the rings (especially the first) in the latter's diffraction pattern. A *comes* much more than 1 mag fainter than its primary will be imperceptible if its image falls on the first ring; with a smaller magnitude difference it may still appear as nothing more than a

*Reference to Bibliography (section 41).
†For work requiring maximum resolving power, increasing the diameter of the central obstruction might profitably be tried. In the case of a refractor, however, it will exaggerate the secondary spectrum, as also the abaxial aberrations. Sir John Herschel long ago reported an improvement in a telescope's resolving power when used with a $D/5$ to $D/6$ central stop, despite the brightening of the rings.

slight local brightening of the ring, and as such be very easily overlooked. Again, a very faint *comes* close to its primary may on occasions be rendered invisible by *increasing* the aperture, it falling in the first interspace of the smaller telescope's image of the primary, but on the first ring of the larger telescope's.

It follows from expression (*c*), p. 15, that the angular size of the spurious disc varies inversely as aperture: a 50-mm objective produces star images four times the angular size of those produced by a 20-cm. The linear size of the image, on the other hand, is a function of the *f*/ratio of the objective (equation (*b*), p. 15). Since it is the angular size of the images that is of importance in the resolution of closely adjacent images, large apertures are the desideratum and focal ratios are irrelevant.

Summarily,

$$\text{linear size} \propto \frac{F}{D}$$

$$\text{angular size} \propto \frac{1}{D}$$

and in addition,

$$\text{apparent size} \propto \text{brightness of image}$$
$$\text{brightness of image} \propto D.$$

2.3 Structure of extended images

All linear and area detail of the image is falsified by diffraction to an extent inversely proportional to the difference between its size and the size of the diffraction pattern of a point image. It varies from a slight distortion of shape, size, or intensity, to complete suppression. Conversely, diffraction can 'create' detail in an image which has no objective counterpart.

Diffraction at the edge of a bright area of the image results in blurring of that edge. Encroachment of light from the bright into the dark causes the dimming of a narrow strip of the former and the 'greying' of an equal strip on the dark side. This blurring of an edge can be observed in the shadow of a ruler (even with parallel light), at both jaws of a spectroscope slit, and among such astronomical objects as planetary limbs and lunar shadows.

A bright line of negligible width, crossing a dark area, may be regarded as consisting of a very large number of contiguous points. Each of these will produce its own diffraction pattern, with the result that the image of the line will be thickened by a fringe on either side.

If the line is turned round upon itself to form a circle, the diffraction band on the inner and outer sides of the image will only be of equal density if the radius of the circle is large compared with r_1. As it approaches r_1 there will be progressive overlapping of the bright rings of the internal fringe, building up a brighter central area, intermediate in intensity between the band itself and the outer fringe. When the radius of the ring is equal to r_1 the first internal rings will all intersect at the centre of the image, with the formation of a bright, star-like point.

The images of small dark spots on bright grounds, and of small bright spots on dark grounds, are modified in respect of both brightness and size: the former being 'greyed' (in extreme cases being rendered invisible through loss of contrast) and reduced, the latter enlarged and brightened centrally. Under certain circumstances a bright patch or point will form at the centre of a small dark spot, and a dark one at the centre of a bright spot on a dark ground.

Astronomical objects which are subject to such diffraction effects include planetary limbs, Martian and lunar 'canali' and fine lunar and planetary detail generally, satellites of sensible disc, shadows of satellites on primaries, inner planets in solar transit (the 'volcano' on Mercury is visible with apertures not larger than about 25 mm), and the 'black drop' (visible with magnifications above about $6D$).

2.4 Resolving power of a telescope

Whereas light grasp and magnification are explicable in terms of geometrical optics, an understanding of resolution thresholds and resolving power involves recognition of the undulatory nature of light.

The separation of the images of a close pair of stars is more than a mere matter of magnification, and is not assisted by increasing F. Since, as has just been seen, these images are not points, but discs, it can happen that they are so large, with a given aperture, as to overlap to an extent that prohibits their visual separation, even though high magnification may have separated their centres to a sufficient distance for them to have become individually visible were they point images. In such a case resolution can only be achieved by reducing the size of the spurious discs, i.e. by increasing D. To increase the magnification alone will achieve nothing, since the size of the discs will increase with their separation, their resolution being therefore unaffected. Increasing M can only result in resolution if the primary images in the focal plane are resolved.

It is generally agreed that the intensity drop midway between the central maxima of two equally bright, overlapping diffraction patterns is relatively deep enough (about 30 per cent) to appear clearly marked to the eye when the distance separating their centres is equal to the radius of the first interspace. The centre of each pattern then lies in the first interspace of the other, and the areas defined by the two first interspaces are overlapping one another by one half (Figure 3 (*b*)). Owing to the character of the light-distribution across the diffraction pattern it is less easy to define precisely the diameter of the 'visible disc' than that of the first minimum, and this has led to some confusion in the literature: some writers speaking of the spurious disc as though bounded by the first interspace, others as though the diameter of the visible disc is only half that of the first interspace.

From equation (*d*), p. 15, we therefore have for the theoretical resolving power, or threshold of resolution, of a telescope:

$$R = \frac{13''84}{D} \quad \dots \dots \dots \dots \dots \dots \quad (f)$$

In practice is it found that a separation of R is clearly perceptible, and that under conditions of good seeing a normally acute eye can just perceive a minimum only 5 per cent less than the intensity of the maxima of the two patterns. This yields a value for R (designated by R' to distinguish it from the theoretical value) which agrees closely with Dawes' empirical criterion,* that a pair of mag 2 stars, separation 4''56, are just resolvable with a 25·4-mm objective:

$$R' = \frac{11''58}{D} \quad \dots \dots \dots \dots \dots \dots \quad (g)$$

This is also in accordance with the results of direct micrometrical measures of spurious discs, e.g. Pickering's value of 0''46 for a 25-cm instrument.

D (cm)	Theoretical resolution threshold	Empirical resolution threshold (Dawes)
	$R = r_1$	R'
2·5	5''54	4''63
5·0	2·77	2·32
7·5	1·85	1·54
10·0	1·38	1·16
12·5	1·11	0·93
15·0	0·92	0·77
20·0	0·69	0·58
25·0	0·55	0·46
30·0	0·46	0·39
35·0	0·40	0·33
40·0	0·35	0·29
45·0	0·31	0·26
50·0	0·28	0·23

*Reached as a result of tests with various apertures, and not, as sometimes stated, based on actual observations with a 25·4-mm OG (B. 3.7).

In practice, the performance of a given instrument and observer may differ from both the theoretical figure and the Dawes criterion, for a number of reasons:

(1) Abnormal acuity or experience of the observer.

(2) The Dawes criterion implies a clearly perceptible intensity minimum between the two maxima. But the existence of two overlapping images may be inferred from the elongation of the residual image although there may be no perceptible intensity drop between two maxima. In this case the pair is not, in the theoretical sense, resolved. Elongation of the image is visible to normal eyes when the separation is not less than about $R'/2$; Burnham was able to detect it with his 15-cm for separations of $0''2$ $(R'/3\cdot8)$. The following separations, achieved by experienced observers, are among the many scattered through the literature:

89-mm	$R'=1''30$	Separation down to $1''16$.
30·5-cm	$R'=0''38$	Separation down to $0''25$.
71-cm (Greenwich)	$R'=0''16$	Separation down to $0''16$.
91-cm (Lick)	$R'=0''13$	Separation down to $0''09$.
102-cm (Yerkes)	$R'=0''11$	Separation clear enough for measurement at $0''2$; estimation of position angle by direction of disc elongation possible below this.
208-cm (McDonald)	$R'=0''06$	Clear separation down to about $0''1$; measurable elongation below this limit.

(3) It might be expected from theoretical considerations—less of the total spurious disc of a faint star being visible than of a bright one*— that faint pairs should be more easily resolved than bright. This is not borne out by experience. From a discussion of a large number of observations, Lewis† found, for example, that whereas bright pairs of roughly equal magnitudes, lying within the mag range 5·7 to 6·4, could be separated at $\dfrac{12''2}{D}$, equal faint pairs (mags between 8·5 and 9·1) could not be resolved unless their separation was $\dfrac{21''6}{D}$. For a very similar magnitude range (8·8 to 9·0) Aitken found that he required a separation of $\dfrac{15''5}{D}$. Similarly, very

*The diameters of a bright and a faint stellar spurious disc, as measured by Pickering with a 25·4-cm objective under good seeing conditions, were $0''60$ and $0''35$ or $\dfrac{15''2}{D}$ and $\dfrac{8''9}{D}$ respectively.

†B. 3.10.

bright pairs are more difficult than those of moderate brightness, as exemplified by Lewis's first group. If one or both components are bright enough to 'dazzle', the effect of irradiation will be added to the already reduced acuity of the eye, and the Dawes criterion will be (in extreme cases, considerably) too low.

(4) The theoretical figure and the Dawes criterion are only strictly applicable to equally bright images;* the greater the magnitude difference between the two stars, the greater will be the discrepancy between either of these figures and the actual telescopic performance. Lewis found that for pairs differing by about 3 mags (the mean mags of the pairs included in his study were 6·2 and 9·5), the resolution threshold was raised to $\dfrac{41\rlap{.}{''}9}{D}$; while for a group of doubles, the mean mags of whose components were 4·7 and 10·4, it had risen to $\dfrac{91\rlap{.}{''}0}{D}$.†

(5) Expression (c), p. 15, shows that the size of the disc varies with the wavelengths of the radiation, the red disc being considerably larger than the blue:

for violet light ($\lambda = 360$nm), $r_1 = \dfrac{9\rlap{.}{''}07}{D}$

for red light ($\lambda = 770$nm), $r_1 = \dfrac{19\rlap{.}{''}4}{D}$.

Hence the resolution of a pair of blue stars may be expected at smaller separations than a pair of red stars.

(6) The resolving power of a reflector is some 5 per cent greater than that of a refractor of the same aperture, owing to the smaller spurious disc of the former (see p. 18), though this depends upon the size of the flat.

(7) It has been suggested that slight residual aberration may account for some instances of the theoretical threshold being passed, in the case of stellar resolution. Atmospheric turbulence would probably, however, more than counteract whatever advantage might be gained in this way.

(8) Even slight atmospheric turbulence will seriously impair the instrument's performance. This, more than any other single factor, is the commonest cause of actual performance falling below that suggested by the Table. Owing to their reduced susceptibility to atmospheric effects,

---

*The figure adopted by Dawes was stated by him to apply to two 6-mag stars seen in a 15-cm telescope. Any departure from this magnitude-aperture relation (even when equally bright images are involved) will, strictly speaking, introduce some modification of the criterion.
†See also B. 3.25.

small instruments generally are more efficient than large—both, of course, in relation to the appropriate value of R'. Thus with a 15-cm Burnham achieved, on the average, $R'/2$, but with a 24-cm less than R', and with apertures between 30 and 90 cm, from $1·15R'$ to $1·6R'$.

(9) The combination of atmospheric turbulence and the low resolution of the emulsion may easily reduce the photographic resolving power by a factor of 20.

The resolving power of prisms and gratings is discussed in section 19.3.2.

2.5 Determination of telescopic resolving power

(1) By the observation at a high altitude of double stars whose separation lies in the neighbourhood of the figures quoted in the Table on p. 22.* The factors listed above must be taken into account, the most important being the state of the atmosphere: no instrument can be fairly tested if the turbulence is marked. Unequal magnitudes and visual inexperience can also combine to make an instrument appear misleadingly inferior.

(2) By the use of a test object. The factor of turbulence can be largely eliminated by reducing the length of the light path, but results so obtained are likely to be misleading for the very reason that conditions do not resemble those under which the instrument is actually used.

(*a*) For stellar sources: a tinfoil sheet, with two minute holes punctured in it, mounted in front of a light source from which it is separated by a sheet of ground glass. Measurement of the distance from the telescope at which the first trace of elongation of the image, the first appearance of a perceptible intensity minimum, and clear separation occur, yields results tolerably near those obtained under normal observational conditions. They tend nevertheless to be on the optimistic side, owing to reduced turbulence.

(*b*) For planetary objects: a fine wire gauze placed in front of a light source and separated from it by a sheet of ground glass. From the distance, a, screen to objective, at which the wires are first separated, and the linear separation of the wires, b, which for a uniform grid can be determined by dividing its overall width by the number of wires it contains, their angular separation can be obtained from

$$R = \frac{b}{a} . 206{,}265$$

in " arc. Alternatively a grid of equal black bars and interspaces can be ruled on a sheet of paper; this is observed with a high-power ocular

*See *B.A.A.H.*, 1922.

and brought towards the telescope until the individual black lines begin to emerge from the uniform grey tone. If each rule and interspace together measure 1 mm, and the grid is placed 10 m from the objective, then from the expression above, the angular separation is approximately $20''$; it is in direct linear proportion for other grid dimensions, and inverse linear proportion for other distances.

As shown by the figures quoted in section 2.6, the resolution threshold obtained in this and similar ways bears little relation to that applicable to stellar images.

2.6 Definition: the resolution of extended detail

The definition—i.e. sharpness, and smallest detail visible—of an extended surface also varies with D, and is heavily dependent on atmospheric state. The image is composed of the overlapping diffraction patterns formed by light emanating from every point in the object; if the radii of these diffraction discs were effectively of the full value r_1 there could be no resolution, however high the power used: but the low intensity of the light, compared with that of stars, reduces their effective size, and the Dawes criterion therefore cannot be applied to the resolution of detail in extended images.

This is clearly demonstrated by the following actual performances of instruments of varied apertures on three types of extended detail:

A. BLACK SPOT ON A WHITE GROUND

(*i*) W. H. Pickering, while testing his 25-cm reflector under good seeing conditions in Jamaica, found that a circular spot became visible when its angular diameter exceeded $0''20$ ($R'/2\cdot3$).*

(*ii*) Experiments by W. H. Steavenson showed that $R'/3$ is a fair average figure.

(*iii*) The naked-eye visibility of sunspots whose diameter is less than R' is well known.

B. SINGLE DARK LINE ON LIGHT GROUND

(*i*) Cassini's division, width about $0''5$, was discovered with about 65-mm aperture ($R'/3\cdot5$).

(*ii*) W. H. Steavenson's experiments yielded about $R'/5$ as an average figure.

---

*The position regarding the visibility of a light disc, as such, on a dark ground is less easy to define, but records of the visibility of Neptune's disc suggest that only about half this aperture is required. See *O.A.A.*, section 10.2.

26

(*iii*) During the testing of a 28-cm refractor at Harvard, a human hair was visible against a light-toned ground at a distance of nearly a quarter of a mile, where it subtended only $0''029$ ($R'/14$).

(*iv*) W. H. Pickering glimpsed a dark line $0''03$ wide with his 25-cm ($R'/15$).

C. PARALLEL DARK LINES ON LIGHT GROUND

(*i*) W. H. Pickering: minimum separation for resolution with a 25-cm reflector was $0''63$ ($1·4R'$).

(*ii*) A similar performance was given by the Arequipa 38-cm, which resolved a pair of parallel lines when their separation was increased past $0''42$ ($1·4R'$), in good seeing; slight atmospheric deterioration immediately raised the threshold to about $2R'$. At less than $0''42$ the lines appeared as a grey band of width about $1\frac{1}{2}$ times their separation.

(*iii*) Resolution of the lines at $12''$ with a 10-mm OG ($1·1R'$). See also sections 2.3, 24.6, 26.7, 26.9.

2.7 Practical demonstration of diffraction and interference effects

Experimental demonstrations of these modifications can be easily staged, and should convince anyone that it is aperture and not magnification that determines the separability or otherwise of a pair of stars.

Make a clean circular hole in two sheets of metal foil, using a fine needle, the diameter of the hole not exceeding about 0·5 mm. If one of the sheets is placed in front of an electric filament and viewed through the second, held close to the eye, a spurious disc and rings will be seen with a perfection seldom provided by a telescope and a star. Increasing the size of the view-hole brightens, and reduces the size of, the disc and rings, while increasing the number of the rings. The dependence of resolution on aperture can be very clearly demonstrated by piercing a close pair of holes in the object-sheet, and observing them through a graded succession of holes in the telescope sheet, ranging in diameter from about 1·5 mm down to as small as the needle can make them.

If a single 'star', or a disc of appreciable diameter, is now viewed through the original 'double star', the action of the interferometer is simulated, the image being traversed by alternate bright bands and dark interspaces. It can be shown by trial that increasing the separation of the view-holes increases both the number and the crowding of the interference fringes.

The effect of diffraction upon extended and linear detail can also be demonstrated. Pierce a white card with two holes and observe them through a view-hole which, at a given distance, is just incapable of resolving them. Then below the two holes rule a black line such that its width is equal to the holes' separation. If the card is reobserved from the same distance it will be found that this line is clearly visible–i.e. R' is too large a figure for a linear image. Lack of resolution of surface detail through insufficient aperture is easily shown by observing any patterned surface, such as a sheet of graph paper, at different distances with the naked eye and with the pinhole.

TELESCOPIC FUNCTION:
MAGNIFICATION AND FIELD SIZE

Since the telescope is a compound optical instrument, in which both the objective and the ocular form images, it is necessary to distinguish between (*a*) the magnifying power of the objective, and (*b*) the magnification of this primary image by the ocular in forming the final image.

3.1 Magnifying power of objective

The size of the image at the primary focus is dependent on F alone. If θ =angular height of object (i.e. the angle subtended by the primary image at the centre of the objective), h=linear height of the primary image, in cm, and F=focal length of objective, also in cm, then

$$h = \frac{\theta F}{k} \quad \dots \dots \dots \dots \dots \dots \dots (a)$$

where k=57·3 (° per radian) when θ is in °, or 206,265 ($''$ per radian) when θ is in $''$.

D, it will be observed, does not figure in the expression for h. For example, substituting $0°53$ for θ in equation (*a*), we have for the linear diameter of the primary image of the Sun in a telescope of 1·2-m focal length (e.g. a typical 75-mm refractor), 11 mm. Similarly, for a 4-m telescope, diameter of primary image of Sun=37 cm.

3.2 Magnification of combined objective and ocular

Any textbook on elementary optics can be referred to for the simple demonstration (it can be inferred from Figure 4) that the magnification of a telescope is given by

$$M = \frac{F}{f} \quad \dots \dots \dots \dots \dots \dots \dots (b)$$

where F and f are the focal lengths of the objective and ocular respectively.

The action of the ocular on the primary image is clearly seen from expression (*b*), where *f* and *M* are inversely proportional. This is confirmed by common sense: as *f* decreases, the distance of the eye from the image decreases, hence the angle it subtends at the eye increases, therefore the magnification of the primary image increases.

The magnification of the final image can therefore be increased in either of two ways, by increasing *F* or by decreasing *f*. Since the former

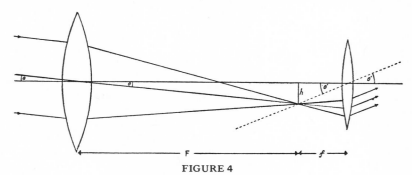

FIGURE 4

The magnification of an eyepiece

involves the use of a different objective (or of a different, e.g. Cassegrain, focus), one is in practice normally confined to the latter. This, however, cannot be carried to indefinite limits: prohibitive aberrations are introduced by the highly curved surfaces of very short focus lenses, and reduction of *f* also reduces the size of the field and the brightness of extended images.

3.3 Lower limit of useful magnification

The size of the exit pupil (section 3.11) also imposes limits, both upper and lower, on the range of magnification possible with a fixed value of *F*. Consider the passage of rays from a star situated on the optical axis: It is clear (Figure 5) that

$$\frac{D}{d} = \frac{F}{f} = M \quad \ldots \ldots \ldots \ldots \ldots \ldots \quad (c)$$

where *D* is the diameter (clear aperture) of the objective, and *d* is the diameter of the exit pupil. (This is the principle behind the use of the Berthon dynameter for measuring the magnification of an ocular.)

It can be seen from equation (*c*) that the size of the exit pupil is inversely proportional to the magnification, or

30

$$d = \frac{D}{M}$$

If the exit pupil is so large as to be incompletely accepted by the pupillary aperture of the dark-adapted eye,* the full light grasp of the objective is not being utilised, and D is in effect being stopped down—not by any diaphragm in front of the objective, but by the iris of the observer's

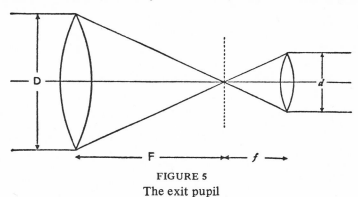

FIGURE 5
The exit pupil

eye. Thus a lower limit is set on the magnification than can be usefully employed.

If we denote this minimal value of M by M', and take the corresponding maximal diameter of the exit pupil as 7·5 mm (equal to a mean night-time value of the diameter of the pupillary aperture), then

$$M' = \frac{D}{0·75} \quad \dots\dots\dots\dots\dots \quad (d)$$

Thus the instrumental factor determining the minimum useful magnification is D, and we have

$$M' \simeq 1·3D$$

In other words, it is useless to employ magnifications lower than about 1·3 times the objective's aperture in cm. Owing to faint illumination from field stars and sky background, the diameter of the pupillary aperture may often be reduced to about 6·5 mm; hence the minimum effective magnification should be taken as about 1·5 per cm of aperture.

*The diameter of the normal pupil in subdued daylight is about 5 mm; in bright light it may contract to less than 2·5 mm. But with the degree of dark adaptation obtained during night-time observation of faint objects it may expand to about 8 mm. When brilliant objects, such as the Moon, are being observed, the pupil will contract again to 2·5 mm or less. It is interesting to note that the older observers were unaware of the extent to which dark adaptation can expand the pupil: J. Herschel, for example, always took its diameter to be 6 mm, and Rosse 2·5 to 5 mm according to the strength of the illumination.

From (*b*) above (the expression for the magnification of a telescope), and putting f' for the ocular focal length giving a magnification of M',

$$f' = \frac{F}{M'} \quad \ldots \ldots \ldots \ldots \ldots \ldots (e)$$

Hence the focal length of the ocular providing the minimum useful magnification depends upon no other instrumental factor than the focal ratio, $\frac{F}{D}$ (from equation (*c*)).*

An entirely different consideration imposes, in certain circumstances, a lower limit on the magnification than can profitably be employed, whose value is about $4M'$. Equation (*g*), p. 22, gives the condition for the resolution of two points in the primary (focal plane) image of a telescope. As emphasised on p. 21, it is hopeless to try to separate two images by increasing the magnification, if they are not resolved in the focal plane. But it is true, conversely, that though they may be resolved by the objective, they will not be individually seen unless a high enough magnification is used to bring them above the resolving threshold of the human eye. No precise or universally applicable value can be assigned to this threshold (see section 24.5), but it is roughly true to say that unless two points subtend an angle of at least $1'$ at the eye (in the case of stellar images it will be nearer $2'$ or $3'$) they will not be separated.

Denoting by M_r the magnification necessary to separate two points, just resolved in the focal plane, to an angular distance of $60''$,

$$R.M_r = 60'' \text{ arc}$$

Also, from (*g*), p. 22,

$$R = \frac{11 \cdot 58''}{D} \text{ arc}$$

(*D*, as always, being expressed in cm). Therefore

$$M_r = \frac{60D}{11 \cdot 58}'' \text{ arc}$$

or

$$M_r \doteqdot 5D$$

Since the selection of M_r is frequently relevant to the observation of close double stars, for which the normal naked-eye limit is $2'$ or more, it

*For example, consider the case of a $f/15$ telescope of 15 cm aperture, 225 cm focal length. Then from (*d*), $M'=19$, and from (*e*), $f'=12$ cm. Hence for a telescope working at $f/15$ the maximum useful ocular focal length is 12 cm, irrespective of the absolute aperture or focal length of the objective.

follows that a magnification *at least* equal to the diameter of the objective expressed in mm will be required if the resolution achieved in the focal plane is to be maintained by the retina. The less optimistic estimates of the eye's resolution threshold for stellar images (2′ and 3′) yield resolving magnifications of 10 and 16 times the aperture in cm, respectively.

With a lower magnification than M_r, there is still resolved detail buried in the focal plane image that has not yet been developed into visibility; increase of magnification beyond M_r will bring out no new detail, merely enlarging the scale of what is already visible. Thus under-magnification may waste some of the objective's resolving power as well as light grasp.

3.4 Upper limit of useful magnification

At the other end of the scale, a decrease of the diameter of the exit pupil below about 1 mm leads to a progressive impairing of vision (as well as to other restricting factors) which imposes an upper limit on the magnification that can be usefully employed—irrespective of atmospheric turbulence and aberrations in oculars of very short focal length.

Using M'' to denote this upper limit, and taking the corresponding (minimal permissible) diameter of the exit pupil as 0·75 mm,

$$M'' = \frac{D}{0·075} \quad \dots\dots\dots\dots\dots\dots (f)$$

or

$$M'' = 13D \quad \dots\dots\dots\dots\dots\dots (g)$$

'Whittaker's rule', which is an empirical expression for the value of M'', states that definition suffers from diffraction effects when the magnification exceeds the number of mm in D, corresponding to the M_r given above; but, as shown below, this rule seems to have little relevance to practical observing experience.

3.5 Discussion of theoretical limits of magnification

It should be emphasised that these 'limits', especially the upper, are not hard and fast. Vision is not affected so adversely as to prohibit observation with magnifications at least double those indicated by equation (g) above—with small instruments, anyway. For this reason, a second set of values of M'', nearer to the practical upper limit of magnification, is included in the Table and graph on p. 35.

Again, certain classes of object stand higher magnification than others;

close doubles, for instance, may react to magnifications of $20D$* or $30D$, providing their angular separation is not below the telescope's resolution threshold and the atmosphere will stand it. The same is true of micrometer measures, where the highest magnification that the atmosphere will permit (seldom above $25D$) must be used.†

The use of lower magnifications than those tabulated under M' will merely involve wastage of the objective's light grasp—a matter of little importance when the object is bright. Use of a lower magnification than M_r wastes the objective's resolving power—which, again, is not vitally important in the observation of comparatively featureless and extended objects, such as some types of nebulosity, or the tails of most comets; more important, in such cases, is the loss of light.

Planets are among the most demanding of objects, since they bear a mass of minor detail, for which the maximum resolution of which the instrument is capable is required, while at the same time being for the most part not so bright that the illumination of the image can be ignored.‡ Since the contrast in planetary detail is much weaker than that of a double star (which is observed against a dark sky background), the planetary observer prefers always to use the lowest magnification consistent with adequate resolution of his subject. This means values near $10D$, with apertures of about 25 cm or more; relatively higher in the case of smaller instruments, where such a magnification would give an uncomfortably small apparent disc diameter.

At the same time it must not be forgotten that the highest magnification that can be used satisfactorily at any time is always dependent to a greater or smaller extent upon the atmosphere (see section 26), which seldom allows more than $20D$ to $25D$. The values given in the Table assume perfect atmospheric conditions. With very large apertures, such as are not found in amateur hands, the quoted values of M'' will become progressively too high, and those of M' too low. A strictly linear relationship between M'' and D is only indicative of the conditions found in practice when D is fairly small. As an example of this, Lewis has quoted the following figures showing the normal range of maximum practicable magnification for double star work with three large telescopes:

Dearborn 47-cm refractor:	390–925	(cf. M'' (pract)=1170)
Greenwich 71-cm refractor:	670–1120	(cf. ” =1780)
Lick 91-cm refractor:	1000–1500	(cf. ” =2290)

*A convenient abbreviation, of a type much used in this section and elsewhere for '25 times the aperture of the objective expressed in cm'.
†Discussed in greater detail, *O.A.A.*, section 18.3.
‡See section 24.6 and *O.A.A.*, section 3.4.

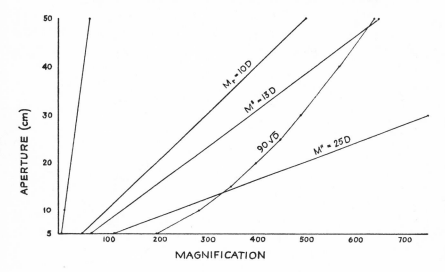

FIGURE 6

Magnifying powers

D (cm)	M'	$M_r = 10D$	$M'' = 13D$	$M'' = 25D$	Lewis, $90\sqrt{D}$
5	6	50	65	125	200
10	12	100	130	250	285
15	19	150	195	375	350
20	25	200	260	500	400
25	31	250	325	625	450
30	38	300	390	750	490
40	50	400	420	1000	570
50	62	500	650	1250	640

As an alternative, he has found that closer correspondence with practice is obtained from

$$M'' = 90\sqrt{D}$$

It will be seen from the graph that the values of this expression approximate to those of $M'' = 25D$ for the smaller apertures of D, and those of $M'' = 12D$ for the larger.

Thus although the size of the primary image is determined by F, and that of the final image by F and f, yet the highest and lowest magnifications that can be usefully employed are functions of D, since both light grasp and resolving power (section 2.4) are determined by D.

35

3.6 Observational characteristics of low and high magnifications

High magnifications suffer from the following drawbacks, as compared with low:

- (*a*) smaller field;
- (*b*) hence more frequent manual adjustment required, in the case of an altazimuth or hand-driven equatorial, with consequent increased difficulty in following, and more disjointed observation;
- (*c*) decreased brightness of extended images;
- (*d*) exaggeration of atmospheric defects;
- (*e*) exaggeration of any defects in mounting, drive, etc.

Perhaps as a counterblast to the beginner's natural instinct, a good deal has been written on the advantages of low powers, with their wider fields. The pendulum may, indeed, have swung too far, for a wide field is in many spheres of work no more than a luxury—provided the mounting is steady, accurately adjusted, and suitably designed. Given these necessary conditions, wide fields are chiefly of interest to beginners 'sightseeing' in the Milky Way, and in certain specialised fields such as variable star work (when suitable comparison stars may be some distance from the variable) or the study of telescopic meteors or large comets. As regards the 'sightseer', it should be pointed out that the type of instrument he frequently uses (a small refractor) is usually designed in such a way as to defeat its own object: the diameter of the drawtube being too small to permit the use of properly designed oculars of the low-magnification range in which the instrument particularly excels. This point is returned to in section 3.10.

3.7 True and apparent angular fields

The apparent angular field (α) is the angular diameter of the field stop in the ocular as seen through the ocular from the observing position, namely, the exit pupil.

The diameter of the stop is limited by three conditions:

- (*a*) it need not be much more than about 45°, the maximum that the average eye can take in at a single glance, i.e. without movement;
- (*b*) it must not be so wide that the extra-axial aberrations are intolerable at its edges;
- (*c*) it must not include more than the field of full illumination (section 3.8).

The true field (θ) is the angular diameter of the area of sky whose image is included in the apparent field. Hence, clearly,

$$\frac{\alpha}{\theta} = M$$

The apparent angular field of different oculars varies somewhat, but an average figure is 40°. Thus,

$$\theta = \frac{40°}{M}$$

This means that the linear diameters of the satisfactory fields of oculars, measured in their focal planes, are commonly of the order of 0·5f to 0·75f.*

3.8 Field of full illumination

When an aperture stop is placed some distance in front of a lens, the total field visible through that lens is not fully illuminated—i.e. illuminated by rays passing through the total aperture of the lens.

In the case of a telescope, the clear aperture of the objective may be regarded as the aperture stop of the ocular, which is in fact so placed that it reveals only the field of full illumination. Figure 7 represents this stop, Ss, located in the focal plane, and (for simplicity) a single-lens ocular, O. S is the image of the furthest object from the optical axis whose divergent cone is received in its entirety by the ocular lens. Thus the whole of the pencil from S passes through the exit pupil ab, while only the shaded section ac of the pencil from S', a point in the focal plane further from the axis, is included in it. The brightness of such images being progressively weakened as the distance FS' increases, the field of vision is restricted to Ss by the diaphragm of the aperture stop.

The size of the fully illuminated field is a function of the aperture, the focal lengths of ocular and objective, and the focal length and aperture of the eye lens.

3.9 True field diameters of different designs of ocular

On page 38 are tabulated the true field diameters (θ) provided by oculars of apparent fields ranging from 30° to 70°, for different magnifications.

The true angular diameter of any field can be determined by clamping the telescope p a star of known Dec, so that the star will trail the field diametrically. Then if ϕ is the diameter of the field in ′ and ″ of arc, t the time of the star's diametrical transit in mins and secs of time, δ the Dec of the star,

$$\phi = 15t \cdot \cos \delta$$

*Taking the linear diameter as 0·7f, $\dfrac{\theta F}{57·3} = 0·7f$, whence $\theta F \simeq 40f$ or $\theta \simeq \dfrac{40°}{M}$.

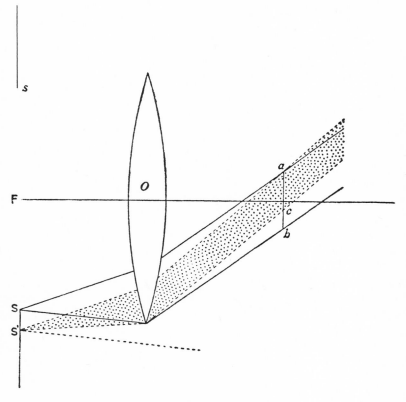

FIGURE 7
Field of full illumination

| Magnification | Apparent diameters | | | | |
	30°	40°	50°	60°	70°
10	3° 0′	4° 0′	5° 0′	6° 0′	7° 0′
25	1 12	1 36	2 0	2 24	2 48
50	36	48	1 0	1 12	1 24
75	24	32	40	48	56
100	18	24	30	36	42
150	12	16	20	24	28
200	9	12	15	18	21
300	6	8	10	12	14
400	$4\frac{1}{2}$	6	$7\frac{1}{2}$	9	$10\frac{1}{2}$
500	$3\frac{1}{2}$	5	6	7	$8\frac{1}{2}$

In the special case of an equatorial star

$$1^h \ 1^m \ 1^s \equiv 15° \ 15' \ 15''$$

3.10 Diameters of eye and field lenses of oculars

Just as the exit pupil must not be larger than the eye's pupillary aperture if the full LP capacity of the telescope is to be utilised, so it must not be larger than the field lens of the eyepiece. In fact, however, this is a condition that is not observed by instruments working at $f/15$ and thereabout (e.g. the typical small refractor) as normally constructed.

We have seen (section 3.7) that the angular diameter of the true field is given by

$$\theta = \frac{\alpha}{M} \quad \dots\dots\dots\dots\dots\dots\dots\dots (i)$$

From (a), p. 29, its linear diameter is

$$h = \frac{\alpha f}{57 \cdot 3}$$

Hence the linear diameter of the objective's image in the focal plane varies directly with f; if the ocular is to take the available field, and the relationship in (i) is to be observed, h may be taken as the diameter of the field lens.

Taking the case of a $f/15$, 75-mm refractor, for which $M'=9$, using an ocular giving $\alpha=40°$,

$$h = \frac{40 \times 12 \cdot 5}{57 \cdot 3} \doteq 87 \text{ mm}$$

i.e. the ocular is larger than the OG; even with a magnification of x11 they would be roughly the same size.

There are two obvious solutions of this difficulty: (a) to modify the eye-end of the telescope to take a larger draw-tube, (b) to use an instrument of smaller focal ratio when using very low magnifications. For by reducing F, while maintaining D constant, a smaller value of f is involved for a given magnification, with consequent reduction of the size of the exit pupil. Suppose, for example, that the telescope is a 75-mm, $f/5$ reflector. Then $F=15$, and f (giving $M=9$) is 4·2. Hence

$$h = \frac{40 \times 4 \cdot 2}{57 \cdot 3} = 29 \text{ mm}$$

instead of 87 mm with the $f/15$ instrument.

This—and not, as is often supposed, increased image brightness or photographic 'speed' as such—is the reason for the small focal ratios of 'comet seekers': only thus can the full extent of the LP field be accepted by the ocular without the latter becoming prohibitively cumbersome.

The condition for full acceptance of the incident pencil by the eye lens is less obvious; nor is the matter one of such practical importance. In fact, the diameter of the eye lens must exceed

$(2 \tan \frac{\alpha}{2} \times$ distance of eye lens to exit pupil$)$ + diameter of exit pupil

In the instance of the 42-mm ocular yielding $M=9$ with the $f/5$, 75-mm reflector the minimum diameter of the eye lens is roughly 18 mm.

3.11 The exit pupil and the position of the eye

The exit pupil, or Ramsden disc, is that section of the emergent pencil, perpendicular to the optical axis, through which rays from every point in the visible field pass. It has the following characteristics:

(a) rays from every point in the field pass through it;
(b) the illumination of the image is there greatest;
(c) rays from the edge of the objective pass through the boundary of the exit pupil;
(d) it is the image of the objective (the entry pupil of the system), when the telescope is pointed at the empty sky;
(e) it is coincident with the field of full illumination;
(f) it is the smallest possible cross-section of the emergent pencil;
(g) the largest field will be seen when the pupil of the eye and the exit pupil coincide.

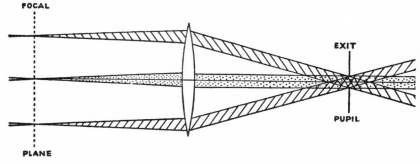

FIGURE 8
Passage of rays through the exit pupil

To ensure that the eye and the exit pupil do coincide during observation, the cap surrounding the eye lens of the ocular is placed a few mm in front of the plane of the exit pupil. With certain types of ocular (see, e.g., section 8.5.1) the exit pupil is so near the surface of the eye lens that there may be an unconscious tendency to hold the head slightly back, thus restricting the field of view and wasting a proportion of the objective's illuminating power. The radius, r, of the field observed is given by

$$\tan r = \frac{\text{(diameter of eye pupil)} - \text{(diameter of exit pupil)}}{2 \times \text{(distance of eye from exit pupil)}}$$

This is a difficulty of which spectacle-wearers should be aware and beware, when using any normally constructed ocular.

The optical axis must also pass through the centre of the pupil of the eye, or part of the exit pupil may be intercepted by the iris and therefore contribute nothing to the retinal image.

From (c) above (section 3.3), it is clear that the equation may be re-written in terms of d, thus:

$$d = f \times \frac{D}{F},$$

D/F being, of course, the reciprocal of the focal ratio. It follows from this that the diameter of the exit pupil formed by any given ocular can be obtained by dividing its focal length by the focal ratio of the objective with which it will be used. The following Table sets out the exit pupil (in mm) for various combinations of eyepiece and objective:

Focal ratio of objective	Focal length of eyepiece (mm)						
	40	25	18	12	9	6	4
4	10·0	6·25	4·5	3·0	2·25	1·5	1·0
6	6·7	4·25	3·0	2·0	1·5	1·0	0·67
8	5·0	3·1	2·25	1·5	1·1	0·75	0·5
10	4·0	2·5	1·8	1·2	0·9	0·6	0·4
12	3·3	2·1	1·5	1·0	0·75	0·5	0·33
15	2·7	1·7	1·2	0·8	0·6	0·4	0·27

ABERRATIONS

4.1 Introduction

Elementary treatment of geometrical optics is concerned with the reflection and refraction of rays which are close to the optical axis.* These conditions are not observed in a telescope since the aperture of the objective is not a negligible fraction of its focal length. The result of this is the breakdown of accurate correspondence between points in the object space and points in the image space. These departures are known as aberrations, and are of two species: those affecting the positions of the images, and those affecting their appearance.

The so-called Seidel errors (or primary aberrations) are five in number; if two types of chromatic aberration, to which refracting systems alone are subject, are added, we have the following seven errors: (1) spherical aberration, (2) coma, (3) astigmatism, (4) field curvature, (5) distortion, (6) longitudinal chromatic aberration, (7) lateral (or transverse) chromatic aberration. The aberrations of a mirror system are, clearly, independent of colour, but those of a refractive system change with wavelength, such changes being called chromatic variation of the aberration concerned.

(1) and (6) are constant over the whole field; (2) and (7) are proportional to the distance from the centre of the field; (3) and (4) vary as the square of that distance; and (5) varies as the cube of that distance. Of the extra-axial aberrations, (2) is generally regarded as the most objectionable.

Diffraction phenomena are also ignored in the geometrical treatment of optical systems; some of the practical implications of diffraction in telescopic image formation are discussed in sections 2.2, 2.3.

---

*A ray can only be considered paraxial if θ, the angle it makes with the optical axis, is small enough to justify the equations $\theta = \sin \theta$, and $\cos \theta = 1$. Or again, the geometrical laws are inapplicable if either θ or $D/2f$ are of such a magnitude that they can no longer be considered negligible when raised to the third power.

4.2 Rayleigh's limit

If the pencil emerging from the objective is free from aberration, and is therefore capable of forming a point image of a point source, its wave front is spherical and all its wave surfaces are sections of concentric spheres; such a pencil, of finite diameter and inclination to the optical axis, is termed stigmatic or homocentric.* The wavefront of an aberrant pencil, on the contrary, is not a simple spherical surface. It has been found empirically that the aberration in the pencil is in general tolerable —i.e. is too small to affect the image appreciably—if its wavefront can be contained between two spheres whose distance apart does not exceed $\lambda/4$. This limit of tolerance is known as Rayleigh's limit.

To achieve this condition in the emergent beam, no local departure from the theoretical surve of the objective's surface or surfaces must exceed $\lambda/2$ in the case of a lens, or $\lambda/8$ in the case of a mirror. This is the justification for the claim that it is easier to make a refractor than a reflector: although there are more surfaces to work, the accuracy to which they are carried need be only one-quarter that of a mirror.

4.3 Spherical aberration

Spherical surfaces are distinguished from other surfaces of revolution in that they do not yield stigmatic reflected or refracted pencils, except in a few special cases which are of no interest to the astronomer. Where the aperture is large, relative to the radius of the curvature, spherical aberration will always occur.

A lens or mirror suffering from spherical aberration is incapable of bringing rays to a focus in the same plane normal to the optical axis if the distances of these rays from the axis are different. If the aberration is under-corrected, the peripheral rays are brought to a focus nearer to the objective than the paraxial rays; if over-corrected, the reverse. In general, spherical mirrors and converging lenses suffer from under-correction; diverging lenses from over-correction.

The reflected rays from the surface of a spherical mirror (Figure 9) define a caustic surface to which they are tangent, whose apex lies at the true focus of the mirror, i.e. the point at which paraxial rays cross the axis. The outer and fainter portions of the caustic are formed by peripheral

---

*The aberrations of inferior objectives may often be greatly reduced by using oculars of similar aberration but opposite sign. This is a makeshift arrangement which cannot be recommended, since it may be required to use the objective for photography at the prime focus, or the oculars with micrometer webs. Both objective and ocular should be stigmatic.

43

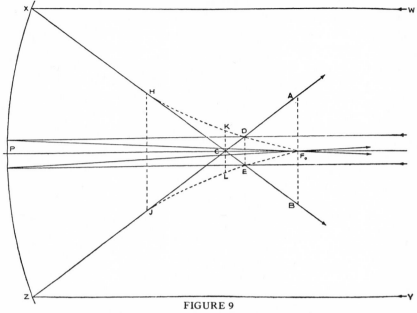

FIGURE 9

The imaging of parallel rays by a spherical mirror

rays, and hence reduction of the aperture leaves a progressively smaller and brighter portion in the vicinity of F_0, approximating to the paraxial image.

Instead of a single point image being formed of a point object on the axis, a series of poorly defined images are strung out along the axis, their edges defining the caustic surface. If the object is a point source (and leaving out of account, for the moment, the effects of diffraction), a screen at *AB* will show a faint disc with a bright central point; as the screen is moved from *AB* towards *DE* the disc will decrease in size and increase in brightness, while the central bright spot will expand and fade; at *DE*, five-eighths of the distance from *AB* to *KL*, the size of the image is minimal: this is the so-called least circle of confusion. From this position inwards the image expands again, the edges becoming markedly brighter than the centre. Thus between *C* and F_0 the brightest region of the image indicates the region of the objective which is focusing long (centre of an under-corrected objective, periphery of an over-corrected one); while between *C* and *P* the brightest region of the image indicates the region of the objective which is focusing short (periphery of an under-corrected objective, central region of an over-corrected one).

In the actual case of a stellar image these effects are superimposed upon the normal diffraction pattern of spurious discs, rings and interspaces.

44

At F_0 the disc is as bright and small as if there were no aberration, and the brightness of the rings diminishes rapidly from the centre outwards; hence spherical aberration tends not to reduce the instrument's resolving power in the case of a pair of equally bright stars. Inside KL the brightness of the rings increases from the centre outwards, the central disc being relatively weak. (This is a convenient and sensitive test for residual spherical aberration—see section 13.9; the effects of a departure from the spherical on the part of the converging wavefront which amounts to $\lambda/8$ are visible in the extrafocal image; this is only half the minimum detectable at the best focus.) The increased brightness of the rings reduces the contrast in an extended image, and hence spherical aberration adversely affects the instrument's resolving power on planetary detail.

In the case illustrated, WX and YZ being the peripheral rays, AF_0 is the extent of the lateral aberration and CF_0 the longitudinal. If we write Δf for CF_0, the difference in focal length as measured by paraxial and peripheral rays, Y for the distance of the peripheral ray from the axis, and take the focal length, f, as unity, then

$$\Delta f = k . Y^2$$

Hence the longitudinal spherical aberration is a function of the square of the distance of the incident ray from the axis.

But $Y = \dfrac{D}{2}$, and hence

$$\Delta f = k . \frac{D^2}{4} \text{ (when } f \text{ is unity)} \quad \ldots\ldots\ldots\ldots (a)$$

k is termed the coefficient of longitudinal spherical aberration, and its value varies with the design of the objective. Thus the longitudinal aberration of a $f/15$ Littrow objective, with, for example, $k = +0\cdot321$, is $+0\cdot00036f$: from (a)

$$\Delta f = 0\cdot321 \frac{D^2}{4} \text{ (when } f = 1)$$

$$= 0\cdot321 \frac{(\frac{1}{15})^2}{4} \text{ (since } D = \tfrac{1}{15} \text{ when } f = 1 \text{ and the focal ratio is } f/15)$$

$$= \frac{0\cdot321}{900}$$

$$= +0\cdot00036 \times f$$

We have also seen that the least circle of confusion lies inside the focus of the paraxial rays and outside that of the peripheral rays (still considering the case of an under-corrected objective) by distances of, respectively,

$$\tfrac{5}{8}.k.\frac{D^2}{4} \quad \text{and} \quad \tfrac{3}{8}.k.\frac{D^2}{4}$$

f being taken as unity in each case.

If the aberration is under-corrected, k is negative; if over-corrected, positive. In the case of concave mirrors there can only be one form obeying the condition that $f=\frac{R}{2}$, and hence one value of k for spherical mirrors generally: namely $-\tfrac{1}{8}$.

The importance of focal ratio in connexion with spherical aberration is revealed by the following expression (arrived at empirically from the examination of the focal and extrafocal images) for the limit of tolerance of longitudinal aberration:

$$\Delta f_{max} = 0.0089 \left(\frac{F}{D}\right)^2 \text{ (all quantities measured in mm).}$$

Hence the precision with which the aberration need be corrected falls off rapidly with decreasing relative aperture:

	mm
$f/1$	$\Delta f_{max}= 0.009$
$f/3$	0.08
$f/5$	0.22
$f/7$	0.44
$f/10$	0.89
$f/15$	2.0
$f/20$	3.6

If we no longer take f as unity, expression (c) becomes

$$\Delta f = k.\frac{D^2}{4f}$$

If ω is the angle ZCP subtended between the peripheral ray ZC and the axis, and ξ the lateral spherical aberration AF_0, then in triangle CF_0A,

$$\omega = \frac{\xi}{\Delta f}$$

and in triangle ZCP,

$$\omega = \frac{r}{f} = \frac{D}{2f}$$

Combining the last three expressions, we have

$$\xi = k.\frac{D^3}{8f^2} \quad \dots\dots\dots\dots\dots\dots (b)$$

46

Correction in the case of a spherical mirror (for which, as we have seen, $k=-\frac{1}{8}$) is obtained by deepening the centre of the curve, a process known

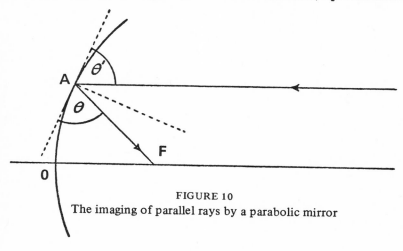

FIGURE 10

The imaging of parallel rays by a parabolic mirror

as figuring, so as to convert the sphere into a paraboloid. For it is clear that the ray, after reflection, must pass through the focus F, and that the curve must be such that θ and θ' (Figure 10) must be equal for any position of the point of incidence, A; the curve defined by this condition is the parabola. The spherical aberration of a parabolic mirror is zero so long as the object is at infinity. For objects nearer than this it is spherically over-corrected. If v is the distance of the image from the mirror, f the focal length, and x the distance of the point of incidence from the pole of the mirror, then the longitudinal spherical aberration is given by

$$\Delta v = \frac{x^2}{2f^2}(v-f)$$

Lenses behave analogously, converging lenses being in general under-corrected and diverging lenses over-corrected; in the latter case the apex of the caustic is towards the objective. By juggling with the radii of curvature of the two surfaces of a lens it is possible to reduce the spherical aberration to a minimum without departing from spherical surfaces. In the case of a converging lens made of glass whose $N=1\cdot50$, the aberration is minimal when $r_2=6r_1$—i.e. one surface is six times more strongly curved than the other, and turned towards the object; such a lens is termed a crossed lens. No manipulation of r_1 and r_2, however, can completely suppress the aberration in a single lens. Complete freedom from spherical aberration involves the combination of two components, one concave and the other convex, the total refraction being shared as equally as possible between the four surfaces. Although the total curvature* of the two lenses is imposed by the necessity of achromatism (see section 4.8), there is an infinite number of individual curvatures to choose from, such that the spherical over-correction of the one is compensated by the under-correction of the other. If the components are in contact, however, this compensation can occur only for a given wavelength. In the case of visual objectives the selected wavelength usually lies near 560 mm in the yellow-green; the red is under-corrected and the violet over-corrected. In the plane containing the best image—namely, that in which the yellow wavelengths are focused—the out-of-focus violet rays of the secondary spectrum will form a halo surrounding the yellow image. It is therefore usual to establish chromatic correction for rays passing through a zone of 0·7 of the full aperture of the objective instead of for the paraxial region. Rays in the paraxial region then exhibit chromatic under-correction, while rays from the marginal zone will be chromatically over-corrected.

The chromatic variation of spherical aberration can be made zero by the two components of the objective assuming a meniscus shape (the Gauss objective), or by introducing a properly-chosen air gap between the two components (as in some Clark objectives, section 5.3). Such a correction is hardly worthwhile in ordinary doublets, where the error will be swamped by the effects of the inevitable secondary spectrum. It can, however, be used to advantage in Cooke photovisual objectives, where the secondary spectrum is (theoretically) zero, leaving only a small tertiary spectrum (section 5.4).

The adequate correction of spherical aberration is of the first importance in the case of telescopic objectives. Photographic lenses, on the other hand, often exhibit a certain amount of residual spherical aberration, since this is less harmful than other aberrations which cannot be removed simultaneously with it.

4.4 Zonal aberration

It has been stated previously (section 4.3) that spherical aberration is a function of the square of the distance of the incident ray from the axis. This is a first approximation (as are, indeed, all the Seidel errors), and in

---

*The total curvature of a lens is given by $C = c_1 - c_2$, where the individual curvatures of the two surfaces are the reciprocals of their radii of curvature, r_1 and r_2. r is positive if the surface is convex to the incident light, negative if concave, and the reciprocals take the same sign. Hence for a biconvex lens $C = c_1 + c_2$ (r_2 and hence c_2 being negative), and for a biconcave, $C = -c_1 - c_2$.

any actual optical system a truer representation of spherical aberration is an infinite series:

$$\Delta F = k_1 Y^2 + k_2 Y^4 + k_3 Y^6 \dots$$

where k_1 is the Seidel coefficient and k_2, k_3, etc. are the coefficients of secondary, tertiary, etc. spherical aberration. Limiting the series to the first two terms, if k_1 and k_2 are of opposite sign, it is always possible to find a value of Y for which $\Delta F = 0$. If this value be Y_0, then

$$Y_0 = \sqrt{(K_1/k_2)}.$$

ΔF will not be zero for other values of Y; hence spherical aberration will be corrected only for a zone of diameter $2Y_0$. The aberration of rays passing through zones other than this corrected zone is referred to as zonal aberration.

Spherical correction is usually designed to be correct for the marginal zone (and in 'brightest light' for a visual objective), and maximum residual aberration is then found at a zone of about 0·7 of the full aperture. In objectives of diameter larger than about 10 cm, the zonal aberration may be corrected by figuring.

This residual aberration should not be confused with the 'zones' of high or low glass produced on the surfaces of objectives by incorrect polishing techniques. This is not a common defect of optical components that have been passed by reputable opticians, though the amateur who makes his own paraboloids will probably be familiar enough with it.

4.5 Coma

Coma is grouped with astigmatism, field curvature, and distortion as the extra-axial aberrations: they all characterise rays which before reflection or refraction are not parallel to the optical axis (i.e. are proceeding from extra-axial objects); hence the greater the angle θ between the axis and the incident ray (consequently the greater the distance H of the image from F, the centre of the focal plane), the greater will be the deformation of the image.

Thus if H is the distance of the image from F (Figure 11), then the extent of the coma increases as H increases. Coma can therefore occur (a) when the aperture is large compared with the focal length—it in fact varies with the square of the relative aperture, (b) when through maladjustment the plane of the objective is not normal to the line joining its centre and the centre of the ocular, or in other words when the optical axes of objective and ocular are not coincident.

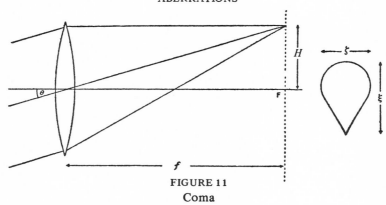

FIGURE 11

Coma

An objective corrected for spherical aberration will focus paraxial rays from a point object not on the axis to a point image, but rays passing through the outer zones will be focused to disc images. The positions of the centres of the discs, and their diameters, will depend upon the diameters of the zones of the objective forming them. The radius of each of these disc images will be half the distance separating its centre from the point image, thus producing the characteristic pear-shaped comatic image, brightest at its 60° apex (the point image) and fading rapidly outward (Figure 11). The major axis of the image will, if produced, pass through F, and ξ is a function of H; if the apex of the image is directed towards F, the condition is termed external coma; if towards the edge of the field, internal coma.

The dimensions of the comatic image of an extra-axial point object are given by

$$\xi = 3B\left(\frac{D}{2f}\right)^2 \tan \theta \quad \ldots \ldots \ldots \ldots \ldots \quad (c)$$

$$\zeta = 2B\left(\frac{D}{2f}\right)^2 \tan \theta \quad \ldots \ldots \ldots \ldots \ldots \quad (d)$$

B = the aberrational coefficient of the objective.

Coma is thus a function of the distance of the image from the optical axis (increasing as $\tan \theta$) and of the relative aperture (increasing as its square). It follows that the coma at a given distance from F can be reduced by stopping down the objective, thus reducing the relative aperture; and, alternatively, that a larger field with the same degree of coma can be obtained.

Coma is not usually observed in its pure form in the telescope. First, because other aberrations, notably astigmatism, are usually present;

50

second, because the incident rays are also inclined at an angle to the axis of the ocular, so that its own coma contributes to the total observed effect (although, in a properly designed objective, eyepiece coma is reduced to become almost insignificant); and third, because in any case the image of a point source is not a point but a diffraction pattern upon which the coma is superimposed. If an extra-axial stellar image is observed with a sufficiently large focal ratio, it will be coma-free; if the aperture is then increased, the following progressive modifications will be observed: (*a*) the rings on one side of the spurious disc will become just perceptibly brighter than those on the other side; this occurs when ξ has increased to about the radius of the spurious disc; (*b*) when ξ is four times the radius of the disc, this inequality of brightness will become most marked in the inner ring, while the outer rings may be reduced to arcs on the bright side of the disc only; (*c*) by the time ξ has increased to ten times the radius of the disc, all the rings will be discontinuous, their remaining arcs outlining the comatic image on one side of the disc, while the region of greatest brightness of the disc will itself be displaced from the centre. In the extrafocal image the rings will still be visible in their entirety, though they will be eccentrically displaced.

Resolution is not affected by coma so long as ξ does not exceed about four times the radius of the spurious disc. Measures of position, on the other hand, are affected as soon as the first trace of asymmetry makes itself perceptible in the image. It is this asymmetrical nature of the comatic image that makes the aberration particularly troublesome (cf. astigmatism, for instance), since it is impossible to determine accurately the true position of the point image within the limits of the comatic image.

Ordinary visual observation is little, if at all, interfered with by the extra-axial aberrations, owing to the comparative restriction of the usual visual field; and providing the objective is at least of average quality, and is correctly squared on, coma will be confined to the outer edges of the field, using normal magnifications. The off-axis aberrations assume great importance in photographic work, on the other hand, as also in micrometer measures; for such work rigorous correction of coma is required.

If an objective is corrected for spherical aberration (Figure 12), the coma coefficient B will be zero for all values of Y when

$$\frac{Y}{\sin \theta} = \text{constant} \quad \dots \dots \dots \dots \dots \dots (e)$$

Whence it follows that points *a*, *b*, *c*, . . . must lie on the surface of a sphere whose centre is the focus F: $aF = bF = cF = dF = eF$. Where the sine condition is not observed, the effect is equivalent to a non-uniform magnification across the object from the centre outwards.

It can be seen, therefore, that the spherical mirror, which has no unique axis if the aperture stop is at its centre of curvature, is coma-free;

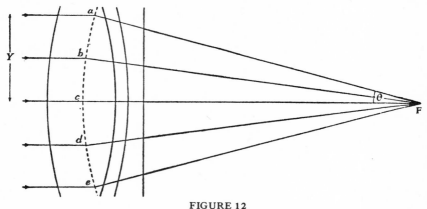

FIGURE 12

The coma-free condition

if its spherical aberration could be eliminated it would be an ideal objective, as was demonstrated by Schmidt (see section 9.12).

The sine condition cannot be rigidly observed by a cemented doublet, which consequently tends to suffer more from extra-axial aberration than a separated-lens doublet. An objective of the latter type, corrected for spherical aberration and at the same time obeying the sine condition, is often termed aplanat.

Neither can the sine condition be fulfilled by a single parabolic mirror, which is therefore subject to coma (Figure 13):

since $$aF \neq bF \neq cF$$

and $$\frac{Y}{\sin \theta} = aF$$

$$\therefore \quad \frac{Y}{\sin \theta} \neq \text{constant}$$

and the sine condition is not fulfilled; it is made possible, however, by the introduction of a second reflecting surface, a device employed in the two-mirror aplanatic telescope (section 9.9). In terms of transverse magnification—which we have seen must be constant if the sine condition is obeyed—the magnification given by the zone of radius x of a paraboloid is a function of aF; also, the magnification given by the axial region of the mirror is a function of PF. But $PF \neq aF$; on the contrary, $PF = aF + a'a$. Since $a'a = \dfrac{x^2}{4f}$, it may be very considerable for large values of x, and the coma correspondingly pronounced.

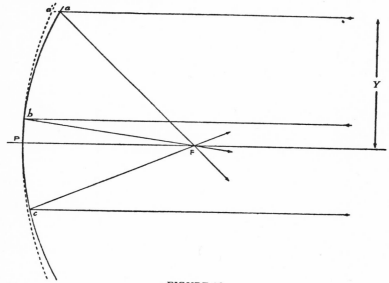

FIGURE 13
Coma in a parabolic mirror

The coma in large modern reflecting telescopes, some of which have aperture ratios as fast as $f/2 \cdot 4$, would limit their photographic fields prohibitively were it not for the use of various types of corrector lenses, placed in the beam near the focus. These lenses can not only correct for coma, but also provide a flat field while introducing negligible spherical and chromatic aberration. Such corrector lenses, pioneered by Ross (B. 3.16), have been developed by Wynne, Gascoigne, and others (B. 15, 17, 26).

The aberrational constant, B is $+0 \cdot 025$ in the case of a Newtonian paraboloid, the coma being external. From (c) we have, therefore,

$$\xi = \tfrac{3}{4}\left(\frac{D}{2f}\right)^2 \tan \theta$$

Coma becomes serious—i.e. visibly distorts images within a few minutes of arc of the axis—at about $f/3$, the practicable limit of relative aperture with a Newtonian.

4.6 Astigmatism and field curvature

An aberration characterising the refracted or reflected pencil from any surface of revolution at which the incident pencil is not normal. Unlike coma and spherical aberration, therefore, the pencils emergent from an objective of small aperture (or from the central region only of a large objective) may be astigmatic, provided only that the incident pencils were oblique. The amount of the aberration will however be greater when the used aperture is large than when it is small.

Astigmatism will also be inherent in an objective whose surface (or surfaces) is not a surface of revolution. If, for example, the rays from the two ends of one diameter of the objective are refracted or reflected to a shorter focus than those from the ends of the diameter perpendicular to the first, then a well-defined image will be found in no single plane.

An astigmatic ray from a point source comes to two foci, neither corresponding geometrically to the object, at different distances from the objective. Midway between them is located the least circle of confusion, or focal circle, but in no position is a geometrical duplicate of the object to be found. If a screen is interposed in the path of the pencil shown in Figure 14, the succession of images caught on it as it is moved away from the lens will be as follows:*

From lens to *T*: an ellipse of decreasing size and increasing eccentricity, its minor axis lying in the vertical plane.
At *T*: the ellipse has fined away to a short straight line, lying in a horizontal plane; it is known as the tangental focal line.

FIGURE 14
Astigmatism

From *T* to *C*: the focal line expands into an ellipse of decreasing eccentricity.
At *C*: it is a small circular disc, the circle of least confusion, whose diameter is half the length of each of the focal lines (*T* and *S*). Though

*The appearance of astigmatic stellar (point) images will be described below.

blurred, the image bears the closest geometrical resemblance to the object.

From C to S: the circle elongates in a direction perpendicular to the tangental focal line.

At S: it has narrowed to a second focal line (the sagittal focal line) lying in a plane perpendicular to that containing the tangental focal line (i.e. vertical).

From S outwards: an increasingly large ellipse, whose major axis is parallel to the sagittal focal line.

The distance ST is termed the astigmatic difference, or longitudinal astigmatism of the lens. In the diagram the lens is represented as being tilted forward in a vertical plane, and the incident pencil as lying in a horizontal plane: then the sagittal focal line, the optical axis of the lens, and the incident ray are contained in a single vertical plane; the tangental focal line and the incident ray are contained in a single horizontal plane.

In Figure 15, OA is, as before, the optical axis of the lens; XY is the axis of an oblique pencil yielding a tangental focal line, a focal circle, and a sagittal focal line at T, C, and S respectively. Then T is contained in a spherical surface whose curvature

FIGURE 15
Field curvature

$(1/r)$ we will call τ, and S in a spherical surface whose curvature will be denoted by σ; C will lie on a sphere of intermediate curvature which is termed the curvature of the field. All these spheres are tangental to the focal plane, GH, at F. Where θ is small, we may write

$$\left.\begin{array}{l} TQ=\dfrac{\tau}{2}\cdot\tan^2\theta \\[2mm] SQ=\dfrac{\sigma}{2}\cdot\tan^2\theta \end{array}\right\} \quad \cdots\cdots\cdots\cdots\cdots \; (f)$$

whence the longitudinal astigmatism

$$ST=\frac{\sigma-\tau}{2}\cdot\tan^2\theta$$

from which it can be seen that it increases very rapidly with increasing inclination of the incident ray. The displacement of the focal circle from the focal plane is

$$CQ=\frac{\sigma+\tau}{2}\cdot\frac{\tan^2\theta}{2}$$

The radius of this circle is given by $r=\frac{D}{2f}\cdot\frac{ST}{2}$, whence its diameter is

$$d=\frac{D}{2f}\cdot\frac{\sigma-\tau}{2}\cdot\tan^2\theta$$

The image formed in the focal plane by the rays passing through this focal circle will be an ellipse whose radial and tangental axes will be respectively (using (f))

$$d_r=\frac{D}{2f}\cdot\tau\cdot\tan^2\theta$$

$$d_t=\frac{D}{2f}\cdot\sigma\cdot\tan^2\theta$$

where d_r is astigmatic and d_t is a function of field curvature, $\frac{\sigma-\tau}{2}$ being termed the coefficient of astigmatism, and $\frac{\sigma+\tau}{2}$ the curvature of the field. If we denote the former by A, and the latter by C, the last five expressions give

$$\left. \begin{aligned} ST&=A\cdot\tan^2\theta\\ CQ&=C\cdot\frac{\tan^2\theta}{2}\\ d&=A\cdot\frac{D}{2f}\cdot\tan^2\theta\\ d_r&=(C-A)\cdot\frac{D}{2f}\cdot\tan^2\theta\\ d_t&=(C+A)\cdot\frac{D}{2f}\cdot\tan^2\theta \end{aligned} \right\} \quad \dots\dots\dots\dots(g)$$

The extrafocal images are distorted but (cf. coma) symmetrical, being ellipses whose major axes are parallel to the nearest focal line. It is for this reason that astigmatism is a less objectionable error than coma. In its pure form, astigmatism can only be observed in aplanatic instruments or when the aperture is heavily stopped down (spherical aberration and coma being functions of D^3 and D^2 respectively).

Below a certain value of ST the focal circle is indistinguishable from a normal spurious disc, and the astigmatism is tolerable. If, in an astigmatic wavefront, the focal plane of the ocular is made to coincide with either of the focal lines, the astigmatism can be detected when the extreme departure (Δx) of the two components $= 0 \cdot 1 \lambda$: the rings will be just perceptibly brighter at the two ends of a diameter, which defines either the tangential or the sagittal axis. When the focal plane of the ocular coincides with that containing the least circle of confusion, Δx can be about $0 \cdot 2 \lambda$ before there is perceptible distortion of the image; the rings will then be brightest at the ends of two mutually perpendicular diameters. If Δx is increased to $0 \cdot 5 \lambda$, the image becomes definitely cruciform; by $\Delta x = 0 \cdot 25 \lambda$ the resolving power of the objective begins to deteriorate.

It can be shown empirically that

$$ST = 8 \cdot \left(\frac{f}{D} \right)^2 . \Delta x$$

and taking $0 \cdot 25 \lambda$ as the limiting value for Xx, we have for the limit of tolerance of longitudinal astigmatism

$$ST_{max} = 2 \cdot \left(\frac{f}{D} \right)^2 . \lambda$$

A range of values of this quantity, taking $\lambda = 560$ nm, is tabulated below:

		mm
$f/1$	$ST_{max} =$	0·0011
$f/3$		0·010
$f/5$		0·028
$f/7$		0·055
$f/10$		0·112
$f/15$		0·252
$f/20$		0·448

Single optical surfaces are incapable of yielding a flat image: i.e. the surface on which the best images lie is a sphere tangent to the focal plane at the point where the optical axis intersects it. In astigmatism-free systems such as the Couder telescope (section 9.9.3), FT and FS are made coincident ($\sigma = \tau$), but the focal surface is still curved.

The condition for minimum field curvature in the case of two lenses in contact is

$$N_1 f_1 = -N_2 f_2$$

57

The coefficients of astigmatism and field curvature, A and C, are independent of the form of the lens, and are the same for all achromatic doublets made of the same types of crown and flint. The values of σ and τ for a typical astronomical doublet are $-1\cdot73$ and $-3\cdot85$ ($A=+1\cdot06$ and $C=-2\cdot79$), the corresponding radii of curvature being $0\cdot578f$ and $0\cdot026f$. The radius of curvature of the surface containing the least circles of confusion is $0\cdot37f$. Hence the best images lie on a spherical surface, concave to the objective, whose radius of curvature is approximately one-third of the focal length.

In the case of a Newtonian reflector (paraboloid), the values of A and C are respectively $+1$ and -1.

Comparing expressions (c) and (d) with (g) it can be seen that the relative importance of astigmatism and coma depends upon the relative aperture of the instrument and the distance of the image from the axis:

(a) As the focal ratio is increased, the distortion of the image due to astigmatism decreases more slowly than that due to coma; hence at large focal ratios coma is less important relative to astigmatism than at small focal ratios, and vice versa.

(b) As the distance of the image from the centre of the field increases, the element of its total distortion that is due to astigmatism will gradually swamp that due to coma.

These relations are illustrated in the following Table:

$\theta=$	15′			30′			1°			2°		
	Astig	Coma	$A+C$	A	C	$A+C$	A	C	$A+C$	A	C	$A+C$
$f/3$	$1\rlap{.}''3$	$18\rlap{.}''7$	$19\rlap{.}''3$	$5\rlap{.}''0$	$37\rlap{.}''4$	$39\rlap{.}''9$	$20\rlap{.}''0$	$74\rlap{.}''9$	$84\rlap{.}''9$	$83\rlap{.}''8$	$140\rlap{.}''8$	$191\rlap{.}''7$
$f/10$	$0\cdot4$	$1\cdot7$	$1\cdot9$	$1\cdot6$	$3\cdot4$	$4\cdot2$	$6\cdot3$	$6\cdot8$	$10\cdot0$	$25\cdot2$	$13\cdot6$	$26\cdot2$

If astigmatism predominates, definition at the edge of the field can be superficially improved by increasing the magnification and hence reducing the field of view; if coma predominates, magnification is no help. But in normal astronomical instruments the two are of roughly equal importance, and the image distorted by both astigmatism and coma is not markedly different from one exhibiting either aberration in the pure form. In Figure 16 are plotted the values of the longitudinal axes (which is never less than $1\frac{1}{2}$ times the minor axis) of the combined astigmatic and comatic images formed by parabolic mirrors of different focal ratios at different distances from their optical axes.

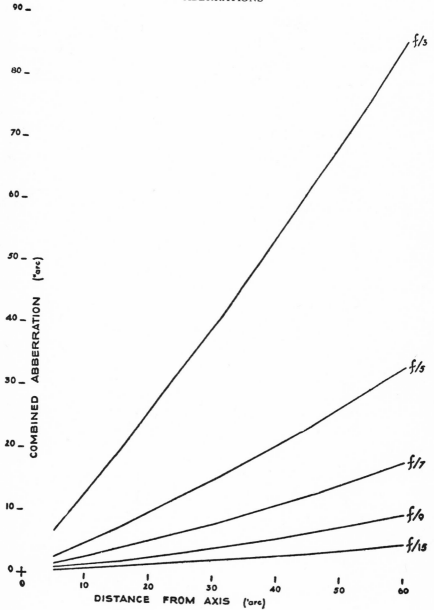

FIGURE 16
Magnitude of combined astigmatism and coma

Neither astigmatism nor field curvature plays an important part in the performance of visual objectives, owing to the comparatively restricted fields that are used; as with coma, however, they are vitally important in the case of wide-field photographic objectives, or those designed for accurate measures of position, and in eyepieces.

4.7 Distortion

If the aperture of a lens system is made very small, all the aforementioned aberrations will have a negligible effect on the image, so that, disregarding diffraction effects, the image will indeed be well defined. It may not, however, be in its ideal position in the field; that is to say, the optical system may suffer from distortion. In a distortion-free system, the magnification M is uniform throughout its field; if an object point is distant H_0 from the optical axis, the distance of the image point from the axis is given by

$$H_0' = MH_0 \quad \ldots\ldots\ldots\ldots\ldots\ldots\ldots \text{ '}(h)$$

and this is true for all points in the field. If, however, distortion is present, the simple expression becomes

$$H_0' + \Delta H_0' = MH_0 + EH_0^3,$$

and, therefore, the displacement due to the distortion term is

$$\Delta H_0' = EH_0^3$$

where E is the distortion coefficient.

If E is positive, a square object will be imaged as a pincushion shape of a size such that it lies outside the ideal square image, but if E is negative the image becomes barrel-shaped and lies wholly within the ideal image.

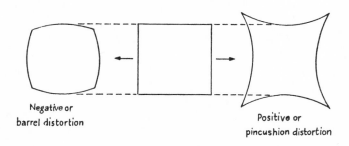

Negative or
barrel distortion

Positive or
pincushion distortion

FIGURE 17
Alternative forms of distortion

60

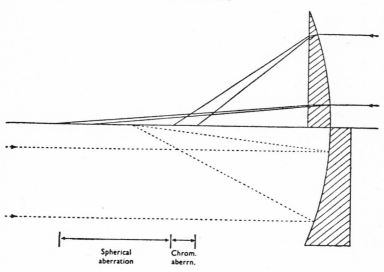

Spherical
aberration

Chrom.
aberrn.

FIGURE 18

Lens and paraboloid compared in respect of spherical
and chromatic aberration (*after F. G. Pease*)

4.8 Chromatic aberration

Chromatic aberration being an effect of refraction, mirrors and all optical
systems consisting wholly of mirrors are free from it. In the case of lenses
it can be effectively reduced by combining components in such a way
that their individual aberrations tend to cancel one another out. It can in
practice never be wholly removed, however, the residual aberration of a
technically 'achromatic' two-lens combination being known as secondary
spectrum.

Since the refractive index of glass increases with the wavelength of the
light with which it is measured, images in different wavelengths are formed
at different positions along the axis of the objective.* Thus a converging
lens (Figure 19) brings the violet rays to a shorter focus than the red; and a
white object will produce a series of sharply focused monochromatic
images strung out along the axis. Thus two dimensions of chromatic
aberration can be distinguished (analogous with longitudinal and lateral
spherical aberration)—aberration of position, and aberration of size; these

---

*The difference of size of the monochromatic images is a matter of less importance
than their formation in different planes.

are known respectively as longitudinal aberration, and transverse chromatic aberration.

The images formed by a fully achromatic objective, on the contrary, are all of the same size and are formed in the same position, irrespective of their wavelength. This condition can only be provided by a mirror, though it can be closely approached by three-component combinations. The importance of chromatic aberration resides in the fact that any residual chromatism sets a limit to the objective's performance even at the centre of the field (cf. the extra-axial aberrations).

Writing f_R, f_B, and f_Y for the focal lengths of a biconvex lens measured in radiations of the wavelengths of the $C, F,$ and D lines* in the red, blue, and yellow respectively, we have as a measure of the longitudinal aberration

$$\frac{f_R-f_B}{f_Y}=\frac{\Delta f}{f} \quad\dots\dots\dots\dots\dots\dots (i)$$

And the dispersive power of the lens for these particular radiations (effectively the visual range) is given by

$$\frac{\Delta f}{f}=\frac{N_B-N_R}{N_Y-1}=\frac{1}{\nu} \quad\dots\dots\dots\dots\dots (j)$$

where $\nu=\frac{f}{\Delta f}$ is known as the constringence of the glass. Its value for crown glasses averages about 60; for flint, about 36.

If C is the total curvature (c_1-c_2) of the lens, then

$$C=\frac{1}{f}\cdot\frac{1}{N-1}$$

whence

$$f=\frac{1}{C(N-1)} \quad\dots\dots\dots\dots\dots (k)$$

Taking f_Y as unity,

$$C=\frac{1}{N_Y-1}$$

whence, taking N_Y for crown glass as 1·5179,

$$C=1\cdot93087 \quad\dots\dots\dots\dots\dots (l)$$

Taking the values of N_R and N_B as 1·5153 and 1·5241, we have for the longitudinal chromatic aberration of a simple biconvex lens, from (i), (k) and (l)

$$f_R-f_B=\frac{1}{C(N_R-1)}-\frac{1}{C(N_B-1)}\cdot f_Y$$

*656·3 nm, 486·1 nm, 589·0 nm.

$$= \frac{1}{1 \cdot 93087(1 \cdot 5153 - 1)} - \frac{1}{1 \cdot 93087(1 \cdot 5241 - 1)} \cdot f_Y$$
$$= 0 \cdot 0169 f_Y*$$

and from (j), $v = 59 \cdot 2$

If a screen is moved along the axis, the succession of images of a poly-chromatic point source will be as follows (Figure 19):

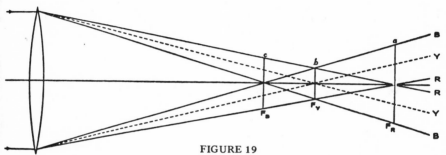

FIGURE 19

Formation of chromatic images by a single lens

At a: The red focus; hence the image will include a central red point. But the out-of-focus blue rays will form a circular blue disc in which it is centrally placed.

From a to b: The whole disc shrinks, position b being the least circle of confusion; at the same time the central point expands and fades as the screen moves away from F_R.

At b: The red and blue discs are here coincident, since F_Y is the focus of neither the red nor the blue rays. The diameter (minimal) of the image is

$$d = \frac{D}{2v}$$

where D is the aperture of the lens, and v the constringence.

From b to c: The disc expands again, the blue component at the same time shrinking, so that the whole image in expanding grows a red border.

$* f_R: +0 \cdot 0050 \brace f_B: -0 \cdot 0120 } \times f_Y$

63

At c: The blue rays being at their focus, the image will consist of a circular disc with a red outer fringe and a central blue point. It is twice as large as the least circle of confusion, or

$$d=\frac{D}{\nu}$$

In the case of a compound optical instrument consisting of an objective and an ocular, the eye automatically selects (by directing the hand focusing the instrument) that image in the yellow to which it is most sensitive; i.e. the ocular is adjusted so that its yellow focus coincides with the yellow focus of the objective. This yellow image will be blurred by the out-of-focus images in all the other colours.

Moving the ocular in from F_Y: blue image, edged with red, most sharply defined when the blue focus of the ocular coincides with F_B.

Moving the ocular out from F_Y: red image, edged with blue, most sharply defined when the red focus of the ocular coincides with F_R.

Hence the longitudinal aberration of the telescope is the sum of the longitudinal aberrations of objective and ocular: that is to say, the distance the ocular must be moved along the optical axis from the position at which the sharpest blue image is seen to that at which the sharpest red image is seen.

The relative contributions of objective and ocular to the total aberration are a function of their focal lengths (measured at the yellow focus, the mean position): writing f'_R and f'_B for the ocular, we have

$$\frac{f_R-f_B}{f'_R-f'_B}=\frac{F}{f}$$

Hence it is that the aberration of the ocular becomes increasingly troublesome when very low magnifications are used.

The possibility of constructing a compound lens that approximates to achromatism springs from the fact that flint and crown glasses are unequally dispersive ($\nu \simeq 60$ and 36 respectively), while their powers of refraction (1·6 and 1·5 respectively) are approximately the same. Hence a flint lens is capable of neutralising to a large extent the colour dispersion of a crown lens (its aberration being of the opposite sign) without similarly neutralising their power of refracting the incident light to a focus (Figure 20). Complete neutralisation of the colour error is not possible with only two components since it would involve the use either of insufficiently durable types of glass or of curvatures of such strength as to make their manufacture very difficult and therefore prohibitively expensive.

Given that the combined dispersions are to be zero, and that the

FIGURE 20
Dispersion and re-combination of light
in an achromatic objective

combined focal length is unity, it is possible to calculate the values of C_1 and C_2, and the powers of the two lenses, in terms of their constringencies and refractive indices. For most astronomical achromatic objectives, C_1 is about $2\frac{1}{2}$ times that of a single lens of the same focal length as the doublet. Also the focal length of the concave component is about $1\cdot7$ times that of the convex, since to secure achromatism the condition must be satisfied that

$$\frac{v_1}{v_2} = \frac{f_1}{f_2}$$

The achromatic doublet does not provide complete colour correction; the dispersion of the combination only approximates to zero. What it does is to bring two colours to the same focus, and the remainder nearly so (Figure 21). The residual colour (about 5 per cent) visible at the best focus is the so-called secondary spectrum. The doublet achieves this by doubling the primary spectrum back on itself, and by correct selection of the coinciding wavelengths it can be arranged that the wavelength selected for use occurs at the point of minimum focus where the spectrum turns over, and where there is consequently a minimum displacement per wavelength along the axis.

With the correction most commonly employed for visual objectives nowadays, the radiations of the F line in the blue and of the C line in the deep red are brought to the same focus, leaving the yellow rays to which the eye is most sensitive (about 560 nm) close to the minimum focus at 555 nm (Figure 21b). In practice, all degrees of correction between $C=F$ and $B=F$ are to be encountered among the best objectives of various makers.

The longitudinal aberration of a typical doublet of this sort is

65

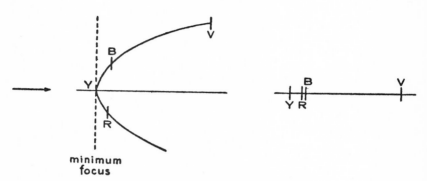

(a) Arrangement of foci along the axis of an
uncorrected biconvex lens

(b) Arrangement of foci of a visual doublet.
(Scale 5 times that of (a))

(c) Alternative colour correction of an achromatic doublet.
(Scale 5 times that of (a))

(d) Photographic correction.
(Scale 5 times that of (a))

(e) Apochromatic correction.
(Scale 25 times that of (a))

FIGURE 21
Different corrections of chromatic aberration

$$\left.\begin{array}{l} f_R: +1 \cdot 00050 \\ f_B: +1 \cdot 00051 \end{array}\right\} \times f_Y$$

and
$$\Delta f = 0 \cdot 0005 f$$

A slightly different correction brings the red focus nearer to, and the blue further from, the minimum focus (Figure 21c).

For a given aperture the trouble experienced from secondary colour increases as the focal ratio is reduced. Taking four times the Rayleigh limit as the greatest tolerable discrepancy between the combined foci and the minimum visible focus (since the former are red and blue, which are little visible compared with the latter, yellow), the minimum permissible focal length of a visual doublet of aperture D is quickly derived:

From p.57,

$$\Delta f = 8 \left(\frac{f}{D}\right)^2 \lambda$$

Also we have just seen that
$$\Delta f = 0 \cdot 0005 f$$

Hence
$$f = 16,000 \left(\frac{f}{D}\right)^2 \lambda$$

Taking $\lambda = 550$ nm $= 0 \cdot 00055$ mm,

$$f \not< 1 \cdot 14 D^2$$

or, roughly,
$$f \not< 1 \cdot 2 D^2$$

Conrady (B. 3.6) gives a rather stricter tolerance, with a recommended minimum focal length of $2D^2$. These formulae therefore give the following limiting focal ratios and lengths:

D (cm)	Sidgwick $(1 \cdot 2D^2)$		Conrady $(2D^2)$	
	$f/$	F (cm)	$f/$	F (cm)
5·0	6	30	10	50
7·5	9	67·5	15	112
10·0	12	120	20	200
12·5	15	188	25	312
15·0	18	270	30	450
20·0	24	480	40	800
25·0	30	750	50	1250
30·0	36	1080	60	1800

*For a single converging lens, primary colour becomes intolerable at about $f = 40D^2$, which was one reason for the enormous length and unwieldiness of the seventeenth-century refractors.

Since for practical reasons (stability, ease of movement, etc) $f/20$ is about the limit for astronomical telescopes, it somewhat surprisingly follows that secondary colour in excess of the tolerable limit will be encountered with all doublets whose aperture is greater than about 15 cm. In this connexion it is of interest that the $H\alpha$ and D_2 foci of the Lick refractor, whose objective is an 'achromatic' doublet, are 81·5 mm apart.

The theoretical distribution of secondary colour in the images of a visually corrected doublet is as follows: at the minimum focus (i.e. *the* focus, visually) a white star will show a yellowish diffraction pattern with little surrounding colour; the image of a very bright object will be surrounded by a halo of out-of-focus secondary colour, bluish or possibly with a tinge of purple; inside the focus a red fringe will develop as the rings proliferate; just outside the focus a minute red point will develop at the centre of the diffraction pattern; further out this vanishes again, and the whole diffraction pattern will be suffused with blue 'flare'. If the objective is under-corrected, the image at the focus will be surrounded by a red fringe which becomes increasingly conspicuous as the ocular is racked in; if it is over-corrected, no red fringe will develop round the intrafocal image.

The curves of object glasses are calculated to give optimum colour correction with a particular magnification, usually in the neighbourhood of $25D$ to $30D$ with instruments of moderate aperture. Since the natural under-correction of the ocular and the eye is exaggerated when the diameter of the pencil with which they are dealing is increased, it follows that the degree of over-correction in the objective necessary to counteract this under-correction at about $25D$ will be insufficient to do so at lower magnifications. With low magnifications, therefore, an objective will appear to be under-corrected, and (though this is much less noticeable or important) with very high magnifications, over-corrected.

The observational effects of secondary colour are important, and not always appreciated. It is not merely a question of falsification of colour, but of light grasp and resolution also:

(*a*) Light grasp is reduced by the secondary spectrum:* light which should be contributing to the intensity of the image on the retina is spread over a wider area. One can easily demonstrate the effect of such dilution of the image by selecting a star in the telescopic field which is just on the threshold of vision, and then throwing it out of focus, when it will disappear altogether.

(*b*) The contrast of an extended image is reduced. Dark detail on a

*See further B. 3.22.

bright ground, for example, is tinted blue by the overlapping unfocused fringes of the images of all the bright adjacent points, and tends to be lost. If there is plenty of illumination, so that a slight loss is of no importance, contrast can be improved by a pale red filter, which absorbs the out-of-focus blue and violet radiations. The freedom of reflectors and photovisuals from this 'image haze' gives them a decided advantage over achromatics in planetary work.

The only way a visually corrected objective can be used for photography is to employ a panchromatic plate at the visual focus, and eliminate the out-of-focus rays with a yellow filter. This involves a considerable light loss, and is, at best, a makeshift arrangement. Objectives for purely photographic use are, therefore, corrected for the colour to which the emulsion is most sensitive: normally blue and violet, but occasionally red. For 'blue' correction, f_B and f_V (the latter about 405 nm, and practically invisible to the eye) are brought into coincidence, leaving the G' line, in the region of greatest sensitivity of ordinary plates (about 435 nm), at the minimum focus (Figure 21(d)); for example:

$$\left. \begin{array}{l} f_R: +1{\cdot}0037 \\ f_Y: +1{\cdot}0024 \\ f_B: +1{\cdot}0005 \\ f_V: +1{\cdot}0005 \end{array} \right\} \times f_{G'}$$

An alternative, giving a compromise between the extremes of 'yellow' and 'blue' correction, would have the minimum focus in the yellow-green, around 490 nm, near the F line. In this case, the focal lengths for the different colours would be approximately:

$$\left. \begin{array}{l} f_R: +1{\cdot}002 \\ f_Y: +1{\cdot}0004 \\ f_V: +1{\cdot}001 \end{array} \right\} \times f_B$$

By using specially-selected combinations of crown and flint glass with fairly small differences of constringence, the focal lengths for the extreme wavelengths of the spectrum can be brought to much closer coincidence with that of the selected colour. Such lenses, known as semi-apochromats, have roughly one-third of the residual colour of a typical achromat; for example:

$$\left. \begin{array}{l} f_R: +1{\cdot}0003 \\ f_B: +1{\cdot}0002 \end{array} \right\} \times f_Y$$

for a visual correction. To be noticeably superior to an ordinary achromat, the glass used in a semi-apochromat must be very carefully selected;

hence, such object glasses are expensive.

Triple apochromats, consisting of three components, each of a different kind of glass, approach very close to the perfectly achromatic object glass, though secondary colour is never completely eliminated; it can, however, be reduced to about one ninth of that of the ordinary achromat. Typical residuals for visual correction are:

$$\left. \begin{array}{l} f_R: +{\cdot}00025 \\ f_B: +{\cdot}0001 \\ f_V: +{\cdot}0004 \end{array} \right\} \times f_Y$$

An apochromat designed for coincidence of the yellow and violet tints is termed 'photovisual', because it can be focused visually (i.e. in yellow light) for use with a blue-sensitive emulsion.

Apochromatic objectives are extremely expensive, and a good doublet will perform better—both regarding its light grasp and resolving power— than the smaller apochromat that the same money will buy, notwithstanding the secondary colour. Secondary colour, indeed, only becomes obtrusive in a doublet of normal focal ratio when its aperture exceeds about 20 cm. For practical reasons, therefore, and whatever the theoretical optician may have to say on the subject of colour correction in apochromats, the reflector offers the only effective escape from chromatic aberration.

4.9 Transverse chromatic aberration

Equation (h) in section 4.7 assumes that M is a constant. This will be so only in a system composed solely of mirrors. For a refracting system, M is a function of wavelength. Calling this new function $M\lambda$,

$$M\lambda = M + \Delta M$$

where ΔM is the change in M due to the change in wavelength; the change in $H_0{}'$ is given by

$$\Delta H_0{}' = \Delta MH.$$

The image of a point source of white light is therefore a coloured line directed towards the centre of the field and of a length varying with H. This image defect, termed transverse, or lateral, chromatic aberration, is of little importance in telescopic objectives which consist of relatively thin components practically in contact and covering small angular fields; but in the case of eyepieces and photographic objectives, both of which consist of widely-spaced, thick components, and cover relatively wide fields of view, correction of this defect is essential. In the well-known Cooke photographic triplet (section 5.4) the two outer convergent components, by reason of their lying on opposite sides of the aperture stop, make contributions to the transverse chromatic aberration that are opposite in sign; it follows that, by choosing appropriate focal lengths for these components, the objective can be rendered free from this defect. In a telephoto lens, which consists basically of a widely-separated positive and negative component, correction of transverse chromatic aberration is obtained by making each component separately corrected for longitudinal chromatic aberration.

TYPES OF OBJECT GLASS

5.1 Introduction

Among the many types of achromatic objective in use, the following are described briefly in order to indicate the dependence of optical characteristics upon the curvatures of the component surfaces. It is convenient to group them as follows:

A. Doublets with lenses in contact, or with very small air spaces.
B. Doublets with separated lenses.
C. Triplets.
D. Anastigmats (photographic objectives).

In the majority of objectives, the crown component is nearer the source of the incident light. Flint outside is theoretically possible, but the advantages of the former arrangement are that the more durable type of glass is exposed to the external air and to the necessity of more frequent cleaning, and that weaker curves are involved when the converging component is placed first, thus facilitating manufacture and reducing costs.

5.2 Contact doublets

It was shown in section 4.8 that, for an object glass to be nominally achromatic (i.e. corrected to give coincidence of foci for two well-separated colours), the focal lengths of its component lenses must be inversely proportional to their dispersive powers as measured for the interval between the two colours. Typically, flint glass has about 1·7 times the dispersive power of crown glass, and the focal lengths of the components will have to be in this proportion. Thus, in an achromatic combination of 150 cm focal length, writing f_c and f_f for the focal length of the crown and flint components respectively:

$$\frac{1}{f_c} - \frac{1}{f_f} = \frac{1}{150}.$$

But $f_c = f_f/1.7$; hence $f_f = 105.0$ cm, $f_c = 61.8$ cm.

Provided these focal lengths are observed, a nominally achromatic combination, of the correct focal length, will result. However, the cross-sectional form of a single lens of given focal length can have an infinite number of forms, all with different aberrational characteristics. The object-glass designer therefore manipulates the curves on his combination of lenses to give, other things being equal, the smallest residual spherical aberration. He may, however, sacrifice a degree of both spherical and chromatic correction in the interests of securing improved correction for some other aberration, particularly coma, if a wide photographic field is required.

The range of possible object-glass forms is very large, but for a particular choice of crown and flint glass there is, in practice, only one optimum prescription for correction of the two most important aberrations. In this sense, therefore, it is a little misleading to speak of 'types' of object glass defined simply by the distribution of curvature on their surfaces, since the choice of curvature is inextricably linked with the selection of the glass types in the first place. In general, however, it can be said that, even when the axial aberrations (spherical and chromatic) are perfectly corrected, the effect of coma will be more marked in object glasses tending in shape towards the left-hand side of Figure 22, and it will be of the internal type (i.e. the head of the comatic image points towards the edge of the field).

FIGURE 22

The possible cross-sectional forms of a contact doublet

A condition of zero coma will occur towards the right-hand end of the series, where the rear surface of the flint lens has a slightly convex shape. External coma (head pointing inwards) appears as the rear surface becomes still more steeply convex.

The study of the change of image characteristics of a lens or lens combination, whose relative surface curvatures remain constant while the overall form is changed (i.e. maintaining the nominal focal length and chromatic correction), is known by designers as 'bending' a lens.

5.2.1. Doublets with 'broken contact': These are not often manufactured nowadays, but were popular in the nineteenth century because the pronounced concavity of the fourth surface gave the optician two concave faces which could be easily checked for sphericity. Objectives by Cooke were commonly of this type, typical curvatures being:

$$c_1 = +2\cdot9$$
$$c_2 = -2\cdot0$$
$$c_3 = -2\cdot1$$
$$c_4 = +0\cdot4$$

Since $c_2 < c_3$, the components could be mounted with glass-to-glass edge contact, the airspace being widest at the centre. Suffering as it does from strong internal coma, squaring-on tolerances are much more critical than with modern objectives.

FIGURE 23
A typical Cooke doublet

5.2.2 Cementable doublets: If $c_2 = c_3$, the components can be cemented rigidly together with balsam. Most binoculars and other small objectives, up to $D = 70$ mm or so, are of this type, for the following reasons:

Advantages:

(*i*) 8 per cent gain in transmission (see section 6.4).

(*ii*) Increased contrast resulting from the elimination of internal reflections.

(*iii*) Centring, once fixed by the makers, cannot get out of adjustment, thus allowing the use of a simpler type of cell.

(*iv*) The effects of local imperfections of figure are minimised.

(*v*) The production of ghosts is avoided; these (usually about 6 mags fainter than the parent image) result from internal reflection in uncemented combinations.

Disadvantages:

The use of hard-setting cement becomes inadvisable with apertures above about 75 mm which will be used for critical work. Differential expansion of the lenses, owing to their different coefficients of expansion (9·5 and $8·0 \times 10^{-6}$ for crown and flint respectively), threatens to crack the balsam; also the different expansions of two cemented lenses sets up strains within them which may cause double refraction. However, the interspace may be nullified optically by applying a thin film of microscope objective immersion oil between the surfaces whose N should, ideally, be intermediate between that of the two glasses used. The oil is retained between the lenses by a sealing peripheral band, e.g. a ring cut from polythene tubing. Any viscous substance of similar consistency (castor or olive oil, for example) will also serve.

The following prescription is for a typical hard-crown/flint astronomical objective; t_1, t_2 refer to the central thicknesses of the crown and flint components respectively, and all quantities are in terms of the finished focal length of the combination. It will be noticed that the rear surface of the flint lens is slightly concave.

Crown lens (Glass-type Schott K5, $N=1·522$, $v=59·5$)

$$c_1 = +2·503$$
$$c_2 = -2·486$$
$$t_1 = 0·008$$

FIGURE 24
A typical hard-crown/dense-flint cementable doublet

Flint lens (Glass-type Schott F4, $N=1·617$, $v=36·6$)

$$c_3 = -2·486$$
$$c_4 = +0·124$$
$$t_2 = 0·004$$

In the so-called Littrow type the crown element is equi-convex, and the choice of glass is more than usually critical:

Crown lens (Glass-type Schott K50, $N=1·523$, $v=60·2$)

$$c_1 = +2·498$$
$$c_2 = -2·498$$
$$t_1 = 0·008$$

Flint lens (Glass-type Schott F3, $N=1\cdot613$, $v=37\cdot0$)

$$c_3=-2\cdot498$$
$$c_4=+0\cdot135$$
$$t_2=\ \ 0\cdot004$$

An alternative design, of similar type, offers the benefit to the maker of a plane rear surface:

Crown lens (Glass-type Chance Hard Crown, $N=1\cdot519$, $v=60\cdot4$)

$$c_1=+2\cdot386$$
$$c_2=-2\cdot386$$
$$t_1=\ \ 0\cdot008$$

Flint lens (Glass-type Chance Dense Flint, $N=1\cdot620$, $v=36\cdot4$)

$$c_3=-2\cdot386$$
$$c_4=\infty$$
$$t_2=\ \ 0\cdot004$$

The Littrow objective offers the following advantages: (*i*) minimum departure from the d'Alembert condition, among the objectives of this series, (*ii*) relative insensitivity to errors of parallelism between the planes of the two lenses, and to distortion near the supporting points (the latter negligible with small objectives, in any case). Its chief disadvantage is that coma is still strong, compared with objectives of types more usually encountered in astronomical telescopes—four times that of the paraboloid ($B=+0\cdot25$), for example—which limits its usefulness for photographic work.

The objectives in modern binoculars and small, mass-produced telescopes are frequently 'bent' still more, having a slightly convex fourth surface and better correction for coma, which is important in wide-field instruments. The following design is for a cementable aplanatic objective:

Crown lens (Glass type Chance Medium Barium Crown, $N=1\cdot572$, $v=57\cdot7$)

$$c_1=+1\cdot923$$
$$c_2=-2\cdot671$$
$$t_1=\ \ 0\cdot008$$

FIGURE 25
An aplanatic
cementable doublet

Flint lens (Glass-type Chance Dense Flint, $N=1\cdot626$, $v=35\cdot7$)

$$c_3 = -2\cdot671$$
$$c_4 = -0\cdot067$$
$$t_2 = 0\cdot004$$

5.2.3 Coma-free doublets: The Fraunhöfer aplanat exemplifies this type. The equiconvex character of the crown lens is departed from, the second surface being about three times as strongly curved as the first; the flint component has become a concave meniscus, though the fourth surface is only weakly curved (in one form, the flint component is plano-concave). The lenses are not cemented, the airspace being narrowest at their centre; relatively thick peripheral spacers are therefore required when the objective is mounted in its cell. A typical Fraunhöfer design, by Dall, is:

Crown lens (Glass-type Schott BaLK 3, $N=1\cdot518$, $v=60\cdot3$)

$$c_1 = +1\cdot657$$
$$c_2 = -3\cdot200$$
$$t_1 = 0\cdot008$$

Flint lens (Glass-type Schott F2, $N=1\cdot620$, $v=36\cdot4$)

$$c_3 = -3\cdot134$$
$$c_4 = -0\cdot684$$
$$t_2 = 0\cdot004$$

$B=0$; hence a wide field of good definition, limited only by astigmatism; also the objective is relatively insensitive to errors of squaring-on (astigmatism only, in the oblique image).

The Fraunhöfer objective is the best of the contact types.

5.3 Separated doublets

The Clark objective—similar to the Littrow and aplanatic types, but with the components separated by a distance equal to about $1\cdot5$ per cent of the focal length—is an example of this class. A large degree of suppression of the normal spherical over-correction of the short wavelengths is obtained. The introduction of the airspace results in increased transverse chromatic aberration, the violet image being about $0\cdot2$ per cent larger than the red. A well-known design of separated doublet is that due to Dr James G. Baker, originally published in B. 7.34:

Crown lens (Glass-type Schott BK7, $N=1\cdot517$, $v=64\cdot2$)

$$c_1 = +1\cdot720$$
$$c_2 = -2\cdot789$$
$$t_1 = 0\cdot0069$$
$$t_2 \text{ (airspace)} = 0\cdot0031$$

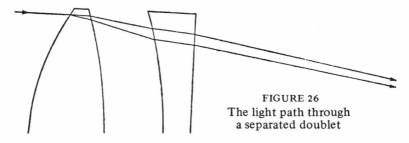

FIGURE 26
The light path through
a separated doublet

Flint lens (Glass type Schott F4, $N=1·617$, $v=36·6$)

$$c_3=-2·789$$
$$c_4=-0·618$$
$$t_3=0·00495$$

5.4 Apochromatic triplets

The Cooke photovisual, exemplifying this class of objective, is a three-lens apochromat of as nearly perfect chromatic correction as is feasible with ordinary glass; its residual colour is from one-eighth to one-tenth that of an ordinary achromatic doublet. Freedom from coma is similar to that of the aplanatic doublet; differential spherical aberration is as small as with the Clark objective. It is comparatively insensitive to errors of flexure and squaring-on.

FIGURE 27
The typical form of a Cooke photovisual triplet

The outer elements are usually of light crown, the central element light flint. The following prescription is from a typical design, but it should be noted that slight variations in the refractive and dispersive constants of the glass can necessitate appreciable readjustment of the curvatures, spacing, or both:

1. Crown lens	*2. Flint lens*	*3. Crown lens*
$N=1·5633$	$N=1·5406$	$N=1·5066$
$v=53·0$	$v=47·0$	$v=62·5$

1. Crown lens	2. Flint lens	3. Crown lens
$c_1 = +4 \cdot 10$	$c_3 = -\ 8 \cdot 43$	$c_5 = +12 \cdot 64$
$c_2 = -8 \cdot 43$	$c_4 = +12 \cdot 64$	$c_6 = -\ 2 \cdot 22$

The components are in contact.

5.5 Anastigmats (photographic)

Lenses designed for photography of wide areas of sky differ in three important aspects from normal visual objectives:

(*i*) A considerably faster focal ratio is required.

(*ii*) Suppression of off-axis aberrations, particularly coma, astigmatism, and field curvature, becomes particularly important.

(*iii*) The resolving power of the photographic emulsion raises the resolving limit of the system to ten or more times the theoretical limit. This means that spherical and chromatic correction can be sacrificed—in order to allow compensation for other factors—to a degree which would be intolerable for visual work.

The anastigmats commonly used for astronomical photography consist of two converging crown elements surrounding a diverging flint element.

FIGURE 28
A wide-field photographic triplet

The arrangement is approximately symmetrical. In some types, one or both of the outer elements is a cemented doublet. Such lenses are usually designed for use in the light of a particular colour, since there is considerable secondary spectrum. The cross-section through a typical modern design of triplet, by Donnel, is shown in Figure 28. For further information on this type of lens, the reader is referred to B. 5.1, 2, 6.

Mention should also be made of the old Petzval design, which gives fine correction of spherical aberration and has a secondary spectrum about half that of a triple anastigmat. Although astigmatism limits its useful field, it can work at a relatively small f/number. The other drawback of

Aperture
stop

FIGURE 29
The Petzval lens

the Petzval and related types, the strong field curvature, can be mitigated by the use of a simple field-flattening lens as described in B. 14.32.

Modern miniature camera lenses are not well suited to astronomical photography at full aperture (usually $f/2$ or $f/2\cdot8$) because of unsuppressed coma and drastic reducation of illumination towards the edges of the field. Performance is greatly improved by a reduction of aperture to $f/4$ or $f/5\cdot6$. (See section 20.14.)

Aperture
stop

FIGURE 30
A modern miniature camera lens

OPTICAL MATERIALS

6.1 Glass: composition and optical properties

Glass consists essentially of fused silicates. Originally, lead and calcium silicates, with sodium or potassium carbonate, were alone used; the former gave the range of flint glasses, the latter of crown. Modern optical instruments owe much to the work of Abbe and Schott, started at Jena about a century ago; as a result of this a very wide range of glasses is nowadays available for the lens designer, with refractive indices varying from 1·45 to 1·96 (N_d), and constringencies (v) from 19·7 to 70. The most important modern varieties are:

light phosphate crown	baryta flint
barium crown	antimony flint
zinc silicate crown	boro-silicate flint
boro-silicate crown	silicate flint
dense baryta crown	dense silicate flint
silicate crown	borate flint

The composition and optical qualities of average crown and average dense flint, together with those of Pyrex glass and silica for comparison, are summarised below:

	Average crown	*Average dense flint*	*Pyrex*	*Silica*
Composition (%)	SiO_2: 70 K_2O: 20 CaO: 10	SiO_2: 45 PbO: 46 K_2O: 9	SiO_2: 81 B_2O_3: 12 Na_2O: 7	SiO_2: 100
Density	2·56	3·60	2·3	2·2
Refractive index (N_d)	1·516	1·618	1·5	1·46
Constringence (v)	59	36	60	67
Linear coefficient of expansion	$9·5 \times 10^{-6}$	$8·0 \times 10^{-6}$	3×10^{-6}	5×10^{-7}

The mean refractive index of a glass (N_d) is taken as its value of N in the wavelength of the helium d line (587·6 nm). N_D, representing the value as measured between the two D lines of sodium (589·0 and 589·6 nm, mean 589·3 nm), was formerly used, and is still quoted in some glass tables. If N_F and N_C are the values of its refractive index at the wavelengths of the F and C lines of hydrogen (486·1 nm, 656·3 nm) respectively, then N_F-N_C is termed its mean dispersion, or partial dispersion for the interval F to C.

The dispersive power of a glass is commonly expressed in the form of its reciprocal ν, known as the constringence:

$$\nu = \frac{N_d - 1}{N_F - N_C}$$

It is in general true that for crown, both mean dispersion and mean refractive index are relatively small; for flint, relatively large. For average crown and dense flint, we have:

	N_d	$N_F - N_C$
crown	1·516	$1·523 - 1·514 = 0·009$
flint	1·618	$1·630 - 1·613 = 0·017$

The distinction is not absolute, however; barium crown, for example, has a large mean refractive index but small mean dispersion.

6.2 Plastics

Plastic optical apparatus is now being produced satisfactorily. Since it is moulded, its manufacture eliminates grinding, figuring and polishing. Polymerisation takes place in the mould at moderate temperatures and without pressure being applied. Cellulose acetate, methyl methacrylate, styrene, and polyvinyl acetate have been in use for a number of years, though only recently have plastics suitable for precision optics been developed.

Two at least are now suitable in this field:

(a) Polycyclohexyl methacrylate (CHM): satisfactory in all respects except low softening temperature (70°C) and the ease with which it is scratched. Optically similar to crown glass.

(b) Polystyrene: chief weakness is its rather large shrinkage, causing loss of figure in the mould; also rather easily scratched. Optically similar to flint glass.

These materials are lighter than glass, and allow of higher curvatures than practicable with glass. Their main drawbacks are softness and mould shrinkage, causing optical imprecision. Their coefficient of thermal expansion is also much higher than that of glass. For astronomical optics, glass, which is far more stable dimensionally and chemically, is unlikely to be superseded.

6.3 Light transmittance of glass

The loss by absorption in a lens depends upon its refractive index and the wavelength of the radiation, and is directly proportional to thickness. Percentage transmissions of a selection of optical glasses, per 25 mm of thickness, are given below:

Glass type	Wavelength (nm)							
	360	380	400	440	500	550	600	700
Borosilicate crown	92	94	98	98	99	99	99	99
Extra-light flint	72	86	95	97	98	99	99	99
Light flint	93	96	99	99	99	99	99	99
Medium barium crown	78	91	97	98	99	100	99	99
Dense barium crown	69	89	96	97	99	99	99	99
Barium flint	62	82	93	96	99	99	99	99
Dense flint	75	89	96	98	99	99	100	100
Extra-dense flint	48	77	91	96	98	99	99	99

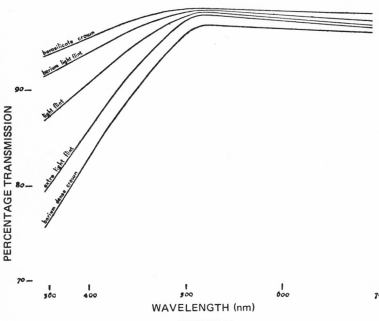

FIGURE 31

Transmission curves for different glasses

83

Taking the light loss resulting from absorption in 25 mm of average optical glass to be 2 per cent in the visual range, we have:

Mean thickness of lens	Transmission factor
25 mm	98%
50 mm	96%
75 mm	94%
100 mm	92%

It will be noticed that crown and flint are, roughly, equally transparent in the visual range (97 to 99 per cent/cm), and both of them considerably less transparent in the ultraviolet. For optical glasses generally it may be said that a very rapid fall-off in transmission sets in at about 500 nm, and below 300 nm they are virtually opaque. Nevertheless it can be seen that, using modern glass types, light absorption is of negligible account in the visual range except in the largest sizes of objective, a mean thickness of 12 cm or so ($D = 100$ cm) being required before even 0·1 mag is lost in transmission.

Various materials are transparent to shorter wavelengths, allowing investigations of the ultraviolet. Schott's 'Uviol' carries a transmission of about 96 per cent per centimetre thickness to 297 nm; Schott's UBK 5 and Vitaglass are also much superior to ordinary optical glass in ultraviolet transparency.

Glass also becomes increasingly opaque in the far infrared. Transmission drops sharply at about

> 2,500 nm with glass
> 4,000 nm with quartz
> 11,000 nm with fluorite (calcium fluoride)
> 18,000 nm with sylvin and rock salt

The last-named, being hygroscopic, requires to be looked after carefully, and kept in a desiccator when not in use, if frequent polishings are to be avoided.

The transmission factors of materials in wavelengths below about 300 nm are of little interest to ground-based astronomers, since this region of stellar and other external radiations is removed by telluric absorption.* Above this threshold, lenses and silver mirrors make a poor showing compared with aluminium films (see below, section 6.6); flint glass generally is less transparent than crown, some flints being opaque almost up to the visible region; even so, 0·2 mm of average crown will completely absorb wavelengths shorter than about 320 nm. The gelatin coating of a photographic plate is strongly absorbent below about 220 nm, while Iceland spar is opaque below about 215 nm, quartz below about 185 nm, and fluorite or fluorspar below about 100 nm.

---

*1 mm of air at 760 mm pressure will absorb all radiations below 170 nm. By means of vacuum spectrographs employing a Rowland grating, laboratory spectra have been obtained as far into the ultraviolet as 13·6 nm.

A further source of loss, which becomes appreciable with lenses of large aperture, is secondary spectrum, into which goes an appreciable fraction of the light transmitted by the objective.* Although the radiation is transmitted by the objective, it does not contribute to the focused image (see section 4.8); a more apt description would, therefore, be light wastage. Wastage due to secondary spectrum is about doubled when both aperture and focal ratio are doubled; trebled when the aperture is doubled without changing the focal ratio.

6.4 Reflection at lens surfaces

A proportion of the light projected at a lens is not refracted through it but is lost by reflection. This occurs at every glass/air and air/glass surface, and causes an integrated light loss of some 4 to 5 per cent.

Where the incident ray is not far from the normal to the surface of the lens we have, putting π = transmission factor,

$$\frac{\pi}{100} = 1 - \left(\frac{N-1}{N+1}\right)^2 \quad \dotfill \quad (a)$$

Putting $N = 1\cdot5$ for the average refractive index of optical glass,

$$\frac{\pi}{100} = 1 - \tfrac{1}{25} = 0\cdot96$$

or
$$\pi = 96\%$$

The precise value of π, the percentage of the incident light of different wavelengths not lost by reflection at the surface of an average crown and flint combination, is given below:

Wavelength (nm)	280	290	300	350	400	450	500	600	700
Crown	0	90·6	90·7	91·1	91·3	91·4	91·6	91·7	91·9
Flint	0	86·2	86·6	87·8	88·4	88·7	88·8	89·1	89·4

It may be noted in passing that the combined transmission of a crown and flint doublet is increased by about 8 per cent if the two lenses are cemented. For optical cementing, Canada balsam is generally used, since it is easily worked, is of satisfactory transparency and stability, and its refractive index is virtually identical with that of crown glass. Thus, if we write $N_c = 1\cdot5$, $N_f = 1\cdot6$, $N_b = 1\cdot5$ for the respective refractive indices

*42 per cent of the light transmitted by a 60-cm object glass, working at $f/15$, is thus lost to the image.

of crown, flint, and balsam, we have from (*a*) above:

for air-spaced combination:

$$\frac{\pi}{100} = \left[1 - \left(\frac{N_c-1}{N_c+1}\right)^2\right]^2 \left[1 - \left(\frac{N_f-1}{N_f+1}\right)^2\right]^2$$
$$= 0 \cdot 826$$
$$\therefore \quad \pi \simeq 83\%$$

for cemented combination:

$$\frac{\pi}{100} = \left[1 - \left(\frac{N_c-1}{N_c+1}\right)^2\right] \left[1 - \left(\frac{N_c-N_b}{N_c+N_b}\right)^2\right] \left[1 - \left(\frac{N_f-N_b}{N_f+N_b}\right)^2\right] \left[1 - \left(\frac{N_f-1}{N_f+1}\right)^2\right]$$
$$= 0 \cdot 908$$
$$\therefore \quad \pi \simeq 91\%$$

showing a saving of 8 per cent.

Various non-reflecting films for coating lenses by the vacuum evaporation method have been devised (see B. 4.2, 3, 4, 12), the best of which virtually eliminate light-loss by reflection at lens surfaces. The principle involved is the suppression of the reflected 5 per cent of the incident light by destructive interference, balanced by constructive interference in the transmitted pencil arising from reflections within the film.

The condition for this to occur is that the film's N^2 should be equal to the glass's N, the optical thickness of the film being $\frac{\lambda}{4}$. Herein lies the first difficulty, for $\sqrt{N}$ for crown and flint are respectively 1·231 and 1·271, and no solids have refractive indices as low as this. Of all the substances used, magnesium fluoride ($N=1\cdot38$) has proved the most satisfactory for single coatings, although calcium fluoride is more frequently used on low-cost optical components. When coated in a high vacuum (at least 10^{-5}mm Hg), with the glass heated to about 200°C, very hard and durable coats are obtained. The light loss through reflection in the chosen waveband is reduced to 1 to 2 per cent, the lower value applying to flint glass (which has a higher value of N). More recent techniques involve double or even multiple coatings, using alternate films of high and low refractive index, to give very low reflection losses over a wide spectral range.

6.5 Percentage reflection at a silver surface

A freshly-deposited silver film of first-class quality reflects about 93 per cent of the incident radiation at the wavelength to which the eye is most sensitive, though this can only be an approximate figure since individual films vary considerably among themselves.

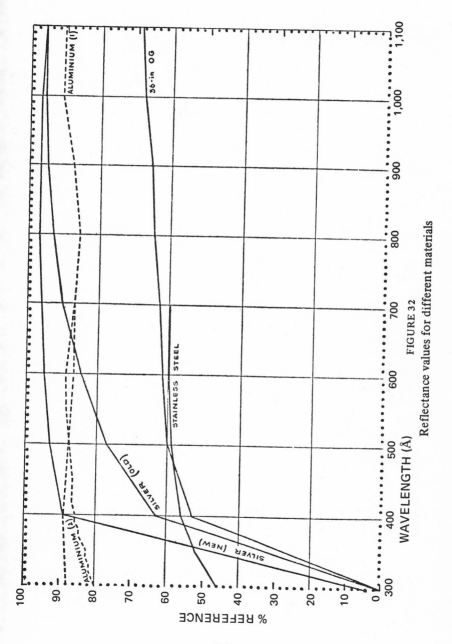

FIGURE 32

Reflectance values for different materials

87

Percentage reflection is also a function of the age of the film and of the wavelength of the incident light. The first two entries in the Table below give figures for silver-film reflectivity at stated wavelengths for a fresh and for an old film; the latter are somewhat arbitrarily chosen, but indicate the sort of effect produced by tarnishing. All the figures, however, must be regarded as to some extent approximate: individual tests have yielded results differing slightly from the average figures quoted here; in particular, the short-wave reflectivity, especially of the old film, may be too optimistic.

Wavelength (nm)	300	400	500	600	700	800	900	1,000	1,100	1,200
Equivalent colour	ultra-violet	violet	blue/green	orange	red	infrared				
Fresh silver	4	89	93	95	96	97	97	97	96	96
Old silver	0	63	77	85	90	93	95	96	96	96
Aluminium	88	89	88	87	87	86	88	91	91	90
Stainless steel	46	56	59	60	60	—	—	—	—	—
91-cm OG (Lick)	0	53	60	62	63	65	66	68	69	70

These figures, plus a second set for aluminium and for fresh silver, are represented graphically in Figure 32.

Beyond the minimum (4 to 6 per cent) at about 316 nm, the reflectivity of silver rises again to about 34 per cent in the region of 215 nm; at the other end of the spectrum it maintains a tolerably constant value of 96 to 98 per cent at least as far as 1500 nm. The progressive loss of the shorter wavelengths as a result of tarnishing should be noted by photographers.

6.6 Percentage reflection at an aluminium surface

It can be seen that, compared with silver films, aluminium is markedly superior in the ultraviolet below about 435 nm,* and inferior to a fresh silver film by 5 to 6 per cent over the visible range, and by 7 to 8 per cent over the infrared as far as 1200 nm.

Its outstanding advantages over silver are its high transmission of the wavelengths to which the photographic plate is sensitive, and its relatively very great freedom from tarnishing.

*The photographic threshold of the 2·5-m Mt Wilson reflector was lowered about 0·5 mag when the present aluminium film was substituted for the original silver film.

6.7 Percentage reflection at other metallic surfaces

The reflectivity of stainless steel is considerably lower than that of even a badly tarnished silver film in all wavelengths, varying with the nature of the alloy and the lapse of time since polished. Where it scores over a silver- or aluminium-on-glass mirror is in its freedom from tarnish, and a high coefficient of expansion which enables it to reach thermal equilibrium after any temperature change more rapidly. During a temperature change, for the same reason, definition will be poorer with a steel than with a glass mirror.

The transmission factor of speculum metal, also a comparatively inefficient reflector, is a direct function of wavelength, giving the image a colour bias of a yellowish-pink tint.

6.8 Telescopic transmission factors

The actual brightness of the telescopic image is always less than the theoretical, owing to light-loss in the optical train. This loss, which varies with wavelength, we have seen to be primarily due to

(a) partial reflection at air/glass and glass/air surfaces in object glass, ocular, and possibly Newtonian prism;
(b) absorption in object glass;
(c) imperfect reflectivity of speculum, and possibly flat;
(d) secondary spectrum of object glass.

Of these, (b) and (d) are negligible in small refractors.

Defining the transmission factor, π, of the instrument as

$$\pi = \frac{\text{actual brightness of image}}{\text{theoretical brightness of image}}$$

the following conclusions can be drawn regarding the efficiency, at visual wavelengths, of reflectors and refractors of about 15-cm aperture (i.e. light wastage through the refractor's secondary spectrum of 5 per cent).

(i) *Refractor with uncoated elements*

Absorption by OG	2%
Reflection losses	18%
Secondary spectrum	5%
	$\pi = 76\%$

(ii) *Refractor with coated elements*

Absorption by OG	2%
Reflection losses	6%
Secondary spectrum	5%
	$\pi = 88\%$

(iii) Refractor with coated and cemented elements

Absorption by OG	2%
Reflection losses	3%
Secondary spectrum	5%
$\pi=$	90%

(iv) Newtonian with aluminised surfaces

Light loss by primary	14%
Light loss by diagonal	14%
Obstruction by diagonal	6%
$\pi=69\%$	

Taking the transmission of the ocular as 82 per cent (uncoated) or 90 per cent (coated) gives the following values of π for various possible combinations:

	π
(*i*) Uncoated objective, uncoated eyepiece	62%
(*ii*) Uncoated objective, coated eyepiece	68%
(*iii*) Coated objective, coated eyepiece	79%
(*iv*) Coated and cemented objective, coated eyepiece	81%
(*v*) Newtonian reflector, uncoated eyepiece	57%
(*vi*) Newtonian reflector, coated eyepiece	63%

This analysis, which admittedly, is somewhat tentative, suggests that in the 15-cm aperture range a refractor is visually more efficient than the equivalent Newtonian reflector by an amount ranging from about 0·1 mag (uncoated objective) to about 0·3 mag (coated and cemented—or oiled—objective). As the aperture is increased, so this advantage is lost, due to the ever more dominant secondary spectrum and possible difficulty in applying a satisfactory anti-reflection coating. A Newtonian with silvered mirrors could transmit as much as 20 per cent more light than aluminised optics (equivalent to about 0·2 mag), but this benefit would rapidly be lost as the film tarnished. See, further, section 23.

Photographically, where sensitivity to light at the blue end of the spectrum is frequently high, the aluminised reflector will enjoy a clear advantage through (*a*) greater efficiency and (*b*) coincidence of foci of all wavelengths.

6.9 Mirror discs

The use of materials other than glass—or, to be more precise, silica alone or in combination—has not, in general, led to satisfactory results. Some mirrors made of pure aluminium have been produced, but this material is too soft to take a blemish-free polish. In addition to its other useful properties, glass has the advantages of hardness (therefore resistance to dust-induced scratches), and an optically homogeneous surface.

The use of plate glass (similar in properties to borosilicate crown) is hallowed by tradition. The 2·5-m reflector at Mount Wilson (1917) has a mirror that was poured using ordinary crown glass. At present, plate glass up to 38 mm in thickness is available in the UK, and, using this easily available material, mirrors of up to 60-cm aperture have been made successfully. The disadvantage of plate glass is its relatively high coefficient of expansion (about $9 \times 10^{-6}/°C$), which leads to a noticeable alteration of focal length and optical figure with quite small temperature change. The development of low-expansion glassware for domestic use has resulted in more satisfactory mirror discs appearing on the market under various trade names (Pyrex, Duran 50, Phoenix, etc.), all with coefficients of expansion of about $3·2 \times 10^{-6}/°C$. For example, Duran 50 discs, manufactured by Schott at Mainz, West Germany, are available in standard sizes ranging from 93·5 mm to 40·6 cm (thicknesses of from 13·5 mm to 76 mm). Further research has resulted in the development of practically zero-expansion ceramic materials, such as Schott's Zerodur, Cer-Vit, etc., the coefficient of expansion of Zerodur being quoted as $0 \pm 0·15 \times 10^{-6}/°C$. Such discs are between two and three times as expensive as ordinary low-expansion discs, but are particularly recommended for duties in solar telescopes, where violent heating effects can be expected.

When glass cools from the molten state, it first loses heat from its surface, so that a temperature gradient is created. Glass cooled too quickly, with too extreme a temperature gradient, is subjected to severe internal strain, and will not be optically stable. The high expansion coefficient of plate glass makes it necessary for the material to be cooled or 'annealed' slowly, to avoid physical fracture; plate glass is, therefore, always practically free from internal strain, and could, indeed, be used in lens manufacture if the green tint were not found objectionable. Low-expansion discs may be cooled much more quickly, and tend to be poorly annealed. Such discs may warp with time, or shatter when perforated for use in Cassegrain-type telescopes, and fine-annealed quality should always be stipulated. See, further, B. 4.8.

In an attempt to reduce the weight of large, thick mirror discs, fabricated blanks using a cellular or ribbed structure between two thin facing discs have been constructed. Accounts of some experiments along these lines will be found in B. 4.5, 9.

METAL-ON-GLASS FILMS

7.1 First-surface mirrors

Astronomical mirrors belong almost always to the type known as first-surface reflectors: i.e. the reflecting film is deposited on the front surface of the glass. The principal advantage of second-surface construction is that the film is protected on one side by the glass, and can be protected on the other by any opaque medium such as an electrolytically deposited film of metal, paint, etc. Such mirrors, however, are useless for any purpose demanding the degree of optical perfection required by an astronomical instrument: the light has to pass twice through the glass in front of the reflecting surface, it undergoes refraction and secondary reflection each time it encounters the glass/air surface, it will be made aberrant if the first surface is not as accurately worked as the second, and the glass itself would have to be of a high grade. A few telescope designs, however, do make use of certain advantages accruing from the use of second-surface or mangin mirrors; see section 9.

With first-surface mirrors, a single surface has to be worked, and the optical quality of the glass is immaterial. The disadvantage of an unprotected film is met in various ways: by the periodical replacement of the film, by the application of a protective layer of some transparent and resistant material, or by the use of a resistant material for the reflecting film itself.

It is probably safe to say that 99 per cent of all amateur-owned mirrors are, these days, given an aluminium film. In practice, bearing in mind the sensitivity of silver to contamination, it will give superior reflecting power over a longer life. Aluminising, once an expensive luxury which drove amateurs to try their own home-silvering methods, is now a relatively low-cost process considering the lifetime of a good film. Home silvering, on the other hand, has become largely uneconomic (the cost of silver nitrate, in particular, being extremely high), while some of the chemicals required are difficult or even impossible for the ordinary amateur to obtain.

Metals suitable for depositing on glass to form first-surface mirrors include:

Silver: Deposited chemically or by evaporation (the surface bloom after chemical deposition must be removed by burnishing); visual reflectivity of a first-class coat, about 93 per cent; its oxide and sulphide, both being dark, cause rather rapid tarnishing; easily scratched; removable by friction when wet; for optical qualities, see section 6.5.

Aluminium: Deposited by evaporation in a vacuum; unaffected by sulphur; on contact with air forms a very thin surface layer of alumina (aluminium oxide, Al_2O_3), which is extremely hard and impervious to water, visual reflectivity about 88 per cent, but superior to silver in the photographic range; for optical qualities generally, see section 6.6.

Rhodium: Highly resistant to weather and physical maltreatment, but has the serious disadvantage that the glass-metal bonding is too strong for easy removal should re-coating be required. Etching of the coat can occur due to electrolytic action with airborne industrial impurities.

Anodised films: Aluminium coats electrically anodised (oxidised), with or without a preliminary overcoat of titanium, are very hard-wearing. They are, however, expensive; much thicker than plain aluminium; and more 'pinholed', which means that unless they are carefully cleaned and kept dry the aluminium corrodes rapidly at the glass surface.

Dielectric films: A small percentage of light is reflected at any sharp boundary between two transmitting surfaces of different refractive index. By arranging a series of very thin transparent layers, such that the light wave-fronts are in constructive or destructive interference (see section 19.2), it is possible to boost reflectance in a selected waveband to a very high figure. Alternate $\lambda/4$ films of high and low refractive index, such as zinc sulphide ($N=2.37$) and magnesium fluoride ($N=1.38$), are used. A reflectivity of 97 per cent will be achieved with a nine-layer coating of this type. Such dielectric mirrors are normally prohibitively expensive, but the principle finds a practical use in the application of narrow-band filters, particularly in solar observation (see section 22.5).

7.2 Length of life

The frequency with which the old film has to be replaced depends upon the local incidence of those substances that attack it:

Aluminium: Sodium chloride; other things being equal, therefore, expectation of life can be related to the distance of the telescope from the sea; in the UK, on the other hand, there is always a measureable amount of salt in the air, since bad storms can carry particles as much as 300 km inland. Undercoats are to be avoided; an inconel undercoat (often used to improve the aluminium-glass bonding) will limit the life of a coat operating in damp and salty conditions to one-tenth of that of a plain aluminium coat.

Silver: Sulphurous acid (H_2SO_3) and hyposulphurous acid ($H_2S_2O_4$), resulting from the dissolving of sulphur dioxide—released into the atmosphere by combustion—in water droplets, are the chief tarnishing agents. The sulphur reacts with the silver to form compounds which reduce the reflectivity, more especially in the shorter (photographically active) wavelengths. A few weeks' exposure to air containing a high enough sulphur content to reduce the visual reflectivity by 10 to 20 per cent will reduce the reflection of the actinic wavelengths by as much as 50 per cent. The length of life of a silver film is, generally speaking, proportional to its distance from the nearest industrial centre or large town, though a single smoke-stack in the immediate vicinity can cause as much trouble as a city 30 km away. Clean-air legislation has, however, led to a considerable reduction of industrial (and domestic) urban air pollution.

If any mirror is allowed to remain dewed, it will quickly develop a matt surface. Industrial impurities in the dew may cause the film to blister or to become pock-marked.

The life of aluminium films is increased by periodical (say, two-monthly) washings with a strong detergent (Dreft, Surf, or liquid Teepol, but *not* commercial washing-up liquids, which contain lanolin). A good aluminium film should, with careful attention, last for years.

When, in 1947, the aluminium films of the 2·5-m and 1·5-m Mt Wilson reflectors were renewed for the first time since before the war, it was found that their reflectivities had dropped by 1 per cent and 3 per cent respectively—an astonishing figure, especially as the observatory is only approximately 25 km from the sea. Both these mirrors had been given six-monthly washdowns with Dreft; the reflectivity of another mirror which had not been so treated had fallen off by 11 per cent during the eight years.

A dulled silver film can be burnished with the minimum of finely-divided polishing powder and cottonwool. The surface of the mirror must first be cleared of every trace of the gritty dust that will have accumulated on it; for this purpose a loose, dry wad of cottonwool should be used as a sweeper. If this precaution is not carefully and thoroughly carried out, there will be a serious risk of scratching both film and glass. It is rarely possible to burnish a silver coat more than twice without seriously reducing its thickness, and, despite the most scrupulous care, the film usually becomes heavily sleeked in the process due to the impossibility of rendering it completely clean.

No rule can be given for the period during which a film will remain serviceable, since too much depends upon the quality of the film and the local conditions. As a very general guide, however, the useful lifetime of an aluminium film should be measured in years rather than months; that of a silver film, in months rather than weeks.

7.3 Care of metallic films

The reflectivity of a film could be maintained indefinitely if it were kept permanently in an airtight container. The best that can be done for a mirror mounted in a telescope is to provide it with some sort of protection against atmospheric sulphur and damp when not actually in use.

Cottonwool, to absorb the inevitable surface moisture that accumulates on a mirror after exposure to air, is probably the most efficient protector. It can simply be in the form of a loose layer about 10 mm thick, backed with a disc of cardboard. A better arrangement would be to use a firmly fitting cap (e.g. plywood with a wide rim of light alloy strip or even heavy linoleum, to fit over the mirror or its cell), faced internally with cottonwool. A pad of chamois leather, as sometimes recommended, will cause smears on the mirror from the oil with which it was impregnated. It is essential that the cap be placed in a dry, warm place while the telescope is in use. If the film has become dewed during an observing session, it should be gently blotted with a paper-tissue handkerchief before the warm cap is replaced.

Even a well-fitting cap, enclosing a very shallow air-space above the film, will materially retard tarnishing, but it must be realised that unless the aluminium film is routinely washed, its surface will be accumulating an ever-increasing contaminating layer from airborne impurities, either solid or dissolved in dew. Upward-facing mirrors are always affected more seriously than face-down ones; for example, the 91-cm primary at the old Greenwich Observatory had its coat ruined by corrosion in less than one year, while the secondary mirror retained a usable film for nearly five years.

7.4 Protective films

Although an aluminium film is relatively resistant to tarnish, partly due to the rapid formation of a hard, colourless oxide layer once it has been exposed to the air, its resistance to corrosion is improved somewhat by applying a layer of silicon dioxide (quartz) to the unoxidised layer, as part of the vacuum process.

The correct thickness is given by $0.25n\lambda$, where n is an odd number (so as to eliminate reflections from the surface of the quartz), and λ is the wavelength of light. The thickness is controlled by observing the colour of a beam of light reflected from the surface of the mirror as the silica is being deposited; as its thickness increases, the colour of the reflected beam passes through all the colours of the spectrum in turn, and the thickness is found to be most satisfactory when it becomes yellow for the second time.

Since minute pinholes in the quartz layer can still allow corrosion to undermine the coat, an overcoated mirror should be protected and washed just as carefully as a bare one. With this precaution, it should be more resistant to the whitish bloom which afflicts ageing mirror coatings.

7.5 Semi-transparent films

With a perfect semi-transparent mirror, the sum of the reflected and transmitted portions of the incident ray would be 100 per cent of the incident light. In fact a certain percentage is always lost by absorption. A compromise has to be found between small absorption and practicable durability of the material of the film.

The following have been used:

(*a*) Thin coating of silver, aluminium, or rhodium. Any degree of transmission relative to reflection can be obtained by varying the thickness of the film.

(*b*) Thin coating of ferrous oxide under controlled conditions. Maximum reflectivity 52 per cent minimum absorption about 7 per cent; resistant to abrasion and water, and does not corrode in air. The transmitted light is strongly orange.

(*c*) Thin coating of titanium dioxide (alternative to (*b*)). Maximum reflectivity 42 per cent, minimum absorption 1 to 2 per cent; transmitted light practically colourless; stability similar to that of (*b*). Both resist vigorous scrubbing with a mixture consisting of equal parts alcohol, water, and 10 per cent ammonia.

(*d*) Duolux.* Absorption higher than with silver or aluminium films (up to about 40 per cent), but more resistant to chemical erosion and physical abrasion. Can be cleaned with detergents, or even nitric acid, and will not abrade appreciably under hard rubbing with a cloth.

(*e*) Inconel (stainless steel). Formerly used as a bonding undercoat between aluminium and glass, inconel can be coated so as to reflect and transmit about 40 per cent each way of the incident light, and will do so substantially evenly throughout the visible spectrum; i.e. its images are colourless.

7.6 Aluminising

Most metals can be evaporated in a vacuum on to glass, forming a thin bright layer. The process is not one that the amateur can undertake unless he has the run of a well-equipped physical laboratory; commercial firms which have facilities for aluminising are listed in section 40.

The aluminising process may be summarised as follows:

(1) thorough cleaning of the glass surface by physical and chemical methods;
(2) loading the vacuum chamber with the mirror and the aluminium pellets;
(3) evacuation of the chamber to 10^{-3} mm;
(4) further cleaning of the glass surface by glow discharge at from 3000 to 5000 volts;
(5) further evacuation to 10^{-6} mm or better;
(6) evaporation of the aluminium;
(7) cooling; destruction of the vacuum; removal and examination of the mirror.

Individual techniques vary slightly in detail. (See further B. 4.10, 4.11.)

7.7 Silvering

Before commercial aluminising firms began offering an easily-available service, amateurs frequently silvered their own telescope mirrors at home, having no practical or economic alternative. Nowadays, it is probably true to say that the cost of the chemicals included in most of the 'classical' recipes would amount at least to the normal aluminising charge. Furthermore, such dangerous chemicals as nitric acid, widely recommended for

*Evaporated Metal Films Corporation, Ithaca, N.Y.

the preliminary cleaning, are unobtainable by members of the public. The days when all the silvering ingredients could be purchased at the local chemist's shop are long past.

A simplified method, using much smaller amounts of silver nitrate than normally necessary, was published by E. E. Hill in B. 4.6. In order to minimise the quantity of solution required, the mirror is set face-up, carefully levelled, and the silvering mixture is carefully poured so as to cover the concave surface with an even layer. The method may be of use where a reasonably reflective film is required at shorter notice than an aluminising firm could offer.

7.1.1 Theory of the precipitation of metallic silver from the solution of a silver salt: All chemical methods of precipitating a film of metallic silver are fundamentally similar, differing among themselves chiefly in the nature of the reducing agent and in the general manipulation of the process.

Two solutions are prepared: one of a silver salt—commonly silver nitrate in ammoniacal solution—and the other an alkaline solution of some reducing agent. The latter, when mixed with the silver solution, reduces the oxygen-containing silver compound, with the precipitation of metallic silver.

Thus silver nitrate plus sodium or potassium hydroxide produces a dark precipitate of silver oxide:

$$2AgNO_3 + 2NaOH = 2NaNO_3 + Ag_2O\downarrow + H_2O$$

This is redissolved by the addition of ammonia, the solution being extremely unstable. The addition of the reducer precipitates metallic silver on any surface that is in contact with (i.e. wetted by—hence the necessity for complete freedom from grease) the solution:

$$Ag_2O + 2H^+ \rightarrow 2Ag\downarrow + H_2O$$

The speed with which the silver is precipitated is a function of temperature, and although different methods work best at different temperatures silvering should never be carried out in a room, or with solutions, whose temperature is below about $13°C$.

7.7.2 Chemicals required: Hill recommends the following ingredients and amounts for silvering a 30-cm diameter mirror:

I   1 gm of silver nitrate dissolved in about 30 ml of distilled water
II   1 gm of potassium hydroxide dissolved in about 30 ml of distilled water
III   0·5 gm of dextrose dissolved in about 30 ml of distilled water

IV About 200 ml of strong ammonia diluted 1:10 with distilled water

V 1 gm of stannous chloride dissolved in 400 ml of tap water. (This solution should not be made up more than one hour beforehand.)

7.7.3 Preliminary cleaning of the glass surface: Absolute cleanliness of mirror and materials is essential. The chief enemy is grease; a carelessly cleaned mirror, or one which has been allowed to become contaminated with grease or chlorides from the fingers after cleaning, is among the commonest sources of failure to obtain a uniform and firm film.

The cleaning process can be done with tap water, but it should have been brought to the temperature of the silvering solutions (about 18°C) beforehand. Swab the mirror, front and edge, with ordinary washing-up detergent and water until all grease has been removed. Wash off the detergent and, using a cottonwool pad, rub over the surface with a paste of whiting and dilute ammonia. Wash again and set the mirror level. Then, using a cottonwool pad held in a plastic clothes peg, swab the mirror all over as the stannous chloride solution (V) is steadily poured over the surface until it has been used up. Wash off, replace the mirror on its levelled support, and cover the surface with distilled water after having swabbed off the last traces of tap water. The surface must not now be allowed to dry until the silvering process has been completed.

7.7.4 The silvering process: Using an eye dropper, carefully add IV to I until the precipitate, which formed with the first drops, has just cleared away. The solution must be stirred with a glass rod while this is being done. Then add all of II. Still stirring, continue adding drops of IV until the solution is of a light straw colour (it should not be completely clear).

Dilute this liquid to 500 ml, using distilled water. Take half of this amount, add half of III, mix vigorously, tip the distilled water off the mirror and immediately fill its concavity with the mixture, using a swab to ensure that the glass is evenly covered. Any loss over the edge should immediately be made good. The film should appear in a minute or two; any areas of the mirror which appear to be progressing slowly should be given extra swabbing. After about five minutes, the solution will be thick with sludge; it is poured off and the second halves of the silvering solutions are mixed together and applied as before. After the process has repeated itself, the mirror is washed under the tap, the excess water is carefully taken up with a paper tissue (remembering that wet silver is extremely fragile), and the mirror is set on its edge in front of a radiant fire to dry as quickly as possible. To prevent the chance of highly unstable silver fulminate being formed, the spent solutions should be washed down the drain as quickly as possible. (See also B. 4.14.)

7.7.5 Post-treatment of the film: As the mirror dries, the yellowish tint that it may have while wet is replaced by a bluish bloom, which has to be removed by burnishing and polishing; it should nevertheless be highly reflective at this stage.

Once thoroughly dry, it is ready for burnishing. This consolidates the film and prepares the way for the polishing, but a preliminary trial should be made on the edge of the mirror, to ensure that the coat really is dry. It should be done with a chamois leather pad stuffed with cottonwool, roughly one-third to one-quarter of the mirror in diameter. Exerting a gentle pressure, cover the whole surface with a circular motion. After several minutes of burnishing, scatter a little finely sifted polishing powder* on the pad, and continue polishing; gradually increase the pressure, which initially must be very light. Best results are obtained during the polishing if the surface of the polisher has first been made smooth and rather shiny by a preliminary rubbing on a piece of perfectly clean glass.

If after several minutes' polishing the quality and reflectivity of the film are unsatisfactory, resilvering is unavoidable.

7.7.6 Determining the thickness of the film: Upon the thickness of the film depends the number of polishings it will stand, and therefore its length of life. 0.1μ to 0.2μ is quite normal; a film of this thickness is not completely opaque, transmitting about 0.01 per cent of the incident light. The filament of an electric-light bulb will therefore be visible through it. If, however, a window frame silhouetting the day sky is visible through it, it is on the thin side. Even if the film is so thick that the Sun cannot be seen through it, there is no cause for worry. By means of its transparency, or otherwise, to light from a window, an electric filament, and the Sun, the thickness of the film can be gauged quickly and precisely enough to decide whether or not a second film must be applied.

The thickness can be determined easily and with great accuracy by means of Nobili's rings, once the mirror is dry. A single minute crystal of iodine is placed on the film. Over this is placed an inverted beaker, to exclude drafts. In a few minutes it will be seen that the crystal is surrounded by a number of concentric yellow rings. These consist of silver iodide, and the ring formation stops when the film immediately beneath the crystal is completely converted to the iodide. The number of rings formed depends directly upon the thickness of the film (see opposite).

7.7.7 Reclamation of silver residues: Unreduced silver solution should never be left standing about; it is unstable, and silver fulminate, an extremely sensitive detonator, may be formed. It can either be thrown

---

*E.g. the dried scum taken from the surface of a jarful of water in which polishing powder has been stirred up.

Number of rings	Thickness of film (μ)	
2	0·04	thin
3	0·07	
4	0·11	
5	0·15	
6	0·18	moderate
7	0·22	
8	0·26	
9	0·29	thick
10	0·33	

away, or the silver precipitated as silver chloride by the addition of sodium chloride solution.

The greyish or black sludge left in the silvering bath is rich in silver—indeed, from 90 per cent upward of the silver that is precipitated is left in this form. Metallic silver can be recovered by washing the precipitate through a filter paper, drying it, and grinding it up with borax; if it is then heated to a bright red heat in a crucible, allowed to cool, and swilled around in a beaker of water to dissolve out the matrix, a bead of metallic silver will be left.

OCULARS

Oculars are rather frequently overlooked as possible sources of inferior performance, which is then blamed on the objective. The general function of the ocular is described in section 3.2; maximum and minimum useful magnification in sections 3.3, 3.4; and field size in sections 3.7–3.9. Here the optical characteristics of oculars in general, and of individual types of ocular in particular, will be dealt with in greater detail.

8.1 Characteristics of the perfect ocular

The perfect ocular exists only in the astronomer's wistful imagination. The following are the main features that are required of an ocular, but no ocular possesses them all, the strength of one type being the weakness of another:

    (*a*) imperceptible aberrations, if not intrinsically, at least with the appropriate objective;

    (*b*) a wide field, with good definition over at least the greater part of it;

    (*c*) a flat field;

    (*d*) a dark field, i.e. freedom from internal reflections which either produce ghosts, or lessen contrast, or both;

    (*e*) bright images, i.e. the reduction to a minimum of light lost by internal reflection and (though this is negligible in most types of ocular) by absorption;

    (*f*) sufficient clearance between the eye lens and the exit pupil to ensure comfortable observing, even when glasses are worn.

8.2 Ocular aberrations

The importance of the aberrations increases with the diameter of the pencil that the ocular has to deal with, and is therefore a function both of the relative aperture of the objective and of its own focal length.

(*a*) The aberrations increase rapidly with increase of the relative aperture, since the latter increases the diameter of the pencil delivered to the field lens of the ocular. It is explained in section 4.3 that spherical aberration is an inverse function of (focal ratio)^2; hence, the reduction of area of this halo will correspond inversely with (focal ratio)^3. To this must be added the effect of residual chromatic aberration, the visibility of which offers a further inverse square effect. The presence of unwanted aberrational halation therefore varies inversely with the fifth power of the focal ratio of the objective, and a poorly-corrected ocular will show an aberrational halo some sixty times more prominent when used with a perfectly-corrected *f*/5 instrument than with an *f*/10 telescope of similar quality. An observer unaware of this high sensitivity might well be led to the unjustified conclusion that his short-focus reflector is faulty, having used an eyepiece that nevertheless performed perfectly adequately in conjuntion with a long-focus refracting telescope.

(*b*) In general the size of the aberrations increases in direct proportion with the focal length of the ocular, and they are therefore more troublesome with low magnifications (when the exit pencil is comparatively large) than with high. The superiority of highly corrected oculars over the commoner types is only apparent with low magnifications, except at small focal ratios when they are essential even with high magnifications.

8.2.1 Spherical aberration: All types of ocular in common use suffer from spherical aberration to a greater or smaller degree. At relatively large focus ratios (around *f*/15) this is not noticeable with high powers, but may also become so with low. Since the residual spherical aberration of both the Ramsden and Huyghenian ocular (the two types in commonest use) is positive, these work better with slightly over-corrected than with perfectly corrected objectives.

We have seen that to increase the relative aperture of an objective by *x* is to increase the illuminated aperture of the ocular in the same proportion, and therefore to increase the spherical aberration by x^2. Hence uncorrected spherical aberration in the ocular only becomes a real menace at small focal ratios, when highly corrected (and therefore expensive) oculars are an absolute necessity if the potential qualities of the objective are to be realised.

8.2.2 Chromatic aberration: Uncorrected oculars suffer from both the longitudinal aberration and chromatic inequality of magnification. The longitudinal aberration introduced by an ocular consisting of simple crown lenses is negligible so long as $f < \frac{1}{500} \cdot \frac{F^2}{D}$ approximately. If *f* increases past this limit, a progressively strong modification of the secondary

103

spectrum of the objective will be observed, becoming really serious when $f > \frac{1}{125} \cdot \frac{F^2}{D}$ approximately. Achromatic oculars are therefore a necessity when the objective has a fairly large relative aperture, though they are less essential with refractors of normal construction. A $f/6$ Newtonian used with, say, a Huyghenian ocular cannot be described as an achromatic instrument.

An ocular may be termed achromatic if it preserves the colour correction of the objective. Hence the term 'achromatic' carries somewhat different connotations when applied to objectives and to oculars. Whereas in the former context it signifies the bringing together of two wavelengths at the same distance from the objective, an ocular is said to be achromatic if it is free from chromatic inequality of magnification, producing an image which is the same size in all colours although the individual monochromatic images are formed in different positions. Images are therefore affected by fringes of out-of-focus colour, though this is uniform over the whole field.

The residual under-correction of the ocular and the human eye in combination is sometimes compensated by making the objective slightly over-corrected. However, such compensation is largely academic, when it is remembered that the error is greatest when it least matters, i.e. at low magnifications. An over-corrected objective may allow an under-corrected LP ocular (e.g. Huyghenian) to perform at least as satisfactorily as a highly corrected orthoscopic type, but may give inferior definition at high magnifications using the more sophisticated types (e.g. achromatic Ramsden) which nowadays are usually offered for high-power duties. Most modern makers correct their objectives for normal achromatism in the focal plane.

Even the condition of chromatic equality of magnification is progressively departed from when the diameter of the pencil becomes large compared with the diameter of the ocular lens through which it is passing. When it becomes marked, its effect is to draw out stellar images in the outer part of the field into short radial spectra, red towards the centre of the field and blue towards the edge, whose lengths are proportional to their distances from the field centre. It is objectionable, therefore, in that it limits the usable field.

The suppression of chromatic inequality of magnification requires that the emergent rays AF_r and BF_v, of different wavelengths, shall be parallel (Figure 33), i.e. that $\theta = \theta'$. The expression for the equivalent focal length of the combination of the two lenses separated by a distance d is

$$\mathcal{J} = \frac{f_1 f_2}{f_1 + f_2 - d} \quad \cdot \quad \cdot \quad \cdot \quad \cdot \quad \cdot \quad \cdot \quad (a)$$

Since both lenses have the same refractive index, for a given $\Delta\lambda$ their focal lengths will both be altered by Δf. Hence

$$\frac{f_1 f_2}{f_1+f_2-d}=\frac{f_1 f_2(1+\Delta f)^2}{f_1(1+\Delta f)+f_2(1+\Delta f)-d}$$

Neglecting powers of Δf and simplifying, we have

$$2d=f_1+f_2$$

which is the condition for achromatism in a pair of separated converging lenses made of the same material. It is observed by the Huyghenian and, approximately, by the Ramsden ocular.

Reflectors and apochromats deserve oculars whose high correction approaches their own perfect achromatism. With ordinary 'achromatics'

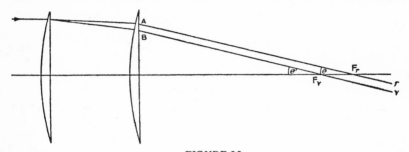

FIGURE 33
Chromatic inequality of magnification

it is less important, since there is inevitably some secondary colour, and an imperfectly corrected ocular may even be discovered that reduces the residual chromatism.

8.2.3 Other aberrations: Coma and astigmatism are normally negligible, though both appear if oculars are used at grossly inappropriate focal ratios. Field curvature is generally present to some extent, its complete suppression being coupled with intolerable astigmatism towards the edges of the field. Distortion, when present, takes the positive form, and is negligible in the central region of the field.

8.3 Single-lens oculars

These have the highest transmission of any, but also the most pronounced aberrations; hence their fields of tolerable definition are very restricted.

The field of full illumination of a single-lens ocular is given by

$$\tan R=\frac{r-\tfrac{1}{2}d}{f}$$

where R = radius of the field, defined by the ocular diaphraghm,
  r = radius of the lens,
  f = its focal length,
  d = diameter of the exit pupil.

The aberrations become intolerable when $2R > 20°$, approximately. The exit pupil lies at a distance from the lens of $f + \left(1 + \dfrac{1}{M}\right)$, or, for practical purposes, f. Good eye clearance is therefore one of the strong points of single-lens oculars.

While the single-lens ocular was recommended by Herschel for all purposes not demanding a large field, and even today has its advocates for such specialised tasks as the delineation of minute planetary detail, the development of a variety of versatile and highly corrected modern oculars has largely superannuated it.

A single plano-convex, mounted with the convex face towards the objective, is sometimes called a Kepler ocular. It possesses the characteristic single-lens features: bright image, small field (not much exceeding about 15°), chromatic inequality, and strong positive distortion.

The use of a single negative lens is confined to the Galilean telescope (see section 9.2).

8.4 Two-lens oculars

These (notably the Huyghenian and Ramsden) were developed in the early days of the telescope to provide an escape from the prohibitive aberrations and restricted field of the single-lens ocular. In particular, the fact that the total refraction is distributed over four surfaces, instead of only two, allows chromatic aberration to be reduced very considerably. As against the Kepler ocular, however, the loss from internal reflections, with resultant loss of contrast and the occurrence of ghosts,* is doubled.

The function of the field lens is to collect a wider cone of rays than would otherwise be received by the eye lens, and it contributes little to the magnification. The eye lens is responsible for the magnification† and

---

*Single reflections brighten the field indirectly, by illuminating the back of the diaphragm. Pencils resulting from two reflections are travelling in the same direction as the incident light, and if their section in the focal plane is small they will be seen as faint ghosts; such ghosts lie on the diameter of the field passing through the parent image, and are at least 7 mags fainter than it. Those too faint or too difuse to be seen as ghost images merely reduce the general contrast and help to destroy the darkness of the field.

†The addition of a field lens actually reduces the magnification that would be yielded by a given eye lens, since it reduces the effective focal length of the objective.

106

for delivering a parallel emergent beam which will come to a focus on the retina of an eye placed at the exit pupil and focused at infinity. A diaphragm in the focal plane of the eye lens defines the field of full illumination.

The exit pupil coincides with the resultant focal plane of the two lenses. This is located between the eye lens and its own focal plane; hence the eye clearance of a single lens is decreased by the addition of a field lens, and the field of full illumination correspondingly increased.

8.4.1 The Huyghenian: Termed 'negative', in contradistinction to the Ramsden and its derivatives. In effect it is a double Kepler, consisting of two plano-convex or convex meniscus lenses, of unequal curvatures, convex towards the objective.

Since the separation of the lenses and their focal lengths are related by $2d=f_1+f_2$ to secure achromatism, and $f_1=3f_2$ is the condition of minimum spherical aberration (in fact f_2 is often as large as $\frac{1}{2}f_1$), it follows that the focal plane lies between the two lenses, since $d=\frac{2}{3}f_1$ and (from (a), p. 104) $\mathcal{J}=\frac{1}{2}f_1$ only. Since the ratio of 3:1 between f_1 and f_2 gives a small eye clearance, the compromise ratio of 2:1 is frequently used, with $\mathcal{J}=\frac{3}{4}f_1$. Hence the Huyghenian cannot be used as a magnifying glass. For the same reason, micrometer webs located in the focal plane will be flanked by fringes of uncorrected colour, since they are being observed with the eye lens alone, and this is not achromatic; the webs and a stellar image, in other words, are observed with different optical systems and are therefore affected by different aberrations.

Though the chromatic aberration of size is thus corrected, that of the field lens being cancelled by that of the eye lens, the longitudinal aberration of the two lenses is additive; there is therefore residual colour, which becomes very noticeable when the ocular is working at small focal ratios. At the same time, strong positive spherical aberration and slight internal coma appear; there is also some distortion towards the edges of the field. For these reasons the Huyghenian is quite unsuitable for use with Newtonians of normal construction, its working limit being in the region of $f/12$.

The field of view, about 40° to 45°, is about the same as that of the Ramsden, though slight positive distortion, astigmatism, and curvature begin to obtrude at more than about 20° to 25° from the centre of the field. The field of full illumination is twice that of a single-lens ocular.

Like the Ramsden, the negative ocular has rather small eye clearance (about 0·3$\mathcal{J}$). It and its derivatives are, however, less troubled by ghosts than the positive eyepieces. Transmission is about 83 per cent of the incident light, increasing to 90 per cent if all air-glass surfaces are coated.

FIGURE 34
The Huyghenian eyepiece

A modification of the Huyghenian, the Mittenzwey, offers good definition across a field of up to $50°$ or so. In this design, the plano-convex field lens is replaced by a convergent meniscus lens.

8.4.2 The Ramsden: The so-called 'positive' ocular, thus named for the not very obvious reason that the focal plane does not lie between the two lenses. It consists of two plano-convex crown lenses of equal radius of curvature, mounted with their convex surfaces facing one another.

If the achromatic condition of $2d = f_1 + f_2$ is exactly obeyed, the focal plane of the eye lens will coincide with the field lens, and the exit pupil with the eye lens; disadvantages of which are, respectively, that any dust particles or defects on the surface of the field lens are visible, magnified, in the field; and that since the eye cannot be placed at the exit pupil, the visible field is decreased and some of the illuminating power of the objective is lost.

Hence d is usually reduced by from 25 to 40 per cent; both focal plane and exit pupil are then external to the system, though at the cost of introducing a sensible degree of chromatic inequality of magnification. Trouble

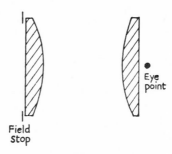

FIGURE 35
The Ramsden eyepiece
108

is also encountered from out-of-focus ghosts due to double reflection in the field lens. Short-sighted persons may find that, despite the reduced d, dust on the field lens is still objectionably in focus. The usual field of view is 40° to 50°, and transmission is similar to that of the Huyghenian.

The Ramsden has considerably less spherical aberration (positive) than the Huyghenian, and less field curvature. It can be used successfully with instruments as fast as $f/6$. It has more colour due to chromatic inequality of magnification, but smaller longitudinal aberration. Slight astigmatism is not troublesome, and there is also slight positive distortion. Useful field in the region of 35° apparent diameter. In the interests of achromatism, the clearance of the exit pupil from the eye lens is not very great—usually about $0.25f$.

8.4.3 **The Kellner and achromatic Ramsden:** The former, though frequently confused with the latter, is of different construction, with a biconvex (usually crossed) field lens and a positive meniscus achromatic eye lens, the concave surface outwards. It is not made nowadays, but was once used in microscopy and (in various forms) as a wide-field or 'comet' eyepiece. The true Kellner has a wide, flat field of view—up to 60°—but a

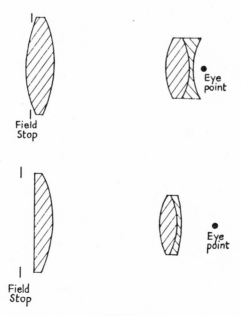

Field Stop

Eye point

Field Stop

Eye point

cIGURE 36

The Kellner (upper) and achromatic Ramsden eyepieces

very small eye clearance. It has two disadvantages: obtrusive ghosts, and a sharply focused field lens. Transmission is about 81 per cent (coated, 89 per cent).

The achromatic Ramsden has become very popular as a fairly low-cost, versatile eyepiece. The plano-convex eye lens is replaced by an achromatic combination, giving the same field of view as the orthodox Ramsden but with improved eye clearance ($\mathcal{J}=0.3$ to 0.45 in different types). Achromatism and spherical aberration permit its use with instruments as fast as $f/4$. The outer component of the eye lens being made of soft flint glass (in common with most oculars containing achromatic components), it is much more vulnerable to scratching from careless cleaning than are the hard crown faces of the Huyghenian and Ramsden types. The achromatic Ramsden transmits 81 per cent of the incident light (coated, 89 per cent). Ghosting is much less obtrusive than in the simple form. It is an ocular commonly used in binoculars.

8.4.4 The Dialsight or Plossl: These eyepieces, consisting of two symmetrical or nearly symmetrical achromatic doublets, their crown surfaces facing inwards and nearly touching, came on to the surplus market after the war in large numbers. Their large eye clearance ($\mathcal{J}=0.8$ in some

FIGURE 37
The Dialsight or Plossl eyepiece

types) and $40°$ flat field make them a suitable eyepiece for sighting telescopes. A development by Zeiss gives a field of view of about $50°$; in this type, the eye lens has a shorter focal length than the field lens, and the eye clearance is reduced. This design is used in the well-known oculars made by the French firm of Clavé. Transmission is 79 per cent (coated, 87 per cent).

8.4.5 The orthoscopics: A family of highly corrected eyepieces. Numerous orthoscopics exist, variants on the original design by Mittenzwey. The eye lens is a plano-convex, and the main function of the field

FIGURE 38

The orthoscopic eyepiece

lens, which is usually a cemented triple achromat, is to supply aberrations opposite to those of the eye lens. Fields of 35° to 40° are normal, but can be as much as 50°, and eye clearance ($f=0.8$ or so) is as generous as in the dialsight type. An additional benefit of this design is that both outer lenses are made of resistant crown glass. Orthoscopic correction permits satisfactory performance down to $f/5$, or even shorter, but some types suffer from ghosting. Transmission is as high as in the simple two-lens types (83 per cent, or, coated, 90 per cent).

8.4.6 **Five- and six-element types:** By increasing the number of lenses to three instead of two, the field of view of the eyepiece can be widened over that normally available with simpler types, although the axial aberrational correction may show no noticeable improvement. From an astronomical point of view, the extra light-loss and general reduction of contrast are high prices to pay, unless wide field is an overwhelming consideration, as may be the case in certain branches of LP observation. These complex types were designed for military use. The five-element tank binocular eyepiece has a field of about 75° and good eye clearance ($f=0.4$ to 0.5), while the well-known Erfle consists of three doublets. The field of view is 65° to 70° in most types; exceptionally, 80°. The eye lenses of most

Field
Stop

FIGURE 39

The Erfle eyepiece

111

samples, which are faced with flint glass, will be found badly abraded from rough usage during their active lifetimes, and will benefit from a re-polish. Transmission figures are rather low (about 70 per cent, or 84 per cent if coated). The Bertele eyepiece, consisting of two separated lenses and a double eye lens, was also made for military use and has similar characteristics.

8.4.7 Other types: In their attempts at improving or modifying the characteristics of standard eyepiece types, designers have produced many variants, and, except perhaps in the case of the simple Huyghenian and Ramsden designs, there is no such thing as a 'pure' type. A number of designs, formerly well-known, have been superseded; these include the Airy (modified Huyghenian), Hastings (achromatic Huyghenian), Gifford or Orthochromat (orthoscopic type), and the Cooke five-element (orthoscopic type). For a general survey of eyepiece designs, see B. 6.4, 8.

8.5 Solid and cemented oculars

In theory these combine the best features of both single-lens and separated-lens oculars; high transmission and a high degree of correction. In practice their superiority over the best oculars with separated lenses is often not spectacular; though between the best of the cemented types, such as the monocentric, and the more primitive types of ocular, there is a gulf fixed.

Generally, they are characterised by dark fields, bright images, and virtually complete freedom from both spherical and chromatic aberration. The lack of scattered light gives them the advantage over separated-lens types in cases where (as in planetary observation) the detection of contrast near the threshold of visibility is a factor of importance.

8.5.1 The Tolles eyepiece: A development of the Coddington lens which may be regarded as a solid Huyghenian: it is cut from a rod of crown glass with spherical ends (Figure 40). The groove defining the image of the objective is cut one-third of the distance from the objective end towards the eye end.

FIGURE 40
The Tolles eyepiece

Chromatic inequality of magnification is suppressed when the overall length of the lens d is related to the refractive index of the glass and the radii of curvature of the surfaces by

$$d = \frac{r_1 - r_2}{1 - \frac{1}{N^2}}$$

and in its usual form, $r_1 = 1 \cdot 5 r_2$.

Its spherical aberration is very considerably less than that of the Huyghenian, and its transmission is higher. For darkness of field and freedom from ghosts the Tolles rivals the monocentric, and its general correction is such that it works well at $f/6$. Unfortunately, it is a very inconvenient eyepiece to use: the field of view is no greater than $30°$, and the exit pupil coincides with the rear surface, so that the whole field cannot normally be seen at the same time. This is not a great disadvantage when the ocular is used for the work to which it is most suited—planetary or double-star observation—with a driven telescope.

8.5.2 **Cemented triplets:** These combine the advantages of single-lens eyepieces (freedom from ghosts and scattered light) with convenient eye relief ($f = 0 \cdot 8$ or so) and reduced field curvature compared with the Tolles type. The well-known monocentric consists of a crown component sandwiched between two flint outer elements, the curves all being struck from a common point, although the eyepiece may not be symmetrical. Steinheil's monocentric (Figure 41) consists of a crown biconvex and two flint meniscuses; all six spherical surfaces are concentric. Spherical aberration restricts its use to objectives of larger focal ratio than $f/6$ approximately. The field of view is about $25°$. The Zeiss monocentric employs slightly larger lenses and a thinner convex component than the Steinheil, giving a rather wider field, and it can be used at about $f/5$. The 'Loups' triplet has similar characteristics, though not a monocentric; it is usable

Field
Stop

Eye
point

FIGURE 41
The Steinheil monocentric eyepiece
113

down to $f/4$ in its best forms. This type of ocular, despite the small field, is the best type available for HP work of a delicate nature: there are no ghosts, practically no light is scattered into the field, and the transmission is about 88 per cent (coated, 92 per cent).

8.6 Terrestrial ocular

Consists of four separated simple lenses, biconvex or plano-convex, of very different curvatures (Figure 42). The rear pair constitutes an ordinary Huyghenian ocular; the other pair constitutes the erector, as shown in the diagram.

The exit pupil of the lens furthest from the eye is exactly defined by a stop between it and the second lens. The field is limited by a diaphragm between the lenses of the Huyghenian pair. The magnification of the pair furthest from the eye—which are together equivalent to a field lens—is inconsiderable, usually x2 or x3.

The usual field is about 35° to 40°, but can be further widened by inserting an equiconvex between the field stop and the eye lens of the Huyghenian pair.

8.7 Dual oculars

The only advantage to be gained by using a dual ocular is reduced eye fatigue during long spells of observation. Neither light grasp nor the resolution of point or extended images is affected when the pencil is divided between the objective and the eye; any parallactic effect that may appear is, likewise, illusory.

Binocular microscopes, complete, have been tried as astronomical oculars, and advocated. Their use, however, is exposed to the following disadvantages:

(*a*) the multiplication of air/glass surfaces;

(*b*) the fact that if the beam is split by a Wenham prism covering half the microscope objective (as is the common practice), half the pencil being received by each eye, then asymmetrical star images will be seen under high magnifications, and the definition of extended images will suffer similarly.

A more satisfactory arrangement would be to split the beam by a half-silvered prism in the neighbourhood of the focal plane of the telescope's objective (Figure 43). The base of the right-angled prism V, which is cemented to the end of the parallelopiped W, is lightly silvered, a film about 0·2λ thick transmitting and reflecting equally. The function of the

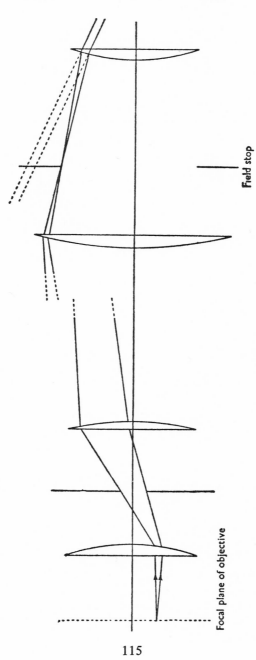

FIGURE 42
The terrestrial ocular

Field stop

Focal plane of objective

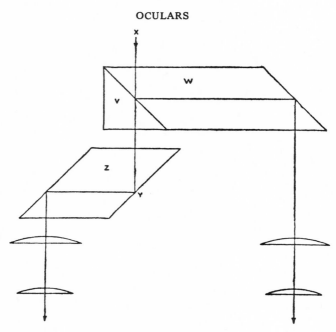

FIGURE 43
A binocular eyepiece

second parallelopiped, Z, is to provide for equal light paths in glass for the two rays, without which provision the two images would be unequally bright. Adjustment of the inter-ocular distance is obtained by rotating Z about XY as axis.

8.8 The Barlow lens

The name Barlow is given to any negative (diverging) lens placed in the convergent cone from the objective, where, by decreasing this convergence, it increases the objective's effective focal length and focal ratio. It usually takes the form of a cemented plano-concave doublet, corrected for incidence at its plane surface. Its secondary spectrum and residual spherical aberration are, or can be, negligible.

Ordinary Barlow lenses of about 20 cm focal length are less satisfactory than the shorter focus telenegative lenses, since the amplification factors possible with the latter are much less limited. Telenegatives were used before the development of the telephoto lens to increase the plate-scale of camera objectives, and are still to be found on the second-hand market. Those of Goerz have been particularly recommended.

Several notable advantages are offered by these extremely useful accessory lenses. By increasing the effective focal length of the objective, thus reducing its relative aperture, it permits the use of relatively poorly corrected oculars—such as the Huyghenians—with Newtonians and similar instruments whose large relative apertures would otherwise render the uncorrected aberrations of the ocular intolerable. Astonishing improvement of definition is from time to time reported (though this should be no matter for surprise) by the addition of a Barlow lens to an optical train consisting of a paraboloid of ordinary Newtonian focal ratio and a Huyghenian ocular. There is a limit to which a Barlow can increase the focal ratio of a given objective, however, since, as will be seen shortly, increase of effective focal length is obtained by moving the lens further from the primary focus, i.e. increasing its own illuminated aperture and therefore bringing its own aberrations into increasing prominence.

A Barlow lens likewise increases the range of magnifications provided by a given set of oculars, and permits high magnifications without employing oculars of impossibly short focal lengths.

It can be seen from equation (a), p. 129, that for a given aperture the size of the flat can be reduced as the focal length of the objective is increased; we have also seen that although the Newtonian's central obstruction may improve some types of stellar resolution, it is a disadvantage in the observation of extended detail owing to the reduced contrast which it causes. Reduction of the size of the diagonal therefore tends to improve, for example, planetary definition. This can be effected by a Barlow, with whose assistance the final image plane can still be brought to the eyepiece position just outside the tube when the flat is moved in nearer to the prime focus, i.e. to a narrower part of the convergent beam.

The operation of the Barlow lens is explained by Figure 44 and the three equations below:

If F = focal length of the objective,
  f_b = focal length of the Barlow,
  $\mathcal{J}$ = effective focal length of the objective and Barlow combined,
  M_b = the Barlow's amplification factor,
  d = distance of the Barlow inside the original focal plane,
  d' = distance of Barlow inside the new focal plane,

then

$$\mathcal{J} = \frac{F \cdot f_b}{f_b - d}$$

$$M_b = \frac{\mathcal{J}}{F} = \frac{f_b}{f_b - d}$$

$$d' = f_b(M_b - 1)$$

FIGURE 44

The Barlow lens

Hence when $d=0$, the Barlow has no effect on F, and $M_b=1$; at the other extreme, when $d=f_b$, $M_b=\infty$, the Barlow having converted the convergent beam into a parallel beam. The range of possible positions for the Barlow therefore stretches from the primary focus toward the objective for a distance equal to its own focal length.

Take, for example, the case of a $f/6$ 30-cm Newtonian ($F=180$) and a 75-mm focus Barlow set 50 mm inside the primary focus. Then from the first of the above equations, the effective focal length of the objective is now increased to 540 cm and its focal ratio from $f/6$ to $f/18$; from the second equation, the Barlow's amplification factor in this position is 3; and from the third, the new focal plane lies 15 cm outside the Barlow.

It will be noticed that d' is extremely sensitive to changes in d. In the case just quoted, where $f_b=75$ mm, we have

d (mm)	25	37·5	50	67·5
M_b	1·5	2	3	6
d' (mm)	37·5	75	150	375

If the Barlow is mounted outside the diagonal (i.e. in the drawtube between the diagonal and the ocular) the inconvenience of widely differing focal extensions is therefore encountered. More conveniently the lens may be cemented to the diagonal (prism); or a combined prism and Barlow could probably be constructed, saving one cemented surface (Figure 45). B. 6.10 describes in detail an arrangement where the drawtube and prism-lens are mounted in one solid piece, adjustable in parallel guides on the telescope tube and held in the required position by a simple locking device.

For a discussion of the Barlow lens and its design, see B. 6.6. The use of a negative lens both to amplify and correct the image produced by a short-focus spherical mirror is described in B. 6.2, 7.

8.9 Polarising eyepiece

A polarising eyepiece can be easily made. A piece of Polaroid, held between microscope slips and mounted between two thin card rings, is fixed against the eye lens of an ordinary ocular. A second piece, similarly

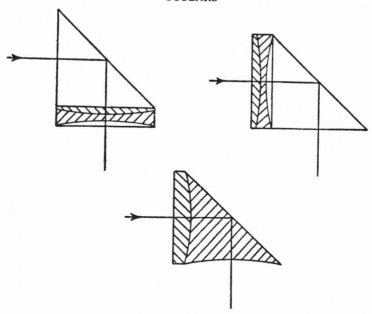

FIGURE 45
Combined Barlow lens and prism arrangements

mounted, is fixed in a screw-on suncap from which the dark glass has been removed; it can thus be rotated relative to the first. Variations of image brightness can be quickly and conveniently made, but there is usually trouble from flare and scattered light. Incomplete polarisation at the extremities of the spectrum produces a purplish tint when the sheets are crossed. Polaroid at the exit pupil is not suitable for solar work owing to the intense concentration of heat at and near this point.

8.10 Diagonals

These may employ total internal reflection in a right-angled prism, a silvered flat, or a prism with hypotenuse face silvered to turn the emergent beam through a right angle. Solar diagonals are unsilvered.*

The reversal of the image is awkward when working with charts, unless the chart is viewed in a mirror; in some cases it is more convenient to take a tracing and use this back to front.

With diagonals of all types it is important that the reflecting surface

---

*Sun diagonals, as well as other types of eyepiece employed in the direct observation of the Sun, are described in *O.A.A.*, section 1.6.

be kept absolutely clean, or scattered light will cause deterioration and dilution of the image.

A zenith prism is a small equilateral prism which turns the emergent beam through 60° (Figure 46). It is mounted in a holder which is usually attached to the ocular by means of a simple clip fitting. Being so close to the exit pupil it has the advantage that it can be quite small.

8.11 Image rotators

Eyepiece attachments which, when rotated about the optical axis, have the effect of rotating the image, are of use in solar work, variable star observation, and other fields. The type shown in Figure 47 consists of a right-angled prism which is mounted in a short tube behind the eyepiece, so that its hypotenuse face is parallel to the optical axis; when, now, this tube is rotated through θ, the image will rotate through 2θ. This type of prism is also sometimes called an erecting prism, since, like a diagonal, it introduces one-plane reversal into the image—i.e. left to right, or up and down. As such it can be used, for example, to give the projected solar image the same

FIGURE 46
A zenith prism

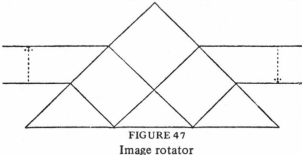

FIGURE 47
Image rotator

orientation as the image viewed direct. This can also be done by the use of an ordinary star diagonal continuing a right-angled prism.

Figure 48 illustrates a similar rotator, but employing reflection at plane mirrors instead of a prism.

8.12 Checking the magnification of an ocular

Since $M=F/f$, there can be no such thing as *the* magnification of an ocular, since M depends in part upon the focal length of the objective

120

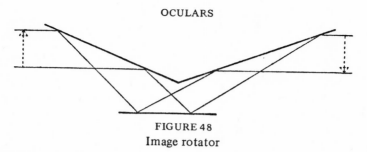

FIGURE 48

Image rotator

with which the ocular is used. The only valid description of an ocular is in terms of its focal length. For this reason the magnification of each newly acquired ocular should be determined with the objective with which it is to be used. Available methods include:

(*a*) Approximate method: An assessment which will be good enough for most practical purposes can be made by comparing the real field of view V_r, as seen through the telescope, with the apparent field of view V_a subtended by the field stop of the eyepiece alone when it is held up to the eye in the normal way. Then,

$$M = \frac{V_a}{V_r}.$$

The diameter of the apparent field can be determined to within a few per cent by holding the ocular to one eye while viewing simultaneously, with the other eye, a scale of known length, e.g. a 50-cm stick. By altering the distance d from the scale to the eye, its length l can be made to coincide with a diameter of the field of view; then,

$$V_a = \frac{l}{d} \times 57°.3.$$

V_r can be determined by allowing a star near the celestial equator (e.g. α Aqr, one of Orion's belt stars, γ Vir, δ Oph) to transit the field. If this takes t seconds, then

$$V_r = \frac{t}{4} \, ' \text{ arc.}$$

(For any other star of Dec δ °, $V_r = t/4 \cos \delta$ ′ arc.)

(*b*) Dynameter: A V-shaped scale graduated to 0·02 mm, described in B. 6.3, is used to measure the linear diameter of the exit pupil. The telescope is directed at any light-toned surface—not a bright one, such as the day sky, or irradiation will falsify the measurement—and the dynameter held against the eye-guard of the ocular in the plane of the exit pupil; it is then viewed from a distance of about 30 cm with the aid

FIGURE 49
The Dynameter

of a magnifying glass. The magnification is then (equation (*c*), p. 30) given by

$$M = \frac{D}{d}$$

where D and d are the diameters of the objective and exit pupil respectively. Care must be taken that the tube stops in a refractor, which define the convergent pupil, are of the correct size and spacing; otherwise the exit pupil may not be an image of the whole aperture of the object glass. Although this method is not highly accurate—owing primarily to the difficulty of reading the scale with sufficient precision*—it is adequate for ordinary purposes. The effect of irradiation can also be eliminated by fixing across, and in contact with, the objective, a card in which two V-shaped notches have been cut at a distance apart equal to about three-quarters of the aperture. The separation of the notches is then measured with the dynameter, this quality replacing d in the formula; their linear separation is substituted for D. An optical scale measuring device, incorporating a finely graduated transparent slip at the focus of a magnifying glass, may also be used.

(*c*) A scale erected at a distance from the telescope is viewed simultaneously with one eye through the telescope and with the other naked; in the case of a Newtonian a prism or plane mirror will have to be used so as to bring the naked-eye image to a position where it can be observed. The distance of the scale is varied until a position is found such that one division of the telescopic image coincides with a convenient number of divisions of the naked-eye image: this number is the magnification. This method is practicable only with very low magnifications, which could be measured sufficiently accurately using method (*b*).

*The fact that it is a chord and not a diameter of the exit pupil that is measured introduces negligible error.

(*d*) The focal lengths engraved on the barrels of most modern oculars being quite trustworthy, the magnification (p. 29) can be derived merely by measuring *F*. Form a sharp image of the Moon on a screen (the Sun's image, unless viewed through a dark filter, will be difficult to focus because of irradiation), and measure the interval between the objective and the screen with a tape rule. In the case of a compound telescope (e.g. Cassegrain), where the focal length is not directly measurable, it can be derived by measuring the focal plane scale of the objective. Selecting a star of known Dec, δ, and using a LP ocular with a well-defined, external field stop, time the diametrical transit of the star *t* across the field. Measure the linear diameter, *d*, of the field stop with the greatest possible accuracy. Then

$$F = \tfrac{1}{2} d \cot \tfrac{1}{2}(15t \cos \delta).$$

8.13 Care and maintenance

Oculars are often subjected to a roughness of handling that would never be meted out to an objective. But it is worth remembering that from the point of view of the final image, the strength of the optical chain is that of its weakest link, not its strongest; and a faulty ocular can ruin the performance of a superlative objective with no difficulty at all.

The precautions to prevent the formation of dew on the lenses, described in section 11.1, should be observed. If condensation does form, it should never be wiped off. Oculars when not in use should at all times be kept in a dust-proof and airtight container of some sort. Lenses should be polished as infrequently as possible, and only when they are perfectly dry. They should first be brushed thoroughly with a fine sable brush (as sold to water-colourists) to ensure that all dust is removed; they can then be polished with lens tissue and the minimum of pressure. Flint is more liable to scratch than crown, and hence those oculars having a flint component outermost (e.g. triplets) should be especially carefully treated.

Faulty oculars should be handed over to an optician, preferably the maker, for repair.

See also section 22.9.

COMPOUND OPTICAL INSTRUMENTS

The images formed at the primary foci of short-focus concave mirrors and convex lenses are too small for effective examination. A telescope is simply an instrument employing what is in effect a magnifying glass to increase the scale of the image formed in the focal plane of the objective, so that resolved points in the focal plane image are separated to an extent which carries them above the eye's resolution threshold. By virtue of the objective the telescope performs two further functions: to reveal fainter objects, and to separate closer objects, than would be possible without it.

9.1 Astronomical refractor

For the purposes of demonstrating the action of a combined image-forming lens (objective) and examining lens (ocular), each may be taken as a simple convex lens. In fact the former is an achromatic combination, residually convex, and the latter usually a combination of two spaced lenses, also residually convex, although the single-lens ocular has its uses in astronomical observation.

Owing to the great distance of astronomical objects, the incident rays are sensibly parallel (Figure 50). Hence they are brought to a focus in the focal plane of the objective, F_{OG}. It is this real, inverted image which is the ocular's object.* The ocular may be regarded as a simple magnifying glass whose position on the optical axis is adjusted so that F_{OG} lies just inside its own focus F_{OC}. Since the eye is focused for infinity when its ciliary muscles are relaxed, it is likely that this adjustment will be unconsciously directed to approximating F_{OG} and F_{OC}. For when the two focal planes coincide, the emergent beam will be parallel and the final image, A, at infinity (Figure 51).

The ocular thus forms a virtual, inverted image which is examined by the eye; in effect the ocular permits the comfortable examination of

---

*Or it may be studied at the prime focus with the aid of a photographic emulsion.

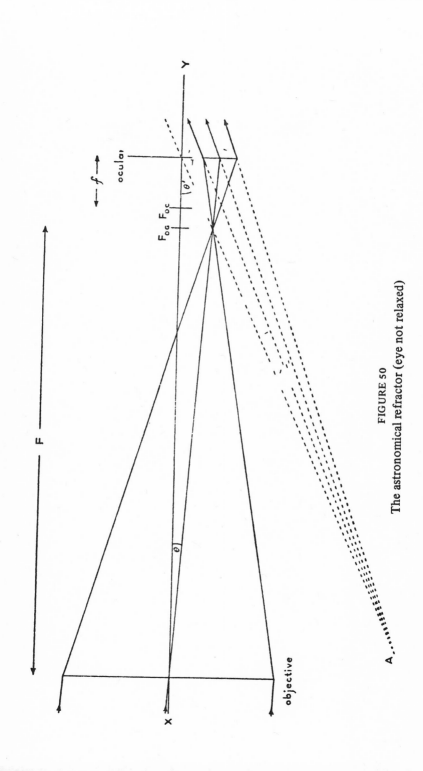

FIGURE 50

The astronomical refractor (eye not relaxed)

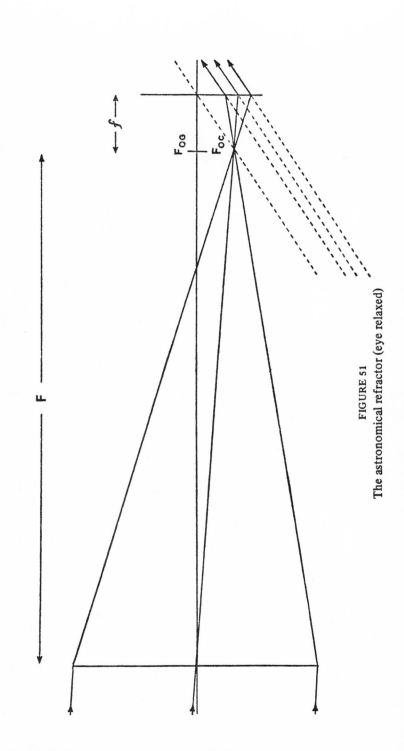

FIGURE 51
The astronomical refractor (eye relaxed)

the primary image from a distance which is much smaller than the distance of distinct vision.

The final magnification, as we have seen, is given by

$$M = \frac{F}{f}$$

Secondary spectrum and the difficulty and expense of manufacturing strongly curved lenses set a limit to the useful focal ratio of about $f/10$ or $f/12$; astronomical refractors are usually built to about $f/15$, and for some types of work (e.g. planetary) even $f/20$ is to be preferred.

9.2 Galilean refractor

Differs from the modern astronomical refractor in using a diverging (negative) system as an ocular. The image is therefore virtual and erect. The magnification, $\frac{F}{f}$, is in practice seldom more than ×5. The ocular is adjusted along the optical axis so that it lies inside the focal plane of the objective by a distance about equal to its own focal length, i.e. to the algebraic sum of the focal lengths, f being negative:

$$d = F + f$$

The OG is usually a Clairaut achromatic, and the ocular a crown equi-concave. (See Figure 52).

The distance of the virtual exit pupil in front of the eye lens is given by

$$a = \frac{df}{d-f} = \frac{d}{M}$$

It being impossible to place the eye at the exit pupil, the whole field can never be viewed simultaneously from a single position. Since the pupillary aperture is smaller than the emergent pencil the magnification is always below M'. The size of the field being, for given values of F and f, a direct function of the aperture,* the Galilean telescope usually has a large relative aperture—$f/3$ is quite normal.

Since the ocular cannot produce a real image from a diverging beam, it cannot be made to focus upon a system of crosswires or any other real object.

The Galilean system is chiefly employed today in opera glasses, where its compactness and lightness are valuable. It has been superseded for astronomical purposes on account of the following disadvantages:

---

*Whence, incidentally, a stop placed over the OG not only reduces the brightness of the image and the telescope's resolving power (as with an astronomical refractor), but also the size of the field.

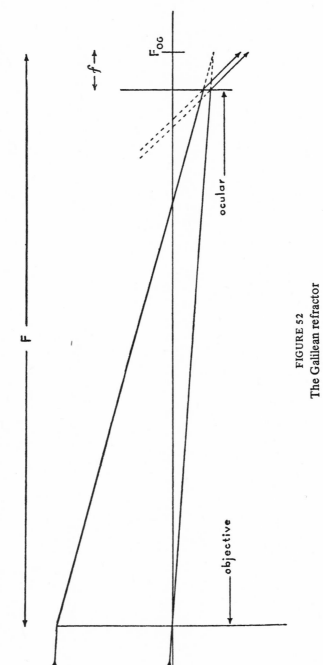

FIGURE 52
The Galilean refractor

(*a*) The full illuminating power of the objective is not used.

(*b*) Any defects in the objective—such as scratches or dust on the surface of the lens, or bubbles in the glass itself—are visible at the eyepiece.

(*c*) The illumination of the field falls off towards the edge.

9.3 Newtonian reflector

The Newtonian employs a concave paraboloid to form the primary image. At a distance inside the focal plane equal to a little more than the radius

of the tube a plane mirror—or, in some small instruments, a totally reflecting prism—is interposed to turn the converging pencil to a position where it is more accessible for examination by the ocular (Figure 53).

Flats are ground elliptical, their axes in the ratio $1:\sqrt{2}$, so that their projection on a plane inclined at $45°$ is circular. The minimum diameter of the flat necessary to reflect the whole pencil to the side of the tube (i.e. when its distance from F is $D/2$), again measured in a plane perpendicular to the optical axis, is given by

$$d=\frac{D^2}{2F}=\frac{D}{2f} \quad\ldots\ldots\ldots\ldots (a)$$

e.g. 15 cm $f/8$, 10 mm; 20 cm $f/6$, 17 mm; 30 cm $f/6$, 25 mm. However, the normal design of a Newtonian requires that the focal point be brought beyond the side of the tube to a point convenient for ocular access. Calling the distance from the focal point to the centre of the diagonal l, then

$$d'=D\times\frac{l}{F}=\frac{l}{f} \quad\ldots\ldots\ldots\ldots (b)$$

Most designs of drawtube impose a focal extension of about 10 cm (although less is desirable). The value of l therefore becomes $(D/2) + 10$; e.g. 15 cm $f/8$, 21·9 mm; 20 cm $f/6$, 33·3 mm; 30 cm $f/6$, 41·7 mm (column i, p. 131).

These formulae represent the geometry of a point image exactly on the axis of the mirror. In other words, the diagonal will not fully intercept the cone of light formed by objects away from the centre of the field, and vignetting will occur. If the diagonal is to be sufficiently large to reflect full illumination

FIGURE 53

The Newtonian reflector

129

everywhere across a field in the focal plane of linear diameter P, then

$$d'' = \frac{l(D-P)}{F} + P \quad \ldots \ldots \ldots \ldots \ldots \ldots \quad (c)$$

Since critical attention is normally applied only to regions near the centre of the field of view, while ocular and objective aberrations may also limit its usable area, it is not necessary to select a value of P that will give full illumination right to the edge of the field stop. At medium magnifications, a value for P of about 10 mm will give the practical appearance of a fully illuminated field, while a HP ocular will not require a value for the minor axis of the flat appreciably greater than d'. Taking the perhaps generous value of $P = 15$ mm as satisfying all wide-field requirements, we have the dimensions given in column ii.

Formula (c) is particularly applicable to instruments designed for prime-focus photography, where it is required to have full illumination across a given diameter on the emulsion. An alternative approach, presented in B. 7.65, is to calculate the minimum size of diagonal which will give tolerable vignetting at the edge of a standard linear field. Column iii gives appropriate sizes of minor axis where a fall-off of 0·5 mag, at the edge of a field 25 mm across, is tolerable. In this case, at the shorter focal ratios, the value corresponds to d', and there is full illumination only on the optical axis.

In general, d should always be kept below $D/4$, resulting in a loss of light, from obstruction, of about 6 per cent; $D/5$ (5 per cent obstruction) is preferable. Light loss due to imperfect reflectivity, assuming a fresh aluminium coat, is about 15 per cent. Light loss by absorption and reflection in a prism is normally in the region of 10 to 15 per cent. On this score, therefore, there is little to choose between prisms and flats.

While prisms do not require periodical re-coating, their use is confined to instruments of small D and not very large relative aperture. If the size of the converging pencil is such as to necessitate the side of the prism exceeding about 60 mm, or the relative aperture is greater than about $f/6$, a flat is preferable to a prism; (a) a prism of such size introduces extra light loss through absorption, and moreover is unlikely to be homogeneous unless very expensive; (b) the incident rays will not strike the prism face even approximately normally; (c) the prism introduces spherical over-correction, and some chromatic aberration, in proportion to the length of its side;* for a given prism it is proportional to the relative aperture.

---

*Many oculars suffer from the opposite errors, and in such cases the aberrations introduced by ocular and diagonal will partially or completely counteract one another, with consequent improvement of definition.

D (cm)	f/10			f/8			f/6			f/5			f/4		
	i	ii	iii	i	ii	iii	i	ii	iii	i	ii	iii	i	ii	iii
10	15·0	27·8	24·5	18·8	30·9	25·4	25·0	36·2	27·1	30·0	40·5	30·0	37·5	46·9	37·5
15	17·5	30·8	27·8	21·9	34·7	29·0	29·2	41·2	31·5	35·0	46·5	35·0	43·8	54·4	43·8
20	20·0	33·5	29·0	25·0	38·1	30·8	33·3	45·8	34·2	40·0	52·0	40·0	50·0	61·2	50·0
25	22·5	36·2	30·5	28·1	41·4	32·5	37·5	50·2	37·5	45·0	57·3	45·0	56·3	67·9	56·3
30	25·0	38·8	31·1	31·3	44·7	33·9	41·7	54·6	41·7	50·0	62·5	50·0	62·5	74·4	62·5
35	27·5	41·3	33·0	34·4	47·9	36·2	45·8	58·9	45·8	55·0	67·6	55·0	68·8	80·8	68·8
40	30·0	43·9	34·2	37·5	51·1	38·0	50·0	63·1	50·0	60·0	72·8	60·0	75·0	87·2	75·0

(Dimensions of minor axis of diagonal in mm)

Hence a prism should not be used with an over-corrected mirror, though it may improve the performance of an under-corrected one. A prism introduces a slight extension of the focal position; a 30-mm crown glass prism used with a 15-cm $f/8$ mirror lengthens F by about 11 mm; a 60-mm prism used with a 30-cm $f/4$ mirror, $\Delta F = 22$ mm. (See B. 7.37.)

The modifications of point and extended images caused by diffraction at the diagonal are discussed in section 2.2; they only attain importance when the diameter of the flat exceeds about $D/5$.

Inferior quality or faulty adjustment of the secondary reflector, especially if it is a prism, are among the commoner causes of disappointing performance by Newtonians. The reflecting surface should at no point depart from planeness by more than $0 \cdot 1\lambda$; nor should the flat be too thick, or the prism too massive, if trouble from differential cooling is to be avoided. The error introduced by a faulty flat is proportional to the linear error of the illuminated area of its surface; hence for a flat of uniform linear error performance will be improved by moving it near to the focus of the primary, where the diameter of the cone is smaller.

In general prisms, unless of the first quality, are likely to give worse performance that a flat, since three optical surfaces are involved instead of one. Small departures from the isoceles figure will prevent the entering and exit rays from being strictly normal to the surfaces, resulting in the distortion of star images into tiny spectra and the ruination of the definition of extended images. Inaccuracy of collimation can introduce extra-axial aberrations—astigmatism, and even coma.

The aberrations of the Newtonian consist of coma ($B=+0\cdot25$), astigmatism ($A=+1\cdot0$), and field curvature ($C=-1\cdot0$). Except for such colour as is contributed by the ocular (and the eye) it is perfectly achromatic; and if correctly figured is free of spherical aberration with parallel incident light. The image at the centre of the field can therefore be as nearly perfect as makes no difference; the field is limited by coma, which sets a limit of about $f/3$ on the relative aperture, above which the field of tolerable definition is hopelessly restricted.

The linear distance from the axis at which coma limits the field, for various focal ratios, is easily derived. Given that $B=+0\cdot25$, then from (c), p. 50, we have

$$\xi = \tfrac{3}{4}\left(\frac{D}{2f}\right)^2 \tan\theta$$

When f is taken as unity, $\tan\theta = H$ (see Figure 11), and

$$\xi = 3H\frac{D^2}{16f^2} \cdot \cdot \cdot \cdot \cdot \cdot \cdot \cdot \quad \cdot \cdot \cdot \cdot \cdot \cdot \cdot \cdot \cdot \quad (b)$$

132

The linear radius of the spurious disc, from equation (b), p. 15, is given by

$$r = 1.22\lambda \frac{f}{D}$$

whence, taking $\lambda = 560$ nm,

$$r = (6.8 \times 10^{-4})\frac{f}{D}\,\text{mm} \quad \dots \dots \dots \dots \quad (c)$$

The distortion of the image becomes prohibitive when ξ exceeds r by a factor of about 3. We may therefore write for the maximum permissible value of ξ

$$\xi'' = 3r$$

Hence from equation (c) above,

$$\xi'' = (2 \times 10^{-3})\frac{f}{D}$$

Substituting this value in (b), we have

$$3Y = (2 \times 10^{-3})\,\frac{f}{D} \cdot \frac{16f^2}{D^2}$$

whence

$$Y = 0.011 \left(\frac{f}{D}\right)^3$$

Y being the distance from the axis, measured in mm, at which coma becomes intolerable. Tabulating Y for various focal ratios, we have:

$\frac{f}{D}$	3	5	7	9	11	13	15
Y (mm)	0.30	1.4	3.7	8.0	14.5	24	37

The particular *advantages* of the Newtonian may be summarised as:

(a) perfect achromatism;
(b) ease of construction;
(c) cheapness, compared with equivalent refractor;
(d) superior light grasp, compared with refractors, when D is large;
(e) comfort and convenience of a horizontal eye position (though this is dependent on the provision of a rotating tube or section);
(f) shorter tube than refractor of same D, and therefore much more easily mounted.

133

And its *disadvantages*:

(a) definition impaired by tube currents (see section 11.2);
(b) definition impaired by diffraction effects at central obstruction (see section 2.2);
(c) observing positions tend to become awkward with some types of mounting when the focal length exceeds 180 cm or so, even with rotating tube or upper section (without the latter, some observing positions may be almost impossible);
(d) limited to relatively short focal ratios.

On balance, the disadvantages are completely eclipsed by the advantages.

9.4 Cassegrain reflector

A design of telescope whose peculiar advantages make its comparative neglect by amateurs a matter for surprise. As normally constructed, it consists of a paraboloidal primary mirror and a coaxially mounted hyperboloidal secondary, located inside the focal plane of the former. The focal ratio of the large mirror commonly lies between $f/4$ and $f/7$: these normal limits are dependent on, as regards the former, practical difficulties connected with the manufacture of the mirrors and their alignment in the instrument, and, as regards the latter, the impossibility of obtaining low enough magnifications since their fields of full illumination would be impossibly restricted.

With small Cassegrains focusing is normally effected by the movement of the secondary mirror, rather than of the ocular, along the optical axis; the focus is extremely sensitive to changes in the distance of the secondary from the focal plane, a small adjustment of the latter making a considerable difference to the former (Figure 54).

If we write:

f_1 for the focal length of the primary (PF),
f_2 for the focal length of the secondary,
s for the distance of the secondary inside the primary focus (ZF),
s' for the distance of the secondary from the final focal plane (ZF'),

then

$$\frac{1}{s'} = \frac{1}{s} + \frac{1}{f_2}$$

Values of these quantities are chosen so that a real image is formed at F', just behind the primary mirror; then $s' > s$. The equivalent focal length of the combination (i.e. the focal length of a single objective giving the same image scale) is given by

FIGURE 54

The Cassegrain reflector

$$\mathcal{J} = \frac{s'}{s} \cdot f_1$$

where s'/s is the magnifying power of the secondary mirror, or $\mathcal{J}/f_1$; in practice this is usually between 2 and 5.

The focal ratio of the primary being given by f_1/D_1, that of the combination is

$$\frac{\mathcal{J}}{D_1} = \frac{s'}{s} \cdot \frac{f_1}{D_1}$$

If the secondary is to be large enough to reflect the whole of the convergent pencil from the primary, the following condition must be satisfied:

$$\frac{D_2}{D_1} > \frac{s}{f_1}$$

The *advantages* of the Cassegrain can be summarised thus:

(*a*) It is the most compact design of visual telescope there is. This, a convenience at any time, particularly commends the Cassegrain to observers engaged on work where a large focal ratio or large linear focal length is desirable.

(*b*) The long equivalent focal length gives a large image-scale.

(*c*) Greatly reduced coma. It is sometimes stated that the Cassegrain is a coma-free system, but this is not so. The coma in the final image is as large as it would be with a single parabolic mirror of aperture D and focal length $\mathcal{J}$. The Cassegrain's freedom from coma, as compared with a Newtonian of the same aperture, derives solely from the fact that $\mathcal{J} > f$, where f is the focal length of a normal Newtonian; the extreme feebleness of the coma in a typical Cassegrain is a direct result of the comparative largeness of its focal ratio $\mathcal{J}/D_1$. The coma in any paraboloidal system (whether Newtonian or Cassegrain) is very weak at $f/15$, and virtually non-existent at $f/20$, over the restricted fields used in visual observation; but is negligible in neither at $f/8$. That the Cassegrain is not an aplanat, whilst being a near approach to one, is demonstrated by the fact that the scale of the image in the focal plane is a function of $\dfrac{XF' \cdot AF}{XF}$ which is not a constant for different positions of A, though it is very nearly so. To render a Cassegrain aplanatic the paraboloid would have to be deepened into a hyperbola, and the curvature of the secondary hyperboloid strengthened even further (see section 9.14).

(*d*) Freedom from spherical aberration. Similarly, the Cassegrain's freedom (compared with the average Newtonian) from spherical aberration is dependent solely upon its greatly reduced relative aperture, and to no miraculous property in the combined action of the two mirrors. In fact the image formed by M_1 is as free from spherical aberration as that of a similarly figured Newtonian paraboloid: to keep the pencil stigmatic after reflection by M_2 it is necessary that the increase in the length of the optical path, due to this reflection, should be a constant for all rays, whether paraxial or peripheral. The curve of M_2 which will maintain $XF'-XF$ a constant is a hyperbola with foci at F and F'—which is why a hyperboloid is chosen for the secondary mirror.

Owing to its comparative freedom from spherical aberration, oculars which are not highly corrected, and which are therefore useless with Newtonians of large relative aperture, perform as well with a Cassegrain as with a refractor. In this respect the Cassegrain combines the advantages of a refractor and Newtonian reflector.

Disadvantages of the Cassegrain:

(*a*) Contrast in the image is apt to be seriously diminished by direct light from the sky falling on it. This, however, can be completely obviated by enclosing the convergent pencil from the secondary mirror in a tube mounted in the aperture of the primary mirror, concentric with the optical axis. When the tube is correctly placed, an eye situated on the optical axis in the focal plane should see nothing but the small mirror and the circular reflection of the primary in it. Thus all light not contributing to the formation of the image is excluded. Alternatively, and more commonly in small instruments, an ocular diaphragm exactly limiting the exit pencil may be employed, though it is rather tricky to locate with precision.

The correct design of a Cassegrain shade tube is a matter of some importance, if the optimum light-sealing effect is to be achieved. The principle is shown in Figure 55, where m_2 represents the actual diameter of the secondary mirror M_2 which is used in forming the image of an object lying on the optical axis of the telescope. The cone m_2-F' must, therefore, be isolated by the shade tube. The outline of M_2 is silhouetted on the face of the primary mirror M_1 as the imaginary shadow M_2'; the intersection of the lines joining the opposite edges of M_2' and F with the edges of the cone m_2-F', indicates the optimum position for the front of the shade tube. If the tube is longer than required, it effectively reduces the aperture; if wider, it increases the diameter of the central obstruction.

Such a design intercepts skylight only at positions on or near the optical axis. Improved sealing is obtained by making M_2 considerably larger than m_2, but at the cost of an increased central obstruction. The radical solution to the problem is to employ a transfer lens (section 9.11 1)

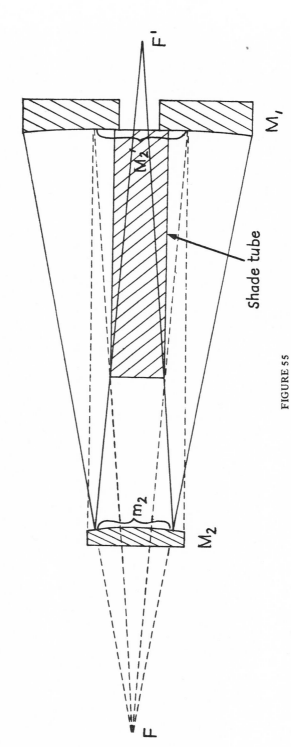

FIGURE 55

Design of shade tube for a Cassegrain-type telescope

(*b*) Small field, the corollary of large image-scale. This should be taken into account when considering the type of instrument best suited to a particular branch of observation—it represents no drawback in planetary work, for example.

(*c*) Astigmatism and field curvature are more marked than with an equivalent Newtonian, though this is to some extent mitigated by the virtual freedom from coma.

(*d*) The hyperbolic secondary is comparatively difficult to figure, a fact which in the past has tended to restrict the Cassegrain design to fairly large instruments.

(*e*) The size of the secondary (rarely less than $D/4$) can introduce undesirable diffraction effects (section 2.2), which tend to impair the definition of critical detail in extended images, and under certain circumstances to interfere with the resolution of close point images. Use of a transfer lens (see above) permits a greatly reduced secondary diameter.

(*f*) Inconvenient observing positions at high altitudes, involving the use of a diagonal.

For a very useful discussion of image-formation in Cassegrain-type telescopes, see B. 7.83, 84.

9.5 Dall-Kirkham reflector

By substituting a spheroidal secondary mirror for the hyperboloid of the classical Cassegrain design, a stigmatic image can be obtained if the primary mirror is given a prolate ellipsoid or 'under-corrected' figure. The fact that outfield coma in the system is considerably greater than that of the Cassegrain is a relatively unimportant consideration when the small visual field of this type of instrument is taken into account; while the curve of the secondary mirror is one which should be produced automatically by the optical polishing process. The cost of a set of Dall-Kirkham mirrors is therefore somewhat less than those for a Cassegrain telescope.

In a typical Dall-Kirkham system, the prolateness of the primary mirror amounts to about 90 per cent of the correction required for a true paraboloid.

9.6 Cassegrain-*coudé* reflector

Like the Cassegrain it employs a coaxial hyperboloid mounted just inside the primary focus; but instead of the converging pencil passing through the centre of the primary, it is reflected to the side of the tube by a diagonal mounted a short distance from the centre of the paraboloid (Figure 56).

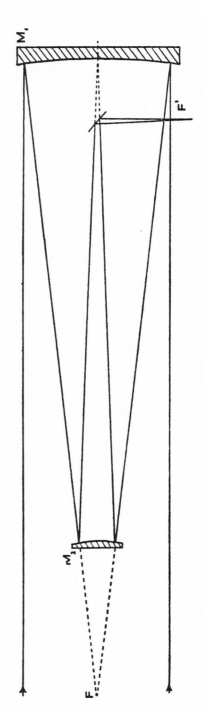

FIGURE 56

The Cassegrain-*coudé*

Advantages:

(*a*) Proximity of the ocular to the Dec axis provides a nearly stationary observing position; it is also nearer to the ground than with the orthodox Newtonian.

(*b*) The distance s' is no longer a function of f_1, since F' does not have to fall behind the primary mirror. By shortening s', the amplification of the secondary can be reduced without having to increase s and, therefore, the diameter of the secondary itself.

(*c*) An existing Newtonian mirror, of suitably short focal ratio, can be converted to Cassegrain use without having to perforate it and, in all probability, spoil the existing figure.

Disadvantages:

(*a*) Reversal of the image.
(*b*) Smaller field of full illumination than with a normal Cassegrain.
(*c*) Small light loss at the additional reflecting surface.

9.7 Cassegrain-Newtonian reflector

Instead of some of the features of each arrangement being combined, as in the last type described, the two arrangements may be incorporated in a single mounting for alternative use. The Cassegrain hyperboloid is interchangeable with a Newtonian flat, and the paraboloid is bored to give passage to the rays converging to the Cassegrain focus. Thus a single mirror can be used at two different focal ratios.

Although a sound scheme in theory, difficulty will probably be experienced in restoring the different optical systems without having to spend time in re-collimation.

9.8 Gregorian reflector

The objective is a concave paraboloid, as in the Newtonian, but instead of a secondary flat or hyperboloid, the convergent pencil is brought to a focus by means of a concave ellipsoid mounted coaxially with the paraboloid; the ocular is mounted on the optical axis, as in the Cassegrain. The focal length of the secondary is considerably shorter than that of the primary; their distance apart is slightly greater than the sum of their focal lengths (Figure 57). The relative sizes of the images at P and F_1 will be as $XP:XF_1$. The conditions for the final image to be stigmatic are that the image formed by M_1 should be so, and that the foci of the ellipsoidal M_2 should lie at F_1 and P.

The Gregorian telescope, as usually constructed, had a fixed ocular,

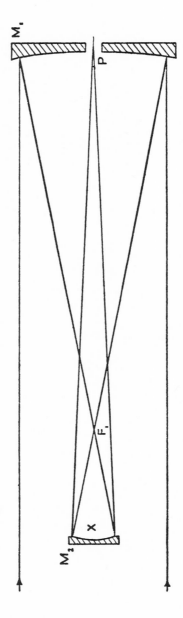

FIGURE 57
The Gregorian reflector

focusing being effected by adjusting the position of the secondary along the optical axis. The image, as seen, is erect.

It possesses the advantage over the Cassegrain-type telescopes that the concave secondary mirror forms a real image of an object and can be tested by orthodox optical techniques. However, the fact that the tube is some 30 per cent longer than that of the equivalent Cassegrain has caused the system to fall into disuse.

9.9 Aplanatic reflecting telescopes

We have seen that the single curved reflecting surface of the Newtonian cannot be rid of coma whilst at the same time escaping spherical aberration; but that by increasing the focal ratio, as in the Cassegrain, it can be reduced to negligible proportions over a restricted field such as is adequate for visual observation. But increasing the focal ratio by increasing F means, with paraboloids, restricted fields; on the other hand, achieving it by decreasing D means reduced resolving power; and increasing it by any method whatever means reduced photographic speed—if it is greater than about $f/8$, prohibitively long exposures are needed for faint objects. These conflicting demands have led to the development, in fairly recent years, of various two-mirror systems and lens-and-mirror systems which combine large relative aperture (and therefore photographic speed) with freedom from spherical aberration, whilst obeying the sine condition giving freedom from coma over wide fields.

Of the remaining aberrations, the effects of field curvature are overcome by distorting the photographic plate to the same curve as that of the focal surface; while the sting is taken out of astigmatism when it is pure, and not combined with coma, since the images are then symmetrical and uniformly illuminated and are therefore as accurately measurable as perfectly circular images. (See further B.7.57.)

9.9.1 Ritchey-Chrétien: The aplanatic form of Cassegrain, using hyperboloidal curves on both mirrors. Designed during the 1920s, it gives wide fields at $f/6$ to $f/10$, from which both coma and spherical aberration are absent. The focus, as in the Cassegrain, is carried just behind the primary mirror, where it is visually accessible. Reduction of field curvature and astigmatism are gained at the expense of light grasp by increasing the size of the seconday mirror; as usually constructed, the compromise arrived at is $D_1/D_2 = 2$ approximately, which gives a fair degree of correction without intolerable light loss.

At $f/8$ the circle of confusion of the astigmatic image of a star lying $30'$ from the axis of a 80-cm Chrétien is about 0·053 mm in diameter. That this is nearly seven times as large as the spurious disc is immaterial so long as the instrument is not used visually, since diffusion in the photographic emulsion will prevent the resolution of two stellar images whose linear separation in the focal plane is less than 0·05 mm. Further than 45′ from the optical axis, the image of the Chrétien is actually larger than that of the equivalent Newtonian would be; but whereas the latter is comatic (therefore asymmetrical as regards both shape and intensity), the former is symmetrical and uniformly illuminated.

As usually constructed, the Ritchey-Chrétien has a primary focal ratio of between 2 and 4·5, and a final focal ratio of about 9. Primary focal ratio and amplification must be very near the optimum relation for the coma-free condition to be satisfied. The focal surface is slightly curved. See, further, B. 7.7.

9.9.2 Schwartzchild reflector: Invented as long ago as 1905, the Schwartzchild is a two-mirror aplanat with aspherical mirrors, giving freedom from coma and spherical aberration over a wide, uncurved field (3° diameter at $f/3\cdot5$, 1°5 at $f/1\cdot2$). Astigmatism is retained.

Hence it combines the photographic desiderata of symmetrical images over a wide field, and extremely short exposures. The focal plane (Figure 58) is located between the two mirrors and nearer to the secondary; the latter diminishes, not magnifies, the primary image, its only function being to correct the coma of the large D/f primary.

The focus is therefore inaccessible for visual inspection, which is disadvantageous since necessitating a separate guide telescope. The Schwartzchild may therefore be considered as having crossed the boundary, ill-defined though it is, between telescopes and cameras.

The heavy light loss through silhouetting (about 25 per cent) is more than compensated for by the phenomenally large relative aperture.

9.9.3 Couder reflector: Generally similar to the Schwartzchild, the Couder not only obeys the sine condition but is anastigmatic as well; freedom from astigmatism, however, is bought at the price of reintroduced field curvature, the radius of the focal

FIGURE 58

The Schwartzchild reflector

144

surface being about $f/2$. Like the Schwartzchild, and unlike the Chrétien, the secondary is concave; it is, in fact, an ellipsoid. The foci of the two mirrors are brought together, so that $f = \frac{s}{2}$, where s is the separation of the two mirrors. It is thus a less compact instrument than the Chrétien, but since it works well at $f/3$ this is no great inconvenience. It does, however, share the more real weakness of the Schwartzchild of having a visually inaccessible focal plane.

9.10 Unobstructed reflecting telescopes

Diffraction effects around the diagonal or secondary, and its supporting vanes, surround the image formed by a reflecting telescope with a certain amount of scattered light which can make low-contrast features (especially planetary detail) appear faint or even invisible. A refracting telescope, on the other hand, although free of such diffraction effects, is afflicted with secondary spectrum whose harm is in some ways analogous. In theory, optimum image contrast will be provided by a reflecting system whose optical path is free from any diffracting elements; hence the attraction, particularly to planetary observers, of the unobstructed reflector. For a general discussion of unobstructed telescopes, see B. 7.21.

9.10.1 Herschelian reflector: Of mainly historical interest, this form was developed by Herschel in the 1780s to avoid a second reflection at the Newtonian diagonal, which in the days of speculum-metal mirrors involved a light loss of some 40 per cent (Figure 59).

By tilting the optical axis of the mirror at a small angle to the incident pencil, the image at the primary focus was directed to the side of the tube,

FIGURE 59
The Herschelian reflector

where it could be conveniently examined by an ocular pointing to the centre of the mirror.

In order to reduce off-axis astigmatism to acceptable limits (i.e. below the resolving power of the aperture), F has to be made very long; for example, $f/23$ if $D=10$ cm; $f/27$ at $D=16$ cm; $f/30$ at $D=20$ cm. The design has therefore been abandoned, except as a curiosity. B. 7.64 describes a folded $f/26$ Herschelian, while B. 7.20 explains how the

145

FIGURE 60
Two forms of folded Herschelian reflector

aberrations of a tilted $f/10$ spherical mirror may be corrected, using two inclined lenses.

9.10.2 Off-axis Newtonian: Unlike the Herschelian, which was a mere expedient, the true off-axis telescope has as its aim the removal of the secondary obstruction from the light path. This effect can be achieved (Figure 61) by placing an eccentric stop of aperture D over a mirror of aperture D', but this is only practical when D' is large. Alternatively, a

FIGURE 61
Effective design of off-axis reflector

mirror can be figured to a special off-axis form, so that its surface resembles the curve to be found inside such an eccentric stop placed over a regular paraboloid. Such off-axis paraboloids are commonly found in industrial collimators, but rarely in astronomical systems because of their great cost.

9.10.3 Schiefspiegler: In its basic form, this instrument resembles an unobstructed Cassegrain-type design (Figure 62). Both mirrors have spherical curvatures, and the system is so designed that the astigmatism introduced by the inclined primary is eliminated by that of the inclined secondary, while the coma introduced by the latter is kept below the resolving limit of the aperture. In order to achieve this, the focal ratio of

146

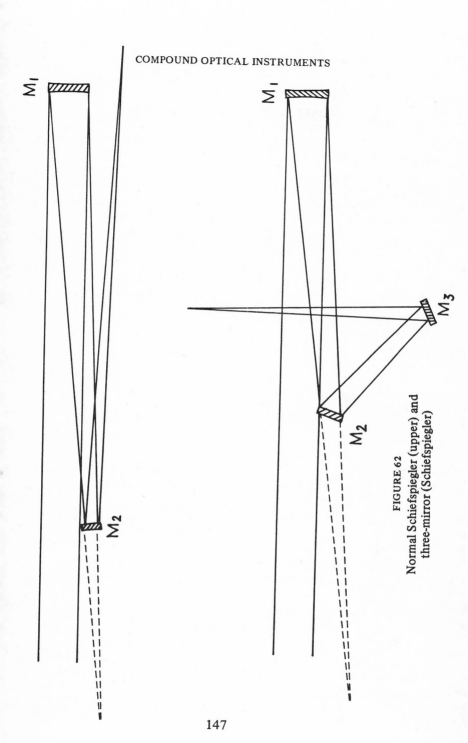

FIGURE 62
Normal Schiefspiegler (upper) and three-mirror (Schiefspiegler)

the primary has to be relatively large (about $f/12$), and this is approximately doubled by the action of the secondary. The restricted field of view and necessarily high magnification of the Schiefspiegler makes it a specialist class of instrument. In a convenient design by Kutter (B. 7.52, 90), both primary and secondary mirrors have the same radius of curvature and can be ground in the same operation.

Residual coma and astigmatism limits the aperture of the two-mirror Schiefspiegler to about 10 cm. In larger apertures, the residual aberration can be suppressed by interposing a corrected lens near the focus, or by interposing a third inclined mirror into the beam (B. 7.10, 53). Typical f/ratios for these systems are from 20 to 30, and the chief practical drawback is the high sensitivity of the mirrors to errors of tilt and position.

9.11 Reflecting systems with relay or compensating lenses

The use of small supplementary lenses, placed in the beam of a Newtonian or Cassegrain-type telescope, can confer either or both of two main benefits: (a) transferring the focal image to a more accessible position; (b) correcting aberrations conferred on the image by the primary optical train.

9.11.1 Cassegrain reflector with transfer lens: A number of advantages accrue from the use of a transfer or relay lens in Cassegrain systems (Figure 63). They include:

(a) By placing a diaphragm exactly isolating the image of M_2 formed by the transfer lens, only the light reflected by M_1 reaches the focus, and sky flooding is entirely eliminated.

(b) The distances s and s' are both much shorter than in the normal Cassegrain system, which means that the diameter of M_2 can be reduced, usually to less than $D/5$.

(c) The final image is erect.

(d) The effective focal length, $\mathcal{F}$, of the system can be varied to some extent by altering the distance between the transfer lens and ocular and, hence, the amplification produced by the lens.

A correctly designed doublet, used in the fairly long-focus beam of a typical Cassegrain, will add negligible secondary colour to the image. The transfer lens is usually designed to amplify the focal length by a factor of 2, and a 15-cm $f/8$-$f/16$ system was described in B. 7.24.

9.11.2 Sampson catadioptric: In 1914, Sampson proposed a lens-and-mirror (catadioptric) system, designed to improve the coma and field curvature of the ordinary Cassegrain (B. 7.67). The figure of the concave primary is intermediate between spherical and paraboloid; the secondary is a convexo-concave lens coated on its rear surface; and about two-thirds

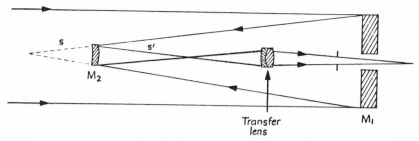

FIGURE 63
Cassegrain-type reflector with transfer lens

of the distance from it to the primary is situated the 'corrector', consisting of a double concave lens and a nearly plano-convex lens mounted coaxially with the mirrors. This corrects coma while introducing negligible chromatic aberration; complete correction of spherical aberration is obtained with negligible field curvature.

Theoretically this is a near approach to an aberration-free system, and the fact that all surfaces except that of the primary are spherical makes for ease of manufacture. The correction of the telescope, however, is extremely sensitive to the centring of the correcting lenses, and in practice this would be the chief drawback of the design.

9.11.3 Buchroeder catadioptric: A modified Cassegrain using, like the Sampson telescope, a mangin secondary, was described by Buchroeder in 1968 (B. 7.9). All surfaces are spherical; the colour introduced by the secondary (which correctes the spherical aberration of the primary mirror) is compensated by a triple erector lens.

The original purpose of the Buchroeder design was to eliminate aspheric reflecting surfaces. As with the Sampson telescope, the image is extremely sensitive to adjustment of the various components. (See Figure 64.)

9.11.4 Dillworth catadioptric: Similar in concept to the Buchroeder design, but requiring only two relay lenses, this aplanatic design was described in B. 7.33.

9.12 Schmidt-type systems

The spherical mirror, whilst being free from the abaxial aberrations (of which the most important is coma), and therefore allowing the use of wide fields, nevertheless suffers from spherical aberration uniformly over the whole field. One method of correcting this is to deepen the curve into a paraboloid; but abaxial rays are then no longer axes of revolution and the

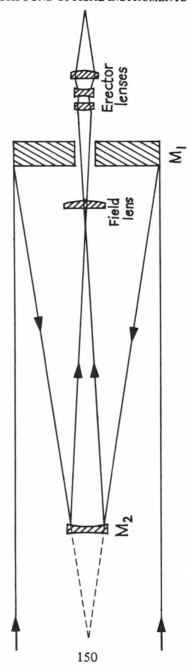

FIGURE 64
Buchroeder catadioptric reflector

coma thus introduced limits the field progressively as the relative aperture is increased. Schmidt's brilliant contribution to telescope design (B. 7.69), made in 1930, consisted in indicating how the spherical aberration of a concave spheroid could be corrected without the introduction of coma in exchange.

The result is an optical system giving good definition over phenomenally wide fields, and at relative apertures providing photographic speeds never before envisaged. For example, a $f/1$ paraboloid would have a well-defined field measurable in $''$ arc, while the coma-free field of the $f/5$ 2·5-m reflector is only $7'.2$, and that of the 5-m about $3'.6$; yet a $f/1$ Schmidt can cover something like $20°$, recording as much detail of extended objects in a few minutes as a first-class lens would require several hours to show.

A concave spheroid possesses the following attributes:

(a) Spherical aberration, which can be reduced to below the Rayleigh limit of $\frac{\lambda}{4}$ by placing a diaphragm of aperture $\not> f/10$ on the optical axis, at the centre of curvature; for its restriction of the diameters of the pencils will mean that the central and peripheral beams are close enough together to be brought to virtually the same focus. This, however, is an impracticable solution, since at $f/10$ insufficient aperture is obtainable unless f is very large.

(b) Field curvature: the focal surface being a section of a sphere concentric with the mirror and of radius equal to $f=R/2$ (the arc FF' in Figure 65).

(c) Freedom from coma by satisfaction of the sine condition, and similarly anastigmatic and undistorted. Since the central ray of any pencil, no matter what its inclination to the axis of the mirror, has to pass through C, it is as much an axis of rotation of the system as AS; i.e. BT could equally well be considered 'the' axis, by tilting the diaphragm until perpendicular to BT. Optically, owing to the sphericity of the mirror and the smallness of the diaphragm, the two pencils centred on AS and BT are indistinguishable.

9.12.1 Schmidt camera: In this instrument (Figure 66), the advantages (c) are retained by employing a spherical mirror.

(b) is retained, whereby the photographic plate has to be warped by pressure to the correct curvature ($r=f$). The curvature of the focal surface can be practically eliminated, for relative apertures less than about $f/5$, by means of a plano-convex lens in contact with the photographic plate.

(a) is removed by means of a thin, aspherical, almost plane parallel correcting plate which replaces the diaphragm of the above account,

FIGURE 65

Principle of the Schmidt reflector

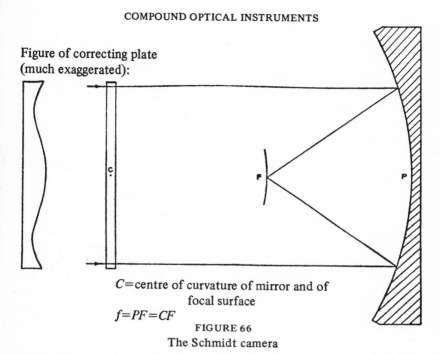

Figure of correcting plate
(much exaggerated):

C=centre of curvature of mirror and of
focal surface

$f=PF=CF$

FIGURE 66

The Schmidt camera

i.e. is situated at the centre of curvature, C (the only point through which all rays pass)*; this introduces into the beam incident to the mirror spherical aberration equal to that of the mirror but of opposite sign. The design of the plate varies,† giving more or less chromatic aberration, but fundamentally it is equivalent to an axial convex lens surrounded by a ring of concave lenses; when its central and peripheral thicknesses are made equal, its power is zero. Being thin, it causes negligible light loss and negligible colour error down to about $f/1$; beyond this the elimination of chromatic aberration requires rather impracticably steep curves for the correcting plate, and it is preferable to make it in the form of an achromatic doublet, or to resort to the thick mirror or solid Schmidt modifications of the basic design. The correcting plate can provide theoretically perfect correction for only a single wave-length, N varying with λ, but the residual chromatic inequality of spherical aberration is comparatively unimportant in a photographic instrument, though it would be intolerable for visual work.

---

*As the plate is moved away from the centre of curvature, the diameter of the corrected field decreases. When a very wide field is not required, this is a convenience which is made use of in some designs of Schmidt.

†Its curvature may be on the outer or inner surface, or distributed between them.

The shortcomings of the Schmidt may be summarised as follows, the first three being important, the last two less so:

(*a*) Tube length is about twice the effective focal length of the system, i.e. a Schmidt is twice as long as a Newtonian of equal *f*.

(*b*) Field curvature: when the plate is flattened again after exposure, distortion occurs.

(*c*) Inaccessible focal surface, making the loading and changing of plates rather awkward.

(*d*) Residual colour: negligible at relative apertures smaller than about $f/1$.

(*e*) Silhouetting (amounting to at least 10 per cent) by the plate holder.

A practical difficulty with all photographic systems where the emulsion has to be placed in the oncoming light path, is the restriction on the available linear field of view to approximately $D/2$. This means that, for a given aperture, angular field coverage varies inversely with F, being theoretically about $25°$ at $f/1$, but only $5°$ to $6°$ at $f/5$. The photographer wishing to cover a large angular field will find an anastigmat of more use if he is troubled with fogging at very large relative apertures. An $f/3$ Schmidt camera, covering a field of up to $10°$ without intolerable inconvenience, is a useful compromise for most purposes.

9.12.2 Lensless Schmidt: It was pointed out above that the achievement of images reaching the Rayleigh limit would require a prohibitively small diaphragm at the centre of curvature of the spherical primary mirror. However, definition of this quality is unnecessary for the small-scale photography at which the Schmidt excels, since the typical resolving power of a photographic emulsion is about 0.05 mm. The approximate diameter d of a star image produced by a spherical mirror is given by

$$d = 0.008 \frac{D}{f/^2};$$

hence, a lenseless Schmidt camera will perform perfectly satisfactorily (i.e. its images will be no larger than the normal developed grain size) when

$$f/ \not> \sqrt{1.6D}$$

giving the following limiting focal ratios for systems of various aperture:

Aperture (cm)	Focal ratio
10	4·0
15	4·9
20	5·7
25	6·3
30	6·9

Fast Schmidt cameras exhibit rapid sky fogging unless the sky is very dark, and the lensless Schmidt would appear to offer useful possibilities to the amateur, since a spherical primary mirror is the only optical component. See B. 7.1.

9.12.3 Schmidt-Cassegrain: The application of the Schmidt plate principle makes possible the manufacture of fast Cassegrain-type instruments for either photographic or visual use with improved field characteristics over the basic two-mirror form. The true Schmidt-Cassegrain is almost flat-fielded and free from both astigmatism and coma, but both mirrors, in addition to the corrector plate, must be figured; the radii of curvature of the two mirrors are similar. At the cost of some compromise in performance, spherically-curved mirrors can be employed. The best-known commericially produced telescope of this type is the Celestron. A Schmidt-Cassegrain system using spherical optics and giving a flat field was described in B. 7.30.

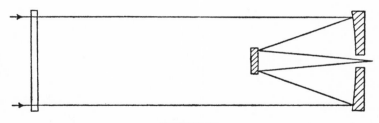

FIGURE 67

The Schmidt-Cassegrain reflector

A further development of the Schmidt-Cassegrain, due to Linfoot, is a family of quadruple systems (two spherical mirrors with two aspherical correcting plates) which combine flat field and anastigmatism with apochromatism: this is made possible by arranging that the aspherical plates have compensating colour errors (Figure 68).

FIGURE 68

Schmidt-Cassegrain reflector (Linfoot)

9.12.4 Baker camera: A family of two-mirror systems described by Baker (B. 7.5), with remarkable performances: aplanatic, anastigmatic, flat-fielded, and free from distortion (Figure 69).

With two spherical mirrors astigmatism is present but negligible; the tube is only three-quarters the length of a Schmidt of the same focal

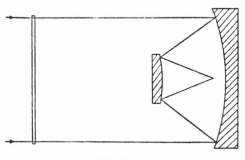

FIGURE 69
The Baker camera

length; in one variant the tube length is reduced to $\frac{1}{3}f$. Figuring one mirror gives complete anastigmatism; figuring both gives anastigmatism and freedom from distortion. The colour error, however, is 3 to 4 times that of the Schmidt; on the other hand it is still very small, even at $f/2$.

9.12.5 Wright camera: Wright has shown (B. 7.86, 7.87) that two important advantages follow from the replacement of Schmidt's spherical mirror by an ellipsoid, the appropriate modifications being made to the correcting plate: the ratio tube-length/focal-length would be reduced from 2 to 1, and the field would be flattened (Figure 70). The focal plane now lies close to the correcting plate, immediately behind which the photographic plate is mounted.

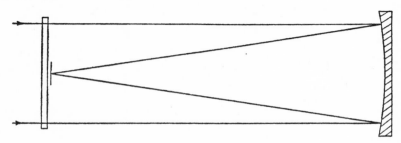

FIGURE 70
The Wright camera

156

The price paid for these advantages is the introduction of astigmatism, which limits the usable field; the colour error is also about twice that of a basic Schmidt of the same focal ratio; on the other hand, it is still very small. The diameter d of the enlarged astigmatic image at an angle θ from the axis, where E is the mirror-corrector plate separation (approximately equal to F), is given by

$$d = \tfrac{1}{4} \frac{E}{f/} \times \tan^2 \theta .$$

The theoretical performance of the Wright camera is superior to that of a Swartzschild aplanat (9.9.2), and it has less astigmatism in a flat field than any other aplanatic telescope with only two components; used as a Newtonian, an $f/3\cdot6$ Wright has a diffraction-limited field about 15 times the diameter of an orthodox $f/4$ paraboloid.

Design and fabrication details for two Wright instruments are given in B. 7.79, 82.

9.13 Maksutov systems

In 1944, D. D. Maksutov described a family of aplanats in which the aspheric corrector plate of the Schmidt is replaced by a steeply-curved, thick, spherical meniscus lens. This meniscus, of almost uniform thickness, is sensibly achromatic, but contributes the negative spherical aberration required to neutralize the positive aberration of the mirror. At faster relative apertures than about $f/3\cdot5$, some apherising of the corrector is required to bring the performance of the system within Rayleigh's limit, but this will be unnecessary if the instrument is designed for prime-focus photography.

camera has some advantages over its Schmidt equivalent: multiple reflections within the corrector plate, and hence ghost images, are avoided, while the tube length is only about two-thirds of that of the equivalent Schmidt for complete suppression of coma (see B. 7.88). The disadvantage, at any rate in large apertures, is the difficulty of manufacturing and mounting the necessarily heavy meniscus corrector, and in these sizes the Schmidt camera reigns supreme.

9.13.2 Meniscus-Schmidt camera: A meniscus which is exactly concentric with the mirror introduces some primary colour, as well as residual spherical aberration. Its errors are, however, virtually uniform over the whole field. Linfoot and Hawkins have shown (B. 7.54) that by combining such a concentric meniscus-mirror system with an aspherical correcting plate at the centre of curvature, a $18°$ field at $f1\cdot3$, at all points in

157

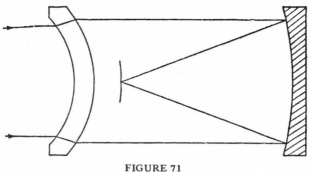

FIGURE 71
The Maksutov camera

which 97 percent of the light of each image is sent into a circle of confusion of smaller diameter than 12″, may be obtained. This is a considerable improvement on the performance of a conventional Schmidt of the same focal ratio. Not only does the correcting plate (which is

FIGURE 72
The Meniscus-Schmidt camera

weaker, and therefore easier to manufacture, than that of the Schmidt) eliminate the residual spherical aberration, but if made as a doublet with a spherical interface it also corrects the primary colour of the concentric meniscus (Figure 72).

9.13.3 Maksutov-Cassegrain: The Maksutov system has achieved considerable popularity in the commercial and home-built telescope fields, not so much as a Newtonian-type instrument (since the work involved in making the meniscus would exceed that of figuring a paraboloid of orthodox focal ratio), but in the Cassegrain form. For full aplanatic performance, a separate secondary mirror is required, but excellent results across typical visual fields can be obtained by bringing the meniscus so close to the primary mirror that the centre of its convex rear surface, suitably

158

aluminised, acts as a spherical secondary mirror (Figure 73). Since the meniscus can be matched with a spherical primary of relative aperture as wide as $f/2 \cdot 5$ without involving more than slight residual figuring, an extremely compact instrument results; e.g. a 15-cm $f/15$ system with a tube only 40 cm long (B. 7.39). Dall describes an erect-image Maksutov-Cassegrain, with reduced central obstruction, in B. 7.25; a design employing low-cost white plate glass for the corrector appears in B. 7.61.

A large literature is growing up around the Maksutov telescope; see, for example, B. 7.26, 39, 91.

FIGURE 73
The Maksutov-Cassegrain reflector

THE MOUNTING OF THE OPTICAL PARTS

10.1 Introduction

Object glasses, mirrors, and secondary reflectors should be mounted so that they are

- (*i*) adequately supported to combat flexure;
- (*ii*) held firmly enough for there to be no movement when the orientation of the telescope is changed in any possible way;
- (*iii*) not so tightly gripped that the glass is subjected to mechanical strain;
- (*iv*) adjustable in every respect required for collimation, centring, and squaring-on;
- (*v*) as free as possible from the effects of differential cooling, dew deposit, and tube currents.

Two accessories of the objective cell may conveniently be mentioned at this point—the protector cover and, in the case of OGs, the dewcap —though there is little to say about them except that the objective must always be kept covered when not in use, but not be covered when dewed-up, and that the dewcaps usually supplied with small refractors are much too short to be any use.

10.2 Mirror cells

While being strong enough to give proper support to the mirror, the cell must not be so massive that it delays unduly the attainment of thermal equilibrium between the mirror and the night air. It must, further, provide uniform support for all parts of the mirror, limit the latter's possible movement to a small fraction of a millimetre, yet not subject the glass to such constraint that thermal expansion and contraction set up strains in it. And this for all positions of the tube.

The detailed construction of cells varies widely, but certain general principles can be stated. For mirrors of $D < 20$ cm approximately, a three-point support is adequate (Figure 74). The supporting screws should lie on a circle, concentric with the mirror, of radius $\frac{1}{2}\sqrt{\dfrac{D^2}{2}}$. Each pair (lateral and supporting) of screws lies on a single radius of the cell, the three radii being separated by 120°. Round-ended brass screws should be used. Between the retaining screws and the rim of the mirror there should be a clearance of a small fraction of a millimetre. There should be a good clearance between the front lip of the mirror and the safety clips, if fitted.

As shown on page 166, the flexibility of mirrors of similar thickness/diameter ratio increases as D^2. When D much exceeds 20 cm, the simple 3-point support ceases to be satisfactory unless the disc is of extraordinary thickness, and some form of balanced multi-point support system must be used. 9- and 18-point systems are advised for mirrors of between about 25 cm and 60 cm aperture; larger sizes still require more sophisticated mounting if the optical surface is to maintain its true figure and alignment at all inclinations of the mirror when used in the telescope.

FIGURE 74
Support system for a small mirror

The principle behind any support system is to arrange for each point of contact with the glass to be supporting an equal weight of material. These points are then linked together so as to provide three fundamental supports to the floor of the cell.

A 9-point support is built up from three triangles, as shown in Figure 75. The adjusting screws through the cell base bear on the centre of gravity of each triangle, which itself carries a small steel ball at each corner. If constructed in accordance with the dimensions given, each ball will be bearing exactly 1/9th of the total weight of the mirror. Each triangle is free to rock on its bearing screw; a loose-fitting pin prevents it from twisting out of the correct orientation. The 18-point support consists of pairs of triangles linked together, and, once again, the couplings between each triangle and its linking bar, and between the bar and its bearing screw, must be free to set themselves in the required plane.

The edge support for a large mirror is a matter of some importance, for as the altitude of the tube lessens the lower rim must take an increasing component of the disc's weight. Altazimuth telescopes always retaining the same configuration, the mirror in such an instrument can be slung within a peripheral band, which provides perfect support around the lower 180° of the disc. An equatorially-mounted mirror must be provided with support all around its edge; a ring of Teflon points, adjustable to leave only a fractional amount of play, will allow the disc to settle back on its supports when the tube is raised from a low elevation. This tendency is helped by restricting the edge support to the zone near the front surface. See, further, B. 7.45.

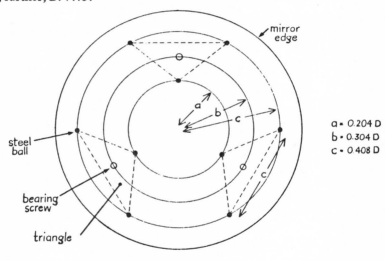

a = 0.204 D
b = 0.304 D
c = 0.408 D

FIGURE 75
Design for a 9-point flotation system

Any inequality in the pressure exerted by the peripheral supports is liable to introduce astigmatism; insufficient central support, allowing the axial region of the mirror to sag, will cause spherical over-correction. Flexure with a three-point support will first be detected in the image by a slight brightening of the rings in three directions equidistant from one another; if gross, it can produce triangular star images.

The disadvantages of these orthodox types of mirror mounting in a solid cell will be referred to in section 10.4.

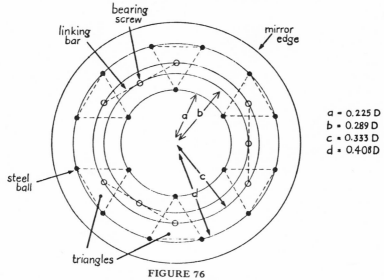

FIGURE 76

Design for an 18-point flotation system

10.3 Mounting the mirror cell

Although a common method is to incorporate squaring-on adjustments directly in the mirror's rear support system, some mirrors, particularly large ones, are mounted permanently in a cell which is itself adjustable in relation to the tube.

The manner in which the cell is attached to the telescope tube depends very much upon the design and shape of the latter. Generally speaking, there must be three points of contact; at least two of these must be adjustable, the third having a little play to allow this adjustment to be made.

Pairs of antagonistic screws are commonly employed (Figure 77), but the disadvantage of these is that their adjustment cannot be made from the observing position. A more convenient arrangement is shown schematically in Figure 78. The rod A is carried up the telescope tube to within reach of the observing position; by turning it the cell flange can be pushed further from, or pulled nearer to, the upper end of the telescope, acting against the spring S.

10.4 Differential cooling of mirror and its surroundings

Ideally, during observation all parts of the objective should be at the same temperature as the surrounding air. In fact the mirror cools more

FIGURE 77

Attachment of mirror cell to telescope tube

FIGURE 78

Remote-adjustment facility for mirror cell

164

slowly than air (and different parts of it at different speeds), so that thermal equilibrium is not attained for a considerable time after it is uncovered to the night air.

Under normal observing conditions the temperature of the mirror falls throughout the night, though not at a steady rate; the result is unequal contraction and loss of the theoretical figure until equilibrium is attained. In a closed cell, and without forced ventilation, the front of the mirror cools more rapidly than the back, the mirror's heat loss being mainly by conduction to the cooler air. Whilst the cooling is in progress the effect upon the figure of a perfect paraboloid is to make it spherically over-corrected; hence the definition of a slightly under-corrected mirror tends more often than not to be superior to that of a theoretically perfect mirror, and markedly superior to that of an over-corrected one.*

Thermal disequilibrium between different parts of the mirror does not attain troublesome proportions with apertures smaller than about 20 cm. With these smaller apertures, however, the temperature lag between them and the surrounding air still remains. A temperature gradient between the mirror and the air of only 2°C can be deduced from the appearance of the extra-focal image,† whilst the rapid development of a 10° difference of temperature can ruin definition completely.

There are various possible situations, usually found in combination:

(*a*) Provided the whole of the mirror and the rest of the instrument are at the same (though changing) temperature, there is a negligible alteration of the focal length:

$$\Delta f = F(\alpha - \alpha')\Delta t$$

where F = focal length,

Δf = increase in focal length,

Δt = decrease in temperature (°C),

α = linear coefficient of expansion of mirror (Pyrex: $3 \cdot 2 \times 10^{-6}$),

α' = linear coefficient of expansion of the tube (steel: $11 \cdot 1 \times 10^{-6}$; wood: negligible).

(*b*) Where there is a perceptible temperature gradient within the mirror (as there invariably is in large mirrors, and usually is in small ones). The difference between the front and back of the mirror is commonly several tenths of a °C. Assuming a linear gradient across the thickness of the mirror,

---

*Conversely, a mirror used only for solar work should be over-corrected.
†Temperature gradients such as are normally encountered in practice have a negligible effect upon the focused image.

$$\Delta f = -2F^2 \frac{\alpha}{e} \Delta t$$

where F, Δf, Δt are as before,

e = thickness of the mirror.

(c) When the two surfaces of the mirror are cooling rapidly and at unequal rates, an effect of astigmatism is produced. This is characteristically a trouble experienced by mirrors in exposed positions (e.g. coelostat). Forced ventilation of the rear surface is the cure.

(d) When the two surfaces are unequally warm across a diameter of the mirror, an effect of spherical aberration is produced: over-correction when the temperature is falling. The magnitude of the effect depends upon the heat capacity of the cell and the thickness of the mirror. Reduction of the latter to the minimum value dictated by considerations of rigidity is therefore indicated.

(e) Short-term, erratic, and unpredictable variations of F, which are probably due, at any rate in part, to wind.

Several steps may be taken to reduce trouble from these causes:

(a) Increasing the contact between the mirror and the air by employing an 'open' cell (e.g. Figure 79*).

(b) Increasing it further by means of a fan (see also section 11.2).

Refractors are comparatively little troubled by temperature differences between the OG and the night air: (a) they are of larger focal ratio than the usual Newtonian; (b) the effects of contraction in a compound lens tend to cancel one another, which is obviously impossible where a single reflecting surface is involved; (c) an object glass being relatively much thinner than a speculum, thermal equilibrium is reached more rapidly. (See further, section 13.1.)

10.5 Flexure

Flexure in object glasses presents no problem with instruments of the size likely to be found in amateur hands. With mirrors, however, it is otherwise, and the manner in which they are supported in the telescope requires careful attention.

The amount of flexure in an objective is a function of four variables: (i) the material of which it is made; (ii) its size; (iii) the manner of its mounting; (iv) the inclination of its axis to the vertical.

(i) The rigidity of a material—i.e. its resistance to flexure under its own weight—is a function of d/E, where d is its density and E is Young's

*Based on Dr Houghton's 20-cm reflector at Teddington.

modulus (related to the elasticity of the material; its value in dynes/cm² varies from 4·9 to 5·9 for flint glass, 5·9 to 7·8 for crown). Optical glass, silica, and steel do not differ from one another significantly in this respect.

FIGURE 79
Design of 'open' cell

(*ii*) Neglecting the other factors, the flexibility of a mirror is a function of r^4/t^2, where $r=D/2$, and $t=$thickness. Hence trouble from this cause varies as D^2 in the case of mirrors whose proportionate apertures and thicknesses are the same.

(*iii*) It has been found empirically that with an unbalanced three-point support, deformation of the image first becomes perceptible when

$$\frac{r^4}{t^2} \cong 1000 \text{ cm}^2$$

If the value exceeds about 1000, a more elaborate form of overall or balanced support must be used. In order to keep the temperature gradients within the mirror at a reasonable level, it is undesirable to have $t>D/8$ for large mirrors, though this may be increased to as much as $D/5$ with small ones. Taking $t=D/6\cdot5$ as a mean figure, the maximum value of D compatible with a tolerable amount of distortion from this cause is obtained by

$$\frac{r^4}{\left(\dfrac{r}{3\cdot25}\right)^2} = 160$$

whence $\qquad\qquad r=9\cdot7$
or $\qquad\qquad D\cong20$ cm

With an effective 18-point flotation system, this thickness ratio can be considerably reduced. This is desirable on several counts, viz.: (*a*) thermal stability; (*b*) reduction of weight; (*c*) cost. Many amateur telescopes in the 30- to 40-cm aperture range have mirrors made from discs of standard 38-mm thick plate glass, and this material has been used successfully for apertures of up to 60 cm, for which a standard mirror blank 80 to 100 mm thick would be prohibitively expensive to purchase, and correspondingly more difficult to mount successfully.

FIGURE 80

Cross-sectional comparison between tapered and
uniform discs of equivalent flexure

Tapered-back or thin-edge mirrors have become popular, particularly in compact instruments where excess weight is particularly undesirable. Many commercial catadioptric telescopes have such mirrors, the disc being mounted on a central boss. This is especially advantageous in small Cassegrain-type instruments, which can then be focused by moving the primary mirror rather than the secondary mirror or eyepiece. Dall (B. 7.27) describes a 15-cm Dall-Kirkham telescope using a tapered-back mirror with a finished central thickness of 10 mm and a peripheral thickness of only 4 mm, weighing 340 grams—about one-third of the weight of a standard 6:1 disc.

10.6 Object glass cells

These must be so constructed as to ensure that there is no possibility of lateral movement of the components, that the separation of the components is maintained at the calculated value, and that the planes of the components are maintained parallel, for all positions of the telescope. At the same time the cell must not subject the components to pressure strong enough to produce mechanical strain.

OG cells present an enormous variety of design, practically every maker having his own individual dodges and devices; those described here illustrate some general principles. Figure 81(a) illustrates schematically the main features of a typical small cell—up to about 10 cm, say. The components are kept in place longitudinally by the annular counter-cell A, which is screwed into the back of the cell until it is in gentle contact with the flint component; the components are separated by tinfoil wedges 120° apart; a lateral clearance of about 0·01 mm is allowed between the edges of the components and the cell. This design has the advantage of simplicity, but it suffers from the drawbacks that it contains no provision for adjustment, and that when screwing in A it is extremely easy to turn the flint component on its axis, thus altering the correct radial correspondence of the two components.

The latter fault is corrected in designs 81(b) and (c); in the former the collar A is machined to the correct depth and held by three equally-spaced retaining screws, while in the latter the setting of the collar is adjustable. Object glasses larger than about 10 cm usually have full three-point support: the components bear fore and aft against three equally-spaced projections on collar A and the retaining ring, are separated by three equally-spaced tinfoil strips, and bear laterally against three equally-spaced projections in the inner cell wall. All these points of support are aligned along three radii of the objective. In instruments larger than about 15 cm these points of support are often adjustable through the side of the cell. Instruments of about 10 cm and upward are also usually fitted with three pairs of antagonistic screws, holding the cell to the tube flange, to facilitate squaring-on (Figure 81(d) and (e)). It is important to remember that one screw of each pair must be slackened off before it is attempted to tighten the other.

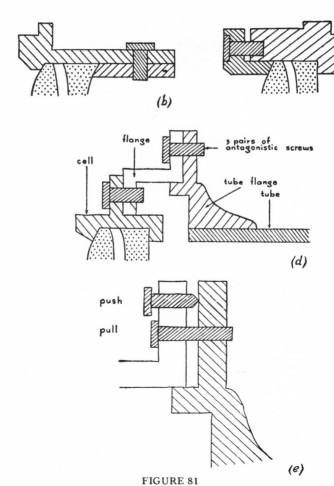

FIGURE 81
Object glass cell designs
170

10.7 Diagonal and secondary mounts

The diagonal, like the primary mirror, must be held securely but without undue pressure in its position in the optical train. The diagonal is normally held in position at the centre of the tube by a three- or four-legged spider. Adjustments must be provided for centring the diagonal in the tube, and also for varying its inclination to the optical axis.

Figure 82 shows a very simple and satisfactory mounting for use in a small home-made telescope with a lightweight diagonal. A round brass strut, a, about 4 to 6 mm thick and threaded for a short distance at each end, passes diametrically through oversize holes through the tube. The diagonal, b, is secured, using a flexible adhesive such as Evo-stik, or double-sided adhesive tape, to a short length of dowel, c, cut off at 45°. A hole for the strut is drilled through the wood, and a slot with a tensioning screw,

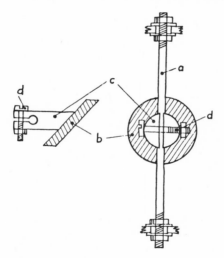

FIGURE 82
Diagonal mounting (diametrical strut)

d, allows the unit to run up and down the strut for correct positioning. If the arrangement is accurately fitted to begin with, a slight adjustment of the strut in its oversize holes will allow the image of the main mirror to be centred.

Diagonals and secondaries whose minor axis is greater than about 60 mm may benefit from the extra rigidity of a three- or four-legged spider. Figure 83 shows a diagonal mount of this type. The diagonal, a, is held at the mouth of a brass tube, b (cut back at an angle of 45°), by four turned-over projections on one side and by a brass plug, *c*, on the other. The central threaded projection of *c* is screwed into the cylinder, *d*, to which the spider is attached. The threaded outer ends of the spider, *e*, pass through slots in the telescope which run parallel with the tube's axis;

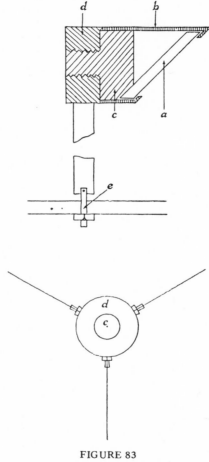

FIGURE 83
Diagonal mounting (spider)

they are secured by nuts on the outer side. These are used for centring the flat and also, by adjustment of *e* along the slot, for adjusting its inclination. The direction in which the reflected cone is turned can also be adjusted by rotating *c* in *d*. This design is simple to construct at home; in professional instruments it is the invariable practice to make provision for all adjustments at the flat mounting, the spider being a fixture.

If a four-legged spider is fitted, it is better that the legs should not converge upon the central line of the flat mount: it is extremely difficult to deaden vibration of a flat suspended in this way, even when considerable tension is put on the supporting struts. A better arrangement is shown in Figure 84, in which the four legs do not radiate from the centre of the flat mounting, but are tangential to a circle concentric with, and slightly smaller than, the mounting. Much greater rigidity is obtained, and the necessity for heavy tensioning of the spider eliminated.

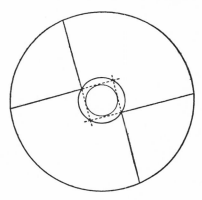

FIGURE 84
Correct design of 4-armed spider

A design which was once common, but is rarely seen today, was employed by the elder Herschel. The diagonal is secured, by a single stout strut, to a rectangular plate into which the eyepiece is fitted. The focus is achieved by sliding the entire unit along a guide fitted to the outside of the tube. An advantage of this system, particularly for photography, is the close proximity of the focal plane to the side of the tube compared with the prolongation required by an orthodox focusing system; its distance from the diagonal at all times approximates to $D/2$, allowing the use of the smallest possible diagonal mirror.

An objection to most forms of diagonal and secondary mounting is that diffraction of light at the edges of the supporting vanes or struts causes faint wings to appear around the image of a star or planet. The thicker the supporting arms, the more noticeable are these wings. By employing a curved support, as in Figure 85, the radial wings are lost, the diffracted light being scattered more evenly around the image. (See, further, B. 7.81.)

173

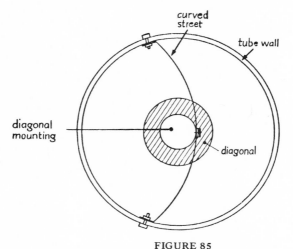

FIGURE 85

Curved-strut design to remove diffraction 'spike'

It is sometimes forgotten that the diagonal, although small, contributes as much to the light-gathering power of the telescope as does the main mirror. Therefore, equal care should be taken to keep its surface in good condition, and a tight-fitting cap, lined with cottonwool, must be used to cover the film when not in use.

If the eyepiece is not at times to assume awkward observing positions with certain types of mounting, there must be provision for rotation of the tube about its axis. With small instruments the whole tube may be rotated in the Dec axis cradle, but if D is more than 30 cm approximately, it will be found more satisfactory to restrict the rotation to a ring in the upper part of the tube on which the drawtube and spider are mounted. The moving parts demand accurate workmanship, since any divergence of the axis of rotation from the optical axis will result in the image moving in the field when the ring or tube is rotated.

DEWING-UP AND TUBE CURRENTS

11.1 Dewing-up

The formation of dew upon any surface whose temperature is below the dewpoint of the surrounding air is explained in section 26.5. The condensation of dew particularly concerns the astronomer when it occurs on object glass, mirror, flat, camera lens, photographic plate, or ocular; such deposits initially reduce light grasp, and only begin to impair definition after a stage is reached at which the light loss has already become prohibitive.

Two conditions are to be distinguished, both of them liable to be encountered during observation:

(a) *Temperature of the air* above its dewpoint:*

Dew will not form on a surface warmer than the air.

Dew will not form on a surface at the same temperature as the air.

Dew may or may not form on a surface colder than the air, according as to whether it is below or above the dewpoint.

(b) *Temperature of the air below its dewpoint:*

Dew may or may not form on a surface warmer than the air, according as to whether it is below or above the dewpoint.

Dew will form on any surface at the same temperature as the air.

Dew will form on any surface cooler than the air.

In case (b) the deposit on exposed surfaces (e.g. object glass) will be heavier than that on surfaces which are relatively protected from the external mass of air (e.g. mirror). This is not so in case (a).

Assuming that during astronomical observation the temperature, if not stationary, is falling, a surface of low thermal conductivity will be less liable to dew-formation than one of high, since its temperature will lag

---

*The air is assumed to be unsaturated but with a finite water content, as non-conditioned air is, even in desert climates.

further behind—i.e. be higher than—that of the air. Hence the aluminium film is less liable to dew than the silver film.

There are thus at least five factors tending to inhibit the formation of dew:

(*a*) removal of the moisture from the air in contact with the surface;
(*b*) the surface being slightly warmer than the air;
(*c*) the surface and its surroundings being of low thermal conductivity;
(*d*) shielding of the surface from the external air as much as possible;
(*e*) absence of dust particles, which act as condensation nuclei, from the surface.

The various methods that have from time to time been proposed for reducing the formation of dew on the optical surfaces of the telescope are applications of one or other of these principles.

(*a*) Replacing the natural moist air above the surface of a lens or mirror by a gentle stream of conditioned air. Franklin-Adams (B. 8.1) describes a somewhat complicated arrangement of this type: air from a pump or pressure bottle is passed through a dust filter, bubbled through pumice impregnated with concentrated sulphuric acid, passed through a tube of phosphorus pentoxide, a tube of soda lime (removing any traces of acid carried over), and a glass-wool plug, checked with litmus paper, and then passed via rubber tubes to jets situated round the objective.

The use of quicklime and calcium chloride has been suggested in two ways: (*i*) as a preventative of dew-formation, the hygroscopic substance being placed in a secondary cell below the speculum cell; this sounds messy and it is unlikely that it would be very effective; (*ii*) as a remover of already formed dew, the dewed part being placed in an airtight desiccator containing the calcium chloride; this is practicable in the case of oculars, less so with flats, and impracticable with objectives, besides being slower than the application of gentle warmth (see below).

(*b*) Keeping the surface at a slightly higher temperature than the air. Dew on a small flat can usually be removed by placing the hand on the supports or against the back of the diagonal for a few moments. An electric hair-dryer is extremely effective.

A flat pad of thoroughly dry, slightly warm flannel in the bottom of the objective's cover, which is then put in place (the telescope being swung downward to avoid contact between the pad and the lens or mirror) will remove dew in a few minutes. This should also be the procedure when closing down for the night. Wet optical surfaces should never be wiped, except, in the case of lenses, when deliberately cleaning them. Insulating the mirror from the surrounding air by enclosing it in a solid cell, desirable from this point of view, is ruled out by the necessity of reducing the temperature gradient between it and the air in the tube as much as possible.

Flats and prime focus plate holders can be maintained at, ideally, a fraction of a degree above the dewpoint by (*i*) a grid of resistance wire, mounted on vulcanite and separated by a small air space from an aluminium plate fixed to the back of the plate holder or flat; a rheostat in the circuit permits fine control of the temperature, which must at all times be the minimum that is effective; (*ii*) a flash-lamp bulb, wrapped in foil and separated from the back of the flat by several layers of tissue paper and cottonwool to ensure even distribution of heat. Similar bulbs mounted in small boxes round the outside of the dewcap of a refractor or camera objective will inhibit the formation of dew, but with a greater risk that the definition will be affected. Many years ago, Steavenson (B. 8.4) reported complete relief from persistent dewing troubles when he mounted a 4-volt bulb, fed by a 3-volt current, inside the base of a 125-mm minor axis diagonal about 22 mm thick.

Oculars, when not in use, can be kept in inside pockets, where they are within reach of the warmth of the observer's body, though they pick up a good deal of fluffy dust; alternatively, but less conveniently, in the nearest warm room. Moisture condenses rapidly on a cold eye lens when the observer's warm face is brought near to it. If, during prolonged observation, an ocular becomes dewed, the simplest and quickest method of clearing it is an airtight jar containing warmed flannel or cottonwool. If excessive trouble from eye-lens condensation is experienced in cold weather when a rotating ocular adaptor is used, some form of gentle electric heating might be tried.*

(*c*) Employing materials of low thermal conductivity, with consequent slowness of cooling. Wooden tubes, for example, carry much greater freedom from dew-formation, both on the flat and the mirror, than metal. Jacketing the outside of the dewcap and upper end of a refractor tube, and the lower end of a reflector tube (the whole tube would be preferable from the point of view of dew elimination, but is impracticable for the reason discussed below), with layers of some non-conducting material such as flannel, or asbestos fibre between layers of flannel. Lining the OG dewcap with a non-conducting and preferably absorbent material, such as black blotting paper, is also a help.

(*d*) Minimum exposure of the surface to the external air, preventing excessive cooling by radiation to the open sky.

*Such a device, giving complete freedom from eye-lens condensation, is described in B. 8.3

Object glasses: a long dewcap, which except in the case of the largest instruments can afford to be as long as $2\frac{1}{2}D$ to $3D$; those supplied by the makers are often too short (about $2D$), but form a convenient base for a homemade extension. On calm nights, however, lengthened dewcaps have been known to suffer from 'tube current' trouble. Long caps are also impracticable with wide-angle cameras.

Flats: the further the diagonal is from the mouth of the tube, the less it will tend to dew up.

Mirrors: a closed tube is the ideal. This, however, is the opposite condition to that dictated by the necessity of reducing the tube currents (section 11.2) to a minimum. The best compromise consists of a tube whose central section is of open or lattice construction, while the upper and lower ends, surrounding the flat and the mirror, are solid. The relatively protected position of the mirror in a partially closed tube, together with the latter's greater mass and therefore temperature lag, are the main reasons for its relative freedom from dew as compared with either OG or flat.

(e) Scrupulous cleanliness of all optical surfaces should be maintained at all times.

11.2 Tube currents

Impairment of definition by currents of unequally warm air in the telescope tube is a characteristic of the open-ended tubes of reflectors, being particularly troublesome when the aperture exceeds about 30 cm. The closed tubes of refractors permit greater homogeneity of the contained air, through which, moreover, the light has to pass only once. Small and moderate-sized refractors are virtually immune from tube currents, though trouble from this cause begins to make itself felt as the aperture is increased; the currents, though less noticeable than those of a reflector of the same size, tend to be more persistent.

It is unfortunate that the conditions militating against thermal air currents are among those that favour the formation of dew on the optical surfaces. The natural night-time temperature lag between the mirror, tube and mounting on the one hand, and the surrounding air on the other, while reducing the likelihood of dew condensation, brings with it the convection currents attendant upon any temperature gradient. These are favoured by the shape of a telescope, and by the fact that its axis is usually not far from the vertical. Since the upper part of the tube is, in an observing hut, more prone to cooling by radiation, the unstable situation arises wherein a mass of cold air is overlying a mass of warmer, and therefore less dense, air lower in the tube. A condition of circulation is thereby set up.

The elimination of tube currents takes precedence over the elimination of dew because their effect on the instruments performance is more harmful, and because dew can be dissipated in various other ways without great difficulty.

Currents of relatively warm air rising from the mirror through colder layers ruin definition by virtue, not of their motion, but of the cause of that motion: their different temperatures, hence densities, hence refractive indices. Similarly, the velocity of the currents is immaterial to their effect upon the image. Any precautions tending to reduce the temperature differences between the telescope and the night air, and between the air in different regions of the tube, are likely to reduce tube currents. These and other helpful measures include:

(a) Keeping the instrument cool during the day. If it is mounted in the open, some sort of screen should be erected to shut off the direct rays of the Sun. An observing hut should be opened up well in advance of the time observing is to begin. The insulation of observing huts is dealt with elsewhere (section 17.2.3).

(b) Reducing the solid tube to an openwork structure. This may take a number of different forms, e.g. a solid tube perforated with as many holes as will not affect its rigidity and firmness; framework or lattice construction; tubular construction (see Figure 79), in which the longitudinal rods are locked in three rings, two of which hold the flat and speculum respectively, the third being bolted to the Dec axis.

Reduction of the solid structure of the tube not only reduces the mass of the material which is at a higher temperature than the air, but by allowing greater circulation accelerates the establishment of thermal equilibrium. While open tubes are perfectly satisfactory in an observing hut, in the open they are liable to meet with an unpleasant amount of trouble from dew.

(c) Lining the tube, if solid, and covering it externally with some material of low heat capacity—cardboard, three-ply, felt, sheet cork, etc. A compromise that may give satisfactory results is the piercing of such a lined tube with ventilation holes. A solid metal tube mounted in the open air is certainly the worst possible combination of circumstances.

(d) Breaking up the convection currents uniformly, and accelerating the cooling of the tube and mirror, by means of an electric fan blowing a steady stream of outside air up the tube from behind the mirror cell, or through the inspection cover just above the mirror. This is the only course open in the case of a solid tube, and proper ventilation can reduce very markedly those faults of performance which are too often assumed to be inherent in the reflector. Indeed, the performance of large refractors is similarly susceptible of improvement—the Greenwich 66-cm, for instance, is fitted with a fan drawing air down the tube from inlets just below the objective.

The disadvantages of a fan (which do not, however, outweigh the advantages) include: it cannot be attached to the tube on account of vibration, and its position has therefore to be constantly changed during observation; even so, it may cause vibration in a lightly constructed hut, especially noticeable when high magnifications are in use; it introduces dust; it introduces sharper vibrations in the effective focal length of the objective than are caused by the normal tube currents, and is hence of less value in photographic work than in visual.

On very still nights in an observing hut, a fan will improve the performance even of a reflector with an open tube.

The rate at which a mirror will cool towards the ambient temperature depends markedly on the thickness of the disc. A study (B. 8.2) produced the following rates for discs of different thickness, under different conditions of ventilation, when $3°C$ above the ambient temperature:

	Mirror thickness		
	30mm	45mm	76mm
One surface exposed	$2·3°/h$	$1·0°/h$	—
Both surfaces exposed	$3·3°/h$	$1·6°/h$	$0·9°/h$
Fans applied to both surfaces	$10·8°/h$	$6·1°/h$	$3·1°/h$

Since the air temperature at nightfall frequently falls at a rate of $3°$ to $4°$ per hour, even a small mirror may be *increasing* in relative warmth if, having achieved the daytime ambient temperature, it is not helped in losing its excess heat.

(e) Tube currents tend to cling to the walls; hence definition is less affected if the diameter of the tube is considerably larger than that of the mirror. For the same reason a square tube, which also presents fewer problems to the amateur telescope builder, is comparatively freer from visible currents. The improvement of definition that often results from stopping down the aperture probably depends to a great extent on the elimination of the affected peripheral rays.

(f) The temperature difference between the observer's body and the air is even greater than that between the instrument and the air. In the Newtonian, particularly, convection currents from the observer's body are liable to cross the path of the incident light. A screen of some light non-conducting material such as cardboard or fibreboard, attached to the upper end of the tube and projecting beyond its mouth, will eliminate most of this trouble; but it is only practicable in an observing hut, where there is no wind to cause unsteadiness. Out of doors, the best palliative is a reserve overcoat to put on as soon as the first has taken up the body heat.

See also section 26.

11.3 Summary

Maximum freedom from dew is provided by a closed tube.

Maximum freedom from tube currents is provided by an open tube (less satisfactorily, by a closed tube and fan).

Maximum freedom from both is provided by an open tube mounted in an observing hut.

NEWTONIAN AND CASSEGRAIN ADJUSTMENTS

12.1 Introduction

Much too great stress has on occasion been laid upon the difficulty of achieving and maintaining correct alignment of the optical components of the Newtonian. It is possible that, as a result, beginners have been scared off, under the impression that it would always be getting out of adjustment.

The collimation of reflectors presents no difficulty if it is tackled systematically, and particularly if the devices mentioned below are incorporated in the telescope—a means of adjusting the set of the mirror from the observing position, and a removable upper section of the tube. Starting from scratch, a couple of hours should suffice for the lining-up of a Newtonian under adverse circumstances; given an assistant, or the devices already mentioned, this time can easily be cut in half. Once adjusted, a Newtonian should not need to be touched again for months, provided it is carefully and reasonably treated.

For a fuller treatment of the subject, see B. 7.41, 7.49.

12.2 Collimation of a Newtonian

The axes of the ocular and the paraboloid must intersect at the centre of the secondary reflecting surface, making equal angles with the normal to this surface at this point.

To achieve this, the following adjustments are available:

  (*a*) of the mounting of the flat relative to the telescope tube;
  (*b*) of the eye-end fitting itself (flat supports and ocular tube);
  (*c*) of the inclination of the flat in two perpendicular planes;
  (*d*) of the paraboloid cell relative to the axis of the telescope tube.

The procedure must be worked through systematically. Point the telescope at the day sky, so that the mirror is filled with fairly bright light. Remove the lenses from a HP ocular and screw it into the draw-tube, where it acts as a stop defining the exit pupil. Then:

(*i*) By means of adjustments (*a*) and (*b*) make the axis of the ocular pass through the centre of the flat. From the eye position at the ocular the outline of the flat is then concentric with the far (i.e. inner) end of the drawtube (see Figure 86).

(*ii*) By means of adjustment (*c*) make the reflection of the primary mirror in the flat concentric with the flat itself—and therefore with the inner end of the drawtube. It is essential that the whole cone of rays

inner end of drawtube

flat mounting

outline of flat

reflection of primary in flat

flat reflected in reflection of primary in flat

Before adjustment

After adjustment

FIGURE 86

Collimating a Newtonian reflector

whose base is the full aperture of the paraboloid be reflected from the flat. With the eye at the lens-less ocular, the edge of the mirror must be visible within the edge of the flat. It is of no importance if the flat is slightly larger than the minimum necessary to ensure this, since the only effect will be a negligible reduction of light grasp. But if the flat is slightly too small, so that a peripheral ring of the convergent cone is not inter-cepted by it, the light loss will be considerable. In the case of a $f/6$ New-tonian, for example, with the flat intercepting the cone at a distance from the prime focus equal to D (thus bringing the Newtonian focus well clear of the tube), a flat diameter one-tenth larger than the minimum necessary to intercept the whole cone will increase the light loss through silhouet-ting by about 0·5 per cent; if, on the other hand, it is one-tenth smaller, the gain of about 0·75 per cent through decreased silhouetting is offset by a loss of 30 per cent due to a peripheral ring of the objective not con-tributing to the image at all.

(*iii*) By means of adjustment (*d*) make the reflection of the flat (a small black disc with the three legs of the spider radiating from it) in the primary mirror concentric with the primary mirror itself. To achieve this final adjustment with economy of time and temper one or other of the following should be provided: (1) an assistant to make the adjustments while their effect on the image is watched from the observing position, (2) means of making the adjustments from the observing position itself.* The visual judgment of the concentricity or otherwise of the primary mirror and the reflection of the flat is made much easier if a circular disc is pasted to the centre of the mirror, of such a size that the reflected flat shows a narrow dark rim round it. (Since this region of the mirror is silhouetted in any case, no damage is done to the usable silver surface.)

(*iv*) Fine adjustment is then carried out on a stellar image. If the displacement of the axis of the primary is slight, a stellar image at the centre of the field will probably exhibit coma only; if the displacement is gross, astigmatism may cloak the coma. Make adjustment (*d*) so as to displace the image at the centre of the field in the direction indicated by the comatic image:

Bring the image back to the centre of the field and repeat the process until the image is symmetrical. The interpretation of an astigmatic image is more difficult, but should not be necessary if the first approximation in daylight has been carried out. The orientation of the sagittal image indicates the diametrically opposed directions in one of which the adjustment must be made:

A ←——————   ——————→ B

but whether towards A or towards B can only be discovered by detecting the eccentricity of the rings. This is much less easy to spot than the shape of the comatic image, but if the drawtube is racked out it should be quite plain in the extrafocal image:

In case of doubt, discover by trial and error which adjustment reduces the astigmatism, and continue the tilting of the mirror in this direction until the image at the centre of the field shows no removable coma, or until the rings of the extrafocal image are as concentric, and as uniformly bright in all position angles, as they can be made.

It must be borne in mind that any alteration of the inclination of the

*See section 10.3.

primary mirror will affect the readings of the position circles. Hence the verniers must be reset after every adjustment of the mirror; so, if the readjustment has been considerable, must the finder.

The so-called 'Cheshire eyepiece' is particularly helpful in achieving critical alignment in any telescope, whether reflector or refractor. The device (Figure 87) consists of a peephole about 1·5 mm across, set on the axis of a brass or aluminium plug whose face has been cut off at 45°. This face is polished to make it reasonably reflective. The device fits into the

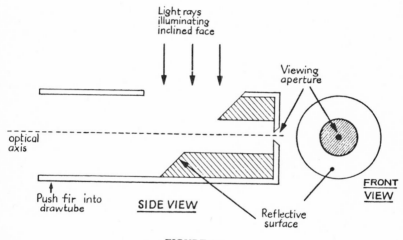

FIGURE 87

The Cheshire collimating eyepiece

ocular tube so that the inclined face is illuminated by skylight (during the day) or artificially. This bright annulus is then made concentric with the other rings and reflections seen on looking along the optical axis.

12.3 Fine adjustment of the secondary (flat)

In the case of instruments of large relative aperture, adjustment (*i*) must be modified. The reason for this is that although a parallel pencil whose axis is incident to a 45° mirror at the centre of the latter will, after reflection, still be concentric with the mirror (Figure 88), a converging beam will no longer be so, since the axis of a cone does not pass centrally through an elliptical section of the cone. The eccentricity of the reflected beam is a function of the steepness of convergence in the incident beam, and therefore of relative aperture; for relative apertures smaller (i.e. focal ratios larger) than about *f*/8 it can be ignored.

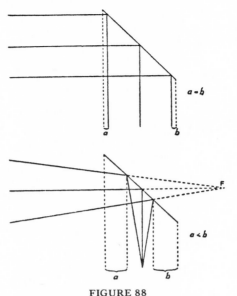

FIGURE 88
Fine adjustment of the diagonal

The new procedure is as follows. A short tube which will fit snugly over the cell of the diagonal is fitted at one end with a pair of cross-wires which exactly define its centre. A hole cut in the side of the tube allows the cone, after reflection, to pass to the drawtube. In this way the true centre of the flat is defined, and it is on this point, rather than on the apparent centre, that the drawtube is aligned. In the remaining adjustments the reflection of the primary in the flat and the reflection of the flat in the primary are to be made concentric, not with the flat, but with the inner end of the drawtube.

12.4 Fine adjustment of the secondary (internally reflecting prism)

If the rays incident to the entrance and exit surfaces of the prism are not normal to these surfaces* (e.g. if the apex is not exactly 90°, or if the prism is tilted in a manner undetectable from the observing position), the procedure already described will not ensure against the introduction of false colour. A further check is required.

---

*Internally reflecting prisms cannot be used as secondary reflectors in a Newtonian whose relative aperture exceeds about $f/6$, for the reason that the convergence of the pencil at the focus is then too steep for this condition to be satisfied.

The inner edge of the drawtube is painted white, and a cloth mask arranged over the mouth of the telescope so that there is just enough light for this white ring to be seen. From the observing position the reflections of two white rings will now be observed in the prism, one from each of the faces bounding the apex. If these two circles are not concentric they must be made so, by means of adjustment (c). This will probably upset the other adjustments, and the two must be repeated alternately, the error being reduced each time, until both are satisfactory. This is liable to be a tedious and rather maddening job, and the device described by Peek in B. 7.41 simplifies and accelerates the adjustment very considerably. The upper section of the telescope tube, carrying the ocular tube and the prism mounting, is made detachable from the main part of the tube, to which it is fixed by bolts engaging in slots running lengthwise (i.e. parallel to the axis of the tube) and secured by wing nuts; they thus allow a margin of tilt adjustment of the upper section of the tube relative to the lower. The upper section is removed from the telescope and the squaring-on of the prism carried out in the manner just described. The section is then returned to the telescope and, by means of the wing nuts, locked in position so that the reflected image of the primary is concentric with the inner end of the drawtube. Finally, adjustment (*iii*) of section 12.2 completes the job.

See also sections 10.3, 10.7.

12.5 Collimation of a Cassegrain

In the case of a Cassegrain the equivalent adjustments are twofold: (*i*) the axis of the secondary must be made to coincide with the axis of the ocular tube (it being assumed both there and in the discussion of the Newtonian adjustments that the axes of the ocular and the ocular tube are coincident); and (*ii*) the axis of the primary must pass through the centre of the secondary:

(*i*) Observe the day sky from the Cassegrain focus through a HP ocular from which the lenses have been removed. The whole of the primary should be reflected in the secondary, this reflection and the secondary being concentric. Any necessary adjustment is made to the secondary.

(*ii*) The image of the secondary in the primary must be concentric with the primary. Fine adjustment of the primary is made by observation of the comatic or astigmatic nature of a centrally placed star image, as described above.

Collimating techniques are discussed in B. 7.18, 19.

TESTING

13.1 Introduction

The most satisfactory method of testing an objective outside a laboratory is by means of the extrafocal images of a star—either natural or artificial (see section 35). This method is both more sensitive and less dependent on absence of atmospheric turbulence than the examination of the image in the focal plane.

If a natural star is used, it should be of moderate brightness (of the 2nd magnitude, say) and must be at or near the zenith. If far from the zenith, differential atmospheric refraction, resulting in dispersion, will introduce false colour: far from the zenith, a star image tends to exhibit a red fringe on its apparent upper side (i.e. the side nearest the horizon), and a greenish or bluish one on the lower. It is also important to keep the eye precisely central behind the ocular—i.e. the optical axis must pass centrally through the eye pupil—or false colour will be introduced. If the eye is displaced,* that side of the image towards which the displacement takes place will become tinged with red. Hence it is possible to suppress the atmospherically produced particoloured fringe surrounding images of objects which are far from the zenith by slightly displacing the eye downward from the centre of the eye lens—'downward' here meaning the direction such that the line from the centre of the eye pupil through the centre of the exit pupil will, if produced, pass through the zenith. For optical methods of surpressing atmospheric dispersion, see B. 19.5, 6.

Turbulence must be very slight, or the image will not be fine enough for the more delicate and sensitive tests to be performed; if the seeing is not satisfactory, testing must be postponed until it is, or an artificial star used.

Magnification should not be so low that the rings are difficult to see and the exit pupil so large that aberrations in the eye and the ocular

---

*Too great a displacement of the pupil will result in effectively cutting down the aperture on one side

become prominent, nor so high that secondary spectrum becomes trouble-some. Where not otherwise stated, about $25D$ is satisfactory.

The telescope, and in particular the objective, must be allowed to reach thermal equilibrium before testing begins.* The length of time required varies with the initial temperature difference between the telescope and the night air, and also with aperture; in the case of a 15-cm refractor, at least half an hour's exposure to the night air should be allowed; an hour would be safer. The outer surface of the crown component of a doublet cools fastest, then the back surface of the flint, and finally the internal surfaces. While these differential contractions are in progress, strains are set up in the objective which simulate the effects of spherical under-correction. Tube currents have been known to introduce effects mistak-able for slight astigmatism. Similarly, when testing with an artificial star in sunlight it is essential that the telescope itself should be completely shaded; thermal effects within the tube, produced by neglecting this pre-caution, can cause great distortion of both focal and extrafocal images.

The order in which the tests are carried out is important. It is a waste of time, for example, trying to apply the very sensitive tests for the geo-metrical aberrations if the correct alignment of the objective has not first been assured. As always, where adjustments in several directions are in-volved, care must be taken to confirm an earlier adjustment if a later one may have disturbed it.

13.2 Telescope tube

Only such characteristics of the tube as affect the optical tests are con-sidered here.

It must be confirmed that the axis of the drawtube of a refractor passes through the centre of the objective (in a Cassegrain it may be assumed that the ocular tube is normal to the mirror cell to which it is fixed). Fix cross-threads over the objective and the mouth of the draw-tube; also fix a slip of plane unsilvered glass across the mouth of the drawtube in contact with the latter set of threads. If the eye is now placed on the optical axis about 25 cm behind the drawtube, its reflection will be seen dimly in the unsilvered slip. If the adjustment of the drawtube is satisfactory, it will be possible to find a position for the eye from which it sees the two sets of threads superimposed upon one another and bi-secting its own reflection.

In the case of a Newtonian, an eye placed at the focus should see only a narrow rim of the tube reflected round the image of mirror in the flat.

*See also section 10.4

189

Finally, remove the ocular and see that none of the diaphragms in the tube is too large or, which comes to the same thing, too near the object glass. That is, the whole of the objective must be visible. To ensure that full illumination is obtained with the lowest powers, see that the whole objective is visible from the extreme edges of the eye-tube.

13.3 Oculars

The oculars used in the testing of an objective must be of established excellence.

With a star central in the field, twist the drawtube or, better, the ocular itself; if it is threaded, this will alter the focus slightly, but this is not important, and the danger of jogging the whole instrument is smaller than in trying to twist the drawtube. Note whether the image of the star is moved in the field as a result of this rotation. Refocus, and see whether the appearance of the image has been modified in any way by the rotation of the ocular. If either has occurred (the alignment of the drawtube having already been established) the ocular should be rejected.

Shift the image to the edge of the field and expand the rings extra-focally; if they are not complete circles, or are brighter on one side than on the other, the ocular diaphragm is not passing the whole of the objective's field of full illumination. With the image still at the edge of the field, refocus the star. A Ramsden will then usually show chromatic inequality of magnification as a fringe of opposite colours on opposite sides of the image where the diameter of the field intersects the image. A Huyghenian showing this should be rejected.

13.4 Flats

Among the commoner causes of disappointing performance with reflectors are the use of unsuitable oculars and of a flat whose optical qualities do not match up to those of the mirror. The flat should therefore be checked before starting on the tests of the primary.

Three methods of testing the planeness of a flat are in common use:

(a) by Newton's rings in monochromatic light, using another flat of known quality;

(b) by knife-edge test of a spheroid of known quality, the flat intercepting the beam at 45°, whence any faults apparently detected in the spherical mirror must in fact be contributed by the flat;

(c) by Rayleigh's water test, which has the advantage that no second glass testing surface is required, though it is by no means easy to operate; it is in fact a variant of (*a*), in which the test flat of known planeness is replaced by a water surface.

The detailed operation of these tests is described in B. 7.48, 7.49.

13.5 Physical character of the object glass

Remove the ocular and point the telescope at the day sky; any physical defects in the object glass will then be strikingly visible to an eye placed at the focus. These can be seen in even greater detail in Foucault focograms (photographs of the appearance of the objective under the Foucault test—see section 13.18).

Bubbles, scratches, and minute particles in the glass which cause local variations of the refractive index and which appear visually as 'threads' are unimportant, since they are small compared with the entry pupil illuminating them; at most they produce an infinitesimal light loss and reduction of contrast through scattering. (The reverse is true of an ocular, where a bubble, for example, is necessarily large compared with the diameter of the pencil in which it is situated.) Large defects of this nature are unlikely to be encountered. If they are, the object glass should be rejected. Covering them with patches of adhesive tape or making irregularly shaped diaphragms to cut them out, as sometimes recommended, is useless.

Internal strain, with variations of refractive index, due to faulty annealing (the outer layers being allowed to set and cool while the interior of the lens still wants to contract further) can, if gross, cause double refraction—strikingly revealed by polarised light—or wings and flares attached to all bright images. No objective from a reputable maker need be suspect.

13.6 Mechanical distortion of the objective*

With a three-point support of the mirror, it may be impossible to eliminate a slight triangular distortion of the stellar image. In any case, the objective must be firmly held without being tightly pinched. An occasional and transient triangular deformation of stellar images has been described from time to time; its cause is unknown, but it is presumably of atmospheric origin; it was noticed in one case that, at an observing station in SE England, this effect only occurred when the wind was in the E or SE.

Excessive and/or unequal pressure from the supports of an object glass, in directions both parallel and perpendicular to the optical axis, is a cause of distorted (typically triangular) focal images and asymmetrical extrafocal

*See also section 10.

rings; even astigmatism may occasionally be traced to unequal pressure on one or both components of an object glass. It must be sufficiently slack in its cell to ensure freedom from undue pressure when the cell is contracted to its minimum size, i.e. at the minimum temperature to which it may be exposed.

Sagging of an object glass under its own weight, at different orientations of the optical axis, will only be encountered in objectives of more than about 15 cm aperture. The flint element is always the more prone to this type of distortion, being both heavier and less robust (owing to its concave section) than the crown.

Symmetrical central sagging produces the effect of spherical aberration, and is allowed for in the calculation of the curves. Sagging between each point of support produces a triangular distortion of the image; if the crown is sagging, light passing through the objective in the vicinity of the supports is brought to a longer focus. The resultant tendency for the two more or less to cancel one another is effective with medium apertures (say, 15 to 25 cm); with larger apertures the only solution is the provision of additional counterpoised points of support, ensuring that the weight of the lenses is distributed evenly over them at all times.

13.7 Centring and squaring-on of object glasses

Squaring-on neither has to be checked as frequently as with a reflector (once adjusted, with a newly acquired instrument, it should not require attention again for years), nor does it so seriously affect the quality of the image if imperfectly carried out. Means of adjustment (see section 10.6) are normally provided when the aperture exceeds 75 or 100 mm only; refractors of 15 cm aperture and larger should be fully adjustable— i.e. both components should be capable of adjustment parallel to, and perpendicular to, the optical axis. The adjustments are tiresome to make, since the screws are not within reach of the observing position; an obliging friend is a great help here.

The following maladjustments may be present singly or in combination:

(a) axes of the components parallel but not coincident: Figure 89 (a);
(b) axes of the components inclined to one another, though the line joining their centres may pass through the centre of the ocular: Figure 89(b);
(c) axes of components coincident, but not passing through the centre of the ocular: Figure 89(c).

The final adjustments in the squaring-on and centring of an object glass

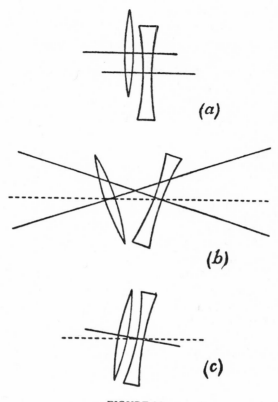

FIGURE 89
The object glass: centring and squaring-on

require an acute and practised eye for the detection of the slight imperfections in the image.

(a) The flint component acts as a prism, the direction of the red dispersion being that in which its centre is displaced from the centre of the crown. The effect is therefore very similar to that produced by differential atmospheric refraction: the image of a bright red star at the centre of the field is fringed with red on one side, while the other is bluish (though this is less easy to detect). In addition to the colour effect, the image will be noticeably elongated if the eccentricity of the components is great (exceeding about 0·1 mm at $f/15$).

If the manufacturer has made no provision for the lateral adjustment of the components, any error in this direction means that the objective must be returned to him.

(*b*) and (*c*) *Approximate checks:* If the telescope passes the following tests, the planes of the lenses are not grossly inclined to each other, nor is their axis grossly divergent from the line joining their centre and the centre of the field of the ocular:

(*i*) Put the telescope tube horizontal, and remove the ocular. Hold a torch bulb between the eye and the front crown surface of the objective. The three images of the filament—reflected from surfaces 1, 2-3, and 4—should coincide when the line through the centre of the drawtube passes through the centre of the flame.

Alternatively, and rather more conveniently, a bright ring (such as the metal frame of a pair of spectacles on the nose of the tester) may be used in place of the flame. The reflected image of the ring and the circular ocular opening should be concentric.

(*ii*) Applicable with objectives whose fourth surface is concave towards the ocular. Cover the object glass; place a white card, illuminated by strong artificial light, about 15 to 25 cm behind the open end of the drawtube; this card is pierced with a 5-mm hole, behind which the eye is placed. The card is moved about until the dark spot (reflection of the hole) on a white circle (reflection of the card, bounded by the edges of the drawtube), seen in the rear surface of the objective, is central within the circle of the drawtube. In this position the white circle and the objective will be concentric if the latter is correctly squared-on. If the circle is nearer one side of the objective, this side must be moved further from the ocular. The test is then repeated until a satisfactory adjustment is obtained.

(*iii*) Cover the object glass. Over the open end of the drawtube place a cardboard mask pierced with two circular holes situated on a radius (Figure 90). The eye is placed behind *A* and shaded from a source of light directed into *B*. When *AB* is vertical, three closely adjacent images of *B* will be seen in the objective, and the line joining these should be vertical. If satisfactory, repeat with *AB* horizontal, when the line joining the images should likewise be horizontal.

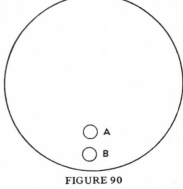

FIGURE 90

Adjustment of an object glass

(*b*) and (*c*) *Fine adjustment:* As usual, the highest degree of precision in the detection and correction of instrumental shortcomings is provided by the examination of a stellar image:

(*i*) Object glass with inherent internal coma: (*c*) is indicated by a comatic image at the centre of the field. Determine in which direction the rings are brightest, and reduce the distance of this side of the objective from the ocular. This will reduce, but may not remove, the coma because (*b*) also causes coma—as well as astigmatism, which may only become visible after the joint squaring-on of the objective has been improved. Of

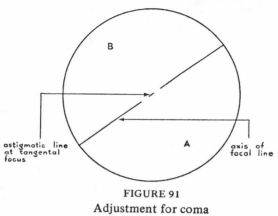

FIGURE 91

Adjustment for coma

the two regions of the objective situated at right angles to the tangental focal line (*A* and *B* in Figure 91), one is too near the ocular, and one too far. The object glass is removed from its cell and the equality of thickness of the inter-lens air-wedges tested by means of Newton's rings, using monochromatic light such as is provided by a mercury-vapour lamp, neon tube, or sodium flame, or alternatively by using an ordinary 100-watt lamp and observing the objective through a monochromatic filter (see section 7.1). The rings are invisible in polychromatic light owing to the relatively great separation of the second and third surfaces. With the source of the illumination held as close to the eye as possible (asbestos sheet between the temple and the flame, if a sodium flame is used), and both source and eye opposite the centre of the crown component, approach the objective until the eye is at the focus of the joint reflecting surface provided by the rear surface of the crown and the front surface of the flint; the objective will then fill with light, although the individual reflections of the source will still be visible. If the Newton's rings are not concentric with the object glass, adjustment is required. Press gently on the edge of the crown component above each of the separating wedges in turn, to discover which decreases the eccentricity of the objective and the system of rings. Then insert a slip of tinfoil between the crown lens and

195

each of the other two spacers, repeating the performance until the concentricity of rings and objective is satisfactory. If c_2 and c_3 are so different as to result in great crowding of the rings towards the edge, the estimation of concentricity of otherwise will be difficult to make; in this case, mark the centre of the crown lens with a minute ink-spot, and judge the concentricity by relation of this to the innermost rings.

The total thickness of the spacers should not be greater than that necessary just to separate the centres of the components if $c_2 > c_3$ (e.g. Fraunhöfer)—in which case the spacers will be of roughly the thickness of a filing card—or just to separate their edges if $c_2 = c_3$ or $c_2 < c_3$—in which case the spacers will be of thin foil. The star test is sensitive to variations in the thickness of the wedges of only 0·01 mm at $f/15$.

When returning the objective to its cell it is advisable to check the tensioning of the counter-cell behind the flint component by the same method, since there is enough 'give' in the spacers completely to wreck the adjustment if undue and unequal pressure is exerted. The necessary pressure to hold the objective firm will inevitably distort the rings somewhat, but a degree of distortion that would involve the complete destruction of the image in a reflector (with its single optical surface) is immaterial in the case of an object glass.

On reassembly, confirm that the joint squaring-on of the components has not been interfered with (it is advisable to mark the setting of the cell relative to the tube before removing it), while improving their relative squaring-on. If reobservation of a star indicates that the astigmatism has only been improved, and not removed, the procedure must be repeated until satisfaction is obtained.

(*ii*) Object glass with non-existent or inherent external coma: (*b*) is indicated by comatic images at the centre of the field. The line joining the points in the field through which the axes of the crown and flint components pass is parallel to the axis of the image, the former lying towards the apex of the image, and the region where the components are too close to each other being indicated by the broad end of the image (Figure 92). The necessary adjustments of the thickness of the wedges are made—by the Newton's rings method described above—and a further test made on a star.

(*c*) Coma having been removed, any trace of astigmatism in the image at the centre of the field indicates inexact squaring-on of the objective as a whole. Slacken the cell-holding screws and pull in the side of the objective marked *A* very slightly (Figure 91); if the condition is aggravated, push *A* further out from the ocular and pull in *B*. By trial and error, the astigmatism may be reduced to zero.

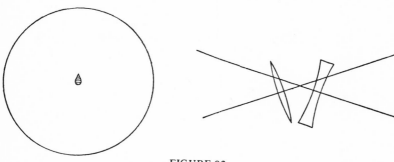

FIGURE 92
Coma due to inclined components

13.8 Alignment of the optical parts of reflectors
See section 12.

13.9 Spherical aberration
The necessity of allowing the objective to attain thermal equilibrium before testing has already been stressed.

A method of detecting residual spherical aberration by the alternate employment of a diaphragm and a central stop, involving a change of focus, is described in the literature.* It is not satisfactory, however, owing to the narrowness of the emergent pencil when all but the central area of the objective is stopped out, which allows an appreciable movement of the drawtube without affecting the sharpness of the image to an extent that is perceptible to the eye. The test is therefore insensitive, and since a simple and extremely sensitive test is available it has nothing to recommend it.

This test involves the comparison of the image of a star inside and outside the focal plane. The latter being the region of uncorrected colour in an 'achromatic', and the extrafocal image being tinged with purplish flare, the use of a green or yellow filter facilitates the observation and increases the accuracy with which the test can be made. The drawtube is racked quickly in and out from the focus, so that the appearance of the image on either side of the focus can be easily compared. Then:

(a) Under-corrected:
Inside focus: light concentrated in the periphery of the image; weak

---

*It was the method used by Herschel for testing his mirrors.

central disc; rings increasing in brightness from the centre outward, being extremely prominent at the outer edges of the diffraction pattern.*

Outside focus: light concentrated at the centre of the image; central disc nearly as small and bright as at the focus; brightness of the rings diminishing from the centre outward, the periphery being faint and ill-defined.

(b) *Over-corrected:*
Inside focus: as for outside focus (above).
Outside focus: as for inside focus (above).

(c) *Perfectly corrected:*
No appreciable difference between the strength of the rings on either side of the focus, the only difference between the two images being (in a refractor) the secondary colour—a greenish or greenish-yellow fringe outside, and maybe a reddish fringe inside, the focus. The concentration of secondary colour at the centre of the image outside focus may produce the impression that the objective is under-corrected; hence the value of a yellow filter.

13.10 Zonal aberration

The test is as for spherical aberration; but even more than the latter it requires an acute and practised eye. Zonal errors are revealed by the anomalous brightness or faintness of one or more rings in the extrafocal image; that is to say, the change in brightness of the rings from the spurious disc to the edge of the image is not uniform. Any anomaly of this nature will obviously be reversed inside and outside focus.

The position in the inside-focus image of an anomalously bright zone corresponds with the position of the zone in the objective that is focusing short; and conversely, that in the outside-focus image, to a zone in the objective that is focusing long.

The test is facilitated by employing a bright star—mag 1, if available—and by a greater displacement of the drawtube than when testing for spherical aberration, a dozen or more rings being expanded. The errors are also more easily detectable on the 'bright' side of the focus—i.e. that on which they show as brighter-than-normal rings, rather than as darker-than-normal.

---

*A spurious effect simulating spherical aberration, due to a temperature gradient in the air immediately surrounding the flat of a Newtonian, is described in B. 9.9.

13.11 Coma

This is unlikely to be encountered in visual instruments, owing to their comparatively restricted fields. The appearance of the comatic image is described in section 4.5. If coma is found to be intrusive, it is more likely that the squaring-on is faulty than that the coma is inherent in the objective.

13.12 Astigmatism

Astigmatism that cannot be removed by adjustment of the squaring-on of the objective (and it is only in the so-called coma-free objectives that faulty squaring-on simulates astigmatism in the extrafocal image on the axis) must be inherent in one or more members of the optical train. If the ocular used for testing has previously been given a clean bill of health it must lie in (*a*) one or more surfaces of the OG (or be caused by varying refractive index across one or both components), (*b*) the main and/or secondary mirror, (*c*) the eye.

(*c*) can be eliminated by changing the orientation of the telescopic field with respect to the line joining the eyes; it is preferable to do this with an image rotator (see section 8.11), since rotation of the head may induce or modify astigmatism in the eye. If this changes the orientation of the astigmatic planes relative to a fixed direction in the objective, the astigmatism lies at least partly in the eye. Try the other eye, which may possibly be free from astigmatism. Changing to a higher magnification will almost certainly reduce or altogether suppress the astigmatism if it is due to the eye (owing to the smaller area of the cornea and crystalline lens being illuminated by the exit pupil), in which case the observer must discover for himself the minimum magnification that his eye will allow him to use, unless the trouble is corrected by spectacles. Astigmatism that remains under high power is invariably inherent in the instrument rather than in the observer.

If the cause is (*a*), and only one component is astigmatic, no improvement of the condition can be made; but if both lenses contribute to the astigmatism, rotation of one of them relative to the other will find an orientation in which the two astigmatisms are opposed, and the trouble minimised, if not entirely suppressed.

Similarly with (*b*), if the mirror alone is astigmatic nothing can be done about it; if it is contributed to by both the primary and the secondary mirror (Newtonian flat or Cassegrain hyperboloid), a relative orientation of the two can be found by trial which gives minimum astigmatism; though in the case of the Newtonian with its elliptical secondary it is easier to scrap it and get another. If it is due to either mirror alone, the direction of the focal line will rotate with the rotation of the mirror.

If the astigmatism is only slight, examination of the focal image, or either the intrafocal or extrafocal image alone, will not easily reveal the error. In the former it will cause a slight blurring of the normal diffraction pattern, and in the latter an elliptical expansion of the rings. As with spherical aberration, the fault is most easily spotted by racking the ocular backwards and forwards across the focus: the expansion of the image in two mutually perpendicular directions is then easily detected.

Any departure whatever from absolute circularity in the extrafocal images, if not due to faulty alignment of the optical parts or to the flat or secondary mirror, justifies the rejection of the objective.

13.13 Chromatic aberration

There remains, in the case of a refractor, the examination of the secondary spectrum, which should be the only residual aberration of any importance. With a magnification of $25D$ to $30D$ examine the focal image and the images inside and outside focus of a white A-type star, about mag 2:

(a) *Well-corrected visual objective·*

At focus: yellowish disc; halo, if any, purplish or indigo (not red or green).

Inside focus (2-3 rings expanded): yellowish disc with faint and narrow reddish-purple border.

Outside focus (2-3 rings expanded): tiny red disc forms at centre.

Outside focus (3-4 rings expanded): mauve centre; no red fringe; greenish or greenish-yellow border.

Outside focus (5-6 rings expanded): bluish flare overlying all the central part of the image.

(b) *Under-corrected:*

Inside focus: white or greenish-white disc; red fringe (scarlet rather than purplish-red) more prominent.

(c) *Over-corrected:*

Inside focus: deep yellow rings; greenish-blue background.

With a blue-white star the reddish-purple fringe inside the focus may be more nearly blue, while with a red star it may be so marked as to compare with that of an under-corrected image of a solar-type star. Further, the lower the magnification (hence the larger the exit pupil) the more the aberration of the eye and ocular will introduce colour under the guise of under-correction in the objective. The choice of test object and magnification is therefore of great importance in the examination of the secondary spectrum. The colour correction of different objectives nevertheless vary between certain accepted limits, while still rating as good, since to this limited extent the nature of the correction is a matter of taste.

13.14 Testing a photovisual objective

(a) *Eccentricity of the component lenses:* The image at the centre of the plate, viewed with a HP ocular mounted axially, should be a brilliantly greenish-yellow point, surrounded by a uniform purplish-red fringe. If this fringe is reddish on one side and bluish on the other, the components of the objective are not concentric. The amateur is advised to return the objective to the makers for adjustment, rather than to attempt it himself. Correct centring is of great importance in a three-component objective.

(b) *Squaring-on of the objective as a whole:* Examine the axial image for astigmatism and coma with a HP ocular mounted at the centre of the film aperture, using a green filter.

(c) *Squaring-on of the film holder:* Insert a plane mirror into the film holder; mark its centre and the centre of the objective with a very small disc of white paper, dampened to make them stick. On looking down the tube, with the eye in front of the objective, the reflection of the pupil of the eye should be in the centre of the mirror.

(d) *Coma:* Mount a HP ocular near the edge of the field, inclined inward at a sufficient angle for its axis to pass through the centre of the objective; if the emergent pencil does not, in this way, pass through the ocular axially, the ocular itself will contribute coma. Using a green filter, look for traces of coma. Astigmatism in its pure state, with no trace of unilateral comatic brightening of the rings, should be seen.

13.15 Squaring-on a camera lens

Since the majority of amateur photographs are taken using standard 35-mm cameras with lenses of various focal lengths, it must be assumed (until proved otherwise) that the fittings are already in accurate adjustment, and that no preliminary attention is required. It may, however, be worth bearing in mind that the 'infinity' focus marked on the lens barrel is not always as precise as it should be.

If a camera is being made at home, whether for miniature film or sheet film, some approximate optical test must be made to ensure that the lens is square on to the film plane, although final adjustment, both of squaring-on and focusing, is usually done by trial exposures (see section 20.9).

Two separate aspects of the lens-film configuration must be checked, and in the following order: (a) optical axis of the lens passing through the centre of the film aperture; (b) film plane at right angles to this axis. To check (a) place a piece of glass or a sheet of semi-transparent paper across the film aperture, with a cross or spot marking the centre, and set the camera on a horizontal support with its back to the light. If the lens is observed through a bright circular ring—the eye being placed directly in front of the lens, with the ring in between—the following will be seen: the mark at the centre of the film aperture, and a series of reflections of the ring at the several surfaces of the lens. If the components of the objective are in correct mutual adjustment, all the reflections of the ring will be concentric (if they are not, the lens must be sent to an optical workshop for adjustment); if, in addition, the lens as a whole is correctly squared on, these concentric reflections will be centred on the mark in the film aperture. If they are not, it is a matter of a few minutes' experimenting with the lens-adjusting screws to make them so.

13.16 Squaring-on a Schmidt-type camera

The adjustments in a prime-focus catadioptric camera being particularly sensitive, the photographer must expect to spend a considerable amount of time (and film) in achieving optimum precision. The task will be lessened considerably by designing the tube and fittings so that (a) all three elements—mirror, film holder, and corrector plate—can be adjusted independently, and (b) the adjustments, once made, are permanent.

Having set the mirror, film holder, and corrector plate at their correct distances apart by ordinary measurement, the corrector is set square-on to the axis of the tube to mechanical accuracy and locked in position. Crossed strings, positioned accurately along two diameters of the plate, are then viewed from a short distance in front of the plate, and are made to coincide with their reflection in the primary mirror by adjusting the inclination of the latter component. The outline of the film holder is then made concentric with the superimposed real and imaged strings.

The final squaring-on and focusing of the film holder must be achieved by experiment. For further details, see B. 9.18.

13.17 Testing resolving performance

See section 2.5.

13.18 Laboratory methods of testing

For simplicity and sensitivity there is nothing to beat the examination of the extrafocal images of point sources for discovering the characteristics of an objective. It is not, however, convenient for the optician who wishes to determine the nature and magnitude of the errors in his objective with a view to correcting them by subsequent work. To meet his requirements a number of laboratory methods have been developed. But since these are primarily of interest to the instrument-maker, and are comparatively unsuitable for the investigation of the performance of a newly acquired telescope, the reader must be content merely with references to other sources.

The Foucault test, for which a knife-edge and an artificial star are set up at the centre of curvature, is probably the most generally used by amateurs. Good general accounts are to be found in B. 7.29, 7.48, 9.21; B. 9.25 describes a convenient form of apparatus for carrying out the test; B. 9.10, 9.11 describe an autocollimating application of Foucault's principle.

The Hartmann test is less convenient to operate than the Foucault, involving photographic work on either side of the focus, and the measurement of the plates. It allows the focal length of any desired zone of the objective to be determined, and is perhaps the most commonly used test where large objectives are concerned. B. 7.29, 7.49 contain good general accounts; B. 9.12 should also be referred to.

The Zernike test is an ingenious use of diffraction to show the high and low regions of the objective's surface in different colours. See B. 7.49, 9.3, 9.26.

Autocollimation methods, in which the error is doubled by arranging for the beam to pass through, or be reflected from, the objective twice, suffer from the disadvantage that a flat as large as the objective is required. Burch has described a method of overcoming this difficulty with quarter-size compensating mirrors in B. 9.4; see also B. 9.5.

The Ronchi test, like the Foucault, requires no complicated equipment. It is an interference method, employing a coarse grating (40 to 200 lines per cm). For general accounts of the method, see B. 7.49, 9.1. Like the Foucault, also, the Ronchi test is qualitative; for a quantitative development of it, see B. 7.49, 9.6. A useful variant of the Ronchi test is described in B. 9.16.

For precise measurement of the departure from accuracy of an optical surface or system, optical workshops use interference techniques, such as are described in B. 7.46. For the amateur, by far the most accurate and convenient test for a paraboloid is the simple null test described by Dall in B. 9.7; for a more detailed treatment of this test, see B. 9.23.

MOUNTINGS

14.1 Introduction

The mounting of the optical parts of a telescope has two functions to perform: (*a*) the maintenance of these parts in their correct, fixed relative positions; (*b*) the provision of means of moving them as a whole in any required direction.

The optical parts themselves, mounted in some form of cell, are maintained in their correct relative positions by mounting them in a single framework (solid or open tube, circular, square, or polygonal in section). In some types of mounting, and particularly those for small instruments, the tube is supported by an altazimuth or equatorial head, permitting its movement about two mutually perpendicular axes; this head is then supported by some form of tripod or column, connecting it with the ground. In some larger mountings the distinction between the head and the support cannot be made, while in other types of mounting in which the movement of the eyepiece is restricted or suppressed (as in the Sun telescope), the tube may be eliminated and the optical parts mounted independently of one another.

14.2 Mountings: general qualifications

(*a*) The primary characteristics of any worthwhile mounting are a rock-like stability and robustness of construction, combined with smoothness, uniformity, and ease of movement about both axes.

(*b*) Overhanging of centres of gravity, involving either the use of counterpoises or excessive strain upon one end of an axis, tends away from stability; other things being equal, a mounting not possessing this characteristic is preferable to one that does.

(*c*) The first point being assured, characteristics of secondary importance may be considered. First among these is the possibility of long uninterrupted runs in RA—preferably from horizon to horizon.

(*d*) Accessibility of all parts of the sky, notably the region of the pole and the zenith.

(*e*) Compactness.

(*f*) Comfortable observing positions in all positions of the telescope.

(*g*) Applicability to as many types of telescope, and arrangement of the optical parts, as possible.

(*h*) Simplicity and cheapness, and/or adaptability to home construc-tion.

B. 10.3 contains useful information on the structural considerations to be borne in mind when designing a mounting; B. 10.8, 9 considers the various aspects to balancing a mounted instrument. Good general accounts appear in the appropriate sections of B. 7.47, 76.

14.3 Altazimuth

If it is to be possible to point a telescope in any direction it must be mov-able in two planes, and these are most conveniently made perpendicular to one another. In the altazimuth mounting they are respectively the horizontal (azimuth being measured westward from S) and the vertical (altitude being measured from the horizontal plane passing through the observer).

The commonest varieties of altazimuth head incorporate two of the following structural features: (*a*) altitude axis reduced to a single pivot situated below the tube at its point of balance; (*b*) altitude axis defined by a pair of trunnions attached to opposite sides of the tube, between which lies the tube's point of balance; (*c*) columnar azimuth axis; (*d*) azi-muth axis in the form of a substantial circular disc running in a ring bearing. (*b*) and (*d*) are illustrated in Figure 94, (*a*) and (*c*) in Figure 93.

FIGURE 93
Basic altazimuth mounting

In the model illustrated in Figure 93 the centre of gravity, C, is not at the point of support, and considerable tensioning of the bolt B is therefore required to keep the tube stable in altitude. This militates against smooth and easy motion. The elongated column supporting the azimuth bearing would also introduce unsteadiness with any but the smallest instruments.

In the model shown in Figure 94, on the other hand, the tube swings between two points of support—a much more satisfactory arrangement.

FIGURE 94
Improved altazimuth mounting

These trunnions, t, t, lie in a horizontal plane, their median line being the altitude axis, aa. The trunnions are supported by a pair of arms, n, n, which are bent back at an angle to the azimuth axis, zz, in order to allow the telescope to be directed at the zenith. The arms are mounted on the horizontal platform, m, which carries the bearings for the short and relatively massive azimuth axis. This bearing is of the plain annular type; freedom of movement about the altitude axis is controlled by means of the screws x, x which bear down upon the metal leaf, y.

The greater tube diameter of the Newtonian, compared with the sort of refractor usually found in amateur hands, necessitates some modification in the design of the head, but the same principles of solidity and lack of vibration, combined with ease of movement, apply (Figure 96).

Steadying rods and slow motions (section 14.25) improve the performance of any altazimuth head by increasing the stability and smoothness of motion, while facilitating the following of the object in its diurnal motion. Indeed, if the altazimuth is to stand up to any sort of comparison with the equatorial, it must be fitted with slow motions, and these must be conveniently placed; the latter point is important, and it is surprising how often it is disregarded.

The Sellers mounting, for a Newtonian, is simple and extremely rigid. Essentially, the telescope tube forms one leg of a tripod, the mirror end resting on the ground, with the two other legs attached to the tube near the eyepiece position. Adjustment in elevation is obtained by a screwed rod passing between the two front legs; in azimuth, by moving the lower

Telescope tube

Vertical slow-motion screw

Adjustable legs

Flat board on ground

Horizontal (twist) motion

FIGURE 95
The Sellers mounting

end of the tube along the ground on small rollers, or, alternatively, by varying the length of the front legs. The tube being supported at each end, the possibility of vibration is completely eliminated; hence its particular attraction for use with telescopes of relatively long F. See B. 10.2, 43.

The advantages and disadvantages of the altazimuth may be sum marised as follows:

Advantages: Simplicity, ease of construction, and therefore cheapness.

Disadvantages: Following an object in its diurnal motion involves continuous readjustment about both axes. While it is true that following soon becomes automatic—and that Denning's planetary observations, for instance, or Markwick's work on Venus, were all carried out with altazimuths—the fact remains that a driven equatorial leaves *all* one's time and attention for observation, drawing, etc.

Two ingenious homemade altazimuths—one using wood and the other gas-piping (a rigid and readily available material)—are described in B. 10.13, 51; a water-pipe mounting for a small refractor is described in B. 10.37, while a design for a simple altazimuth, with slow motions, appears in B. 7.61.

14.4 Altazimuth modifications

Most types of altazimuth head can be more or less easily converted into equatorials by tilting the azimuth axis in the meridian until it is parallel to the earth's axis, and then securing it in that position. Such conversions, though cheap and easily contrived, are at best only makeshifts, and in view of the ease with which a solid and well-designed equatorial head can be made (if a professionally manufactured one cannot be afforded) they have little to commend them.

If the head is of the pillar type, it can be locked in a groove made in the surface of a hardwood prism whose base angle, ϕ, is equal to the latitude of the observing site, and whose side *SN* is horizontal and oriented south and north (Figure 97). The head is then mounted on a tripod or pier.

If the head is of the fork and base-plate type, the base can be countersunk and screwed to the face of a truncated hardwood prism whose apex angle, ϕ, is equal to the latitude of the observing site and whose side *SN* is oriented south and north (Figure 98). A counterpoise in RA is required, and this, unless it is forked, will interfere with the observation of low Decs.

The modification known as the Earl of Crawford's equatorial is illustrated in Figure 99. The rod *XY* lies in the meridian; the point of attachment of the wire, *B*, is such that the angle *CBX* equals the latitude of the observing site; the length of the wire, *AB*, is adjusted according to the Declination of the object; the wire is kept taut by a counterpoise at *W*. A force applied horizontally to the eye end of the telescope will then

FIGURE 96
Altazimuth fork mounting
for Newtonian

FIGURE 97
Equatorial modification
of basic altazimuth

make it follow the diurnal motion of a star. This method is only applicable to objects south of the zenith.

14.5 Equatorial mountings

Like altazimuths, equatorials provide for motion of the telescope about two mutually perpendicular axes; these are, however, no longer horizontal and vertical, but parallel to the earth's axis of rotation (the polar axis) and at right angles to it (Declination axis).

FIGURE 98
Equatorial modification of altazimuth fork mounting

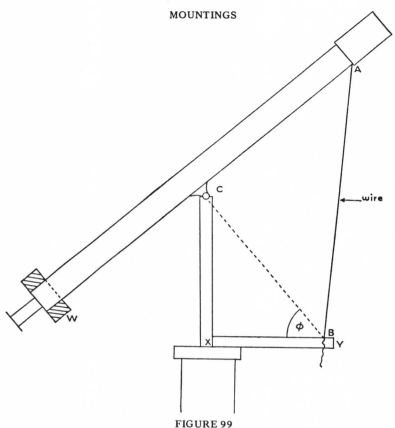

FIGURE 99

The Earl of Crawford's equatorial

The specific *advantages* of the equatorial are:

(a) ease of following;
(b) ease of re-finding (even without a driving clock) when the observation is interrupted by cloud or ocular changing, or the eye is taken from the telescope for whatever reason;
(c) possibility of a mechanical drive, applied to one axis only;
(d) possibility of finding and locating approximate positions by means of graduated circles.

It is even more a *sine qua non* of photographic work than of visual observation and though not strictly essential to the latter, its advantages over the altazimuth are so overwhelming that for all serious work two thoughts will not be given to the choice.

14.6 The German mounting

Invented by Fraunhöfer. The telescope is carried at one end of the Dec axis, which is approximately bisected by the polar axis; thus the centre of gravity is offset with respect to the polar axis, and the weight of the telescope has to be balanced by a counterpoise at the other end of the Dec axis. This is also a very convenient place to mount subsidiary apparatus such as cameras, since no increase in the total counterpoise weight is then involved.

A typical German head, suitable for small instruments, is illustrated in Figure 100. The polar axis is carried on bearings inside the housing, a; the Dec axis is similarly mounted in b. The cradle, c, carrying the telescope, must be of sufficient length to give adequate support to the tube, and to ensure that it is always normal to the Dec axis. Declination and hour circles (see section 14.24) are at d and e respectively. The counterpoise f balances the weight of the telescope and cradle; accurate counterpoising is important, especially if the polar axis is clock-driven. Slow motion (see section 14.25) in RA is applied by the rod g, through a universal joint and worm, to the driving circle h. Platform i is for a driving clock (see section 15). Slow motion in Dec is applied by the rod j. Both axes are supplied with clamps. Adjustment of the elevation of the polar axis for any latitude from $0°$ to $70°$ can be made by slackening off the bolt k and reading the angle between the polar axis and the horizontal from scale l; most heads also provide for $5°$ to $10°$ of adjustment in azimuth. Levels (m, n) are also provided to facilitate the setting up of the head (see section 16).

The whole sky is visible from the two positions of the telescope (E and W of the pier) but not from each. With normally designed mountings, a refractor placed W of the polar axis cannot pass the meridian when directed at altitudes greater than a few degrees N of the zenith; reflectors can generally pass the meridian at higher altitudes than a refractor. At least half an hour's run past the meridian at all but very high Decs is desirable, but cannot always be provided; the operative factor, for altitudes N of the zenith, is the clearance between the eye end of the telescope and the mounting; for other altitudes it is often the obstruction caused by the presence of the driving gear.

Reversing the telescope involves a rotation of the polar axis through 12^h and of the Dec axis through $180°$. The tube and its parts—e.g. field diaphragm or photographic film—are therefore inverted relative to the field. The interruption to observation is an inconvenience in visual work, but is even more serious for the photographer, since the exposure cannot

FIGURE 100

A well-equipped German equatorial mounting

be continued on the other side of the mounting owing to the 180° rotation of the film.

Objects E of the meridian should be observed with the tube W of the mounting. Observation can then be continued till the object is on or a little past the meridian; the telescope is then moved back eastwards 6^h, turned over in Dec, and brought once more to the meridian by a further 6^h eastward rotation of the polar axis; it is then reset in Declination, and the observation continued.

Among the *advantages* of the German mounting are:

(*a*) easily portable with small instruments (cf. English mounting);

(*b*) only a single supporting pier or tripod is required (cf. English mounting);

FIGURE 101

The author's clock-driven German equatorial,
mounting an 11·5-cm refractor and camera

213

(c) compact (cf. English mounting);

(d) visibility of the whole hemisphere of the sky.

Its *disadvantages* are:

(a) necessity of reversing when passing the meridian;

(b) overhung centres of gravity, involving a counterpoise on the Dec axis and overloading of the upper end of the polar axis.

Numerous descriptions of home-made German equatorial heads appear in the literature; see, for example B. 10.33, 39, and appropriate sections in B. 7.47, 48.

14.7 The English mounting

Sometimes also called the Yoke mounting (Figure 102). The points of support of the tube are on either side of it, and the Dec axis is borne in a cradle which is supported at both ends; this cradle, or yoke, constitutes

FIGURE 102
The English mounting

the polar axis. Thus no counterpoise is required, and both ends of the polar axis take the weight of the telescope, making for greater stability and freedom from flexure in the axes. It is particularly suitable, therefore, for large instruments (e.g. the 2·5-m Hooker telescope at Mt Wilson).

While reversal at the meridian is eliminated, the region round the pole is inaccessible*—a drawback which has led to the developement of the

*If Airy's 32-cm at Greenwich is regarded as an English equatorial, then this statement requires modification, since the pole is accessible with it. Airy's mounting may, however, perhaps more correctly be regarded as a halfway stage to the Horseshoe mounting.

Horseshoe mounting (see section 14.11). Also it is not a design that is well adapted to use with small refractors or Cassegrains, owing to the great height involved. An English mount with sector drive was described in B. 10.40.

14.8 The modified English mounting

Also known as the English, or Cross-axis, mounting (Figure 103). The yoke of the English mounting has been here reduced to a single beam, at whose centre lies the Dec axis bearing; as in the German mounting,

FIGURE 103
The modified English mounting

the Dec axis must be counterpoised. The 183-cm reflector of the Dominion Astrophysical Observatory (Victoria, B.C.) is a well-known example; one of the earliest was Admiral Smyth's equatorial at Bedford (*c.* 1830).

Advantages:

(*a*) two-point support of the polar axis reduces the heavy load on its upper end (cf. German);

(*b*) polar region accessible (cf. English);

(*c*) equally suitable for refractor and Newtonian (cf. English).

Disadvantages: Dec axis counterpoise (cf. English).

For a description of a modified English mounting, see B. 10.21.

14.9 The Astrographic* mounting

Designed to combine the advantages of the English and the German mountings by means of the overhanging polar axis:

*The term is used here in its wider sense. The mountings of the photographic telescopes used for the International Astrographic Chart (Carte du Ciel) were not generally of this form.

(*a*) no reversal at the meridian, which involves not merely the interruption but the termination of exposures;

(*b*) compactness.

Condition (*a*) is satisfied if the upper part, A, of the pier (Figure 104) is longer than the half-length of the guide telescope. The astrographic is not, therefore, a design which is generally applicable to telescopes of normal focal length.

14.10 The Fork mounting

Whereas the modified English mounting consists in effect of one side of a Yoke mounting, the Fork is in effect the lower half of an English mounting; it resembles the German mounting as regards its polar axis bearings, and the English as regards its Declination bearings (Figure 105). To avoid flexure, the polar axis, its bearings, and the fork have to be massive. Given the requisite solidity of construction, this is one of the most useful forms of mountings for Newtonians. The 122-cm Schmidt at Mt Palomar, the Mt Wilson 150-cm, and the Lick 3-m are all mounted in this way.

Advantages:

(*a*) pole accessible to Newtonian or camera;

(*b*) no meridian trouble;

(*c*) no Dec axis counterpoise.

Disadvantages:

(*a*) weight of axis and fork bear outside the points of support of the polar axis; flexure of the fork difficult to overcome with large instruments;

(*b*) unsuitability for refractors and small Cassegrains, there being a wide region, centred on the pole, which is inaccessible.

An interesting Fork mounting, homemade out of reinforced concrete, is described in B. 10.20.

14.11 The Horseshoe mounting

Also known as the Open Yoke type. The mounting designed by Russell Porter for the Mt Palomar 5-m is the prototype. It aims at combining the advantages of the German and English types, while avoiding the weakness of the Fork (Figure 106).

The first telescope to be mounted in this form was the Palomar 5-m reflector, but the design has been employed with most of the very large

FIGURE 104
The Astrographic mounting

FIGURE 105
The Fork mounting

View from South
(telescope tube omitted
to simplify the diagram)

View from East

FIGURE 106
The Horseshoe mounting

217

modern reflectors (e.g. Kitt Peak 4-m, Anglo-Australian 4-m). The upper bearing ring of the polar axis is in the form of a horseshoe, which is connected to the lower bearing by a short polar frame within which the telescope swings.

Advantages:

(*a*) no meridian reversal;
(*b*) polar region accessible;
(*c*) all masses lie between their points of support;
(*d*) the three-point support of the yoke provides great rigidity in all positions of the telescope.

Disadvantage: Unsuitable for other than very compact instruments.

An interesting homemade mounting of this type is described in B. 10.26.

The modern tendency has been to abbreviate the polar frame as much as possible, mounting the tube actually within the horseshoe in the form advocated by Russell Porter in his 'split-ring' mounting (see B. 10.18, 42).

14.12 Summary of equatorials and their characteristics

The characteristic strengths and weaknesses of the types of equatorial mentioned are summarised below under five heads:

A = *Newtonian* B = *Cassegrain* *or refr.*	*Accessibility of pole*	*No reversal at meridian*	*Tube inside Dec axis supports*	*Dec axis inside polar axis supports*	*Compactness*	*Total*
German: A	*	—	—	—	*	2
B	*	—	—	—	*	2
English: A	—	*	*	*	—	3
Modified English: A	*	—	—	*	—	2
B	*	—	—	*	—	2
Astrographic	*	*	—	—	*	3
Fork: A	*	*	*	—	*	4
B	—	*	*	—	*	3
Horseshoe: A	*	*	*	*	—	4

14.13 Two simple photographic mountings

Although amateur ingenuity has led to much sophistication in mounting design, it has also been directed to obtaining useful results from the simplest of materials, as in the following two examples of equatorial mountings for small cameras.

14.13.1 'Scotch' mounting: Described in B. 10.14, 15 by G. Y. Haig, this consists (Figure 107) of two boards hinged together along one side. The lower board is secured to a wooden block cut away at a base angle

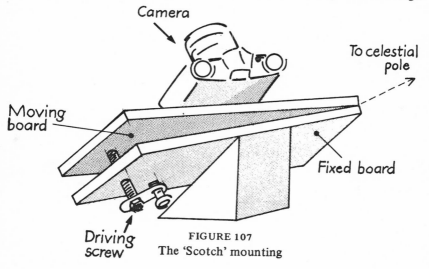

FIGURE 107
The 'Scotch' mounting

ϕ°, and is set in azimuth so that the axis of the hinges is directed towards the celestial pole. A length of screwed rod passing through this lower board drives the upper board, to which the camera is attached, at an appropriate rate (for a further discussion of tangent drives of this type, see section 15.3.1).

14.13.2 Poncet mounting: Also usable for a short Newtonian or Cassegrain, the Poncet has the advantage of extreme rigidity. It consists of an approximately horizontal board, pivoted loosely near one edge, while the opposite edge is free to slide over a surface set in the Earth's equatorial plane (Figure 108). Under these conditions, the motion of the board, and anything attached to it, becomes an equatorial one. The board can be driven by a tangent arm or similar arrangement. (See B. 10.35.)

A drawback common to both these simple mountings—though far outweighed by the ease with which they can be constructed—is the relatively restricted run available in RA before they have to be re-set.

VIEW FROM EAST

FIGURE 108
The Poncet mounting

14.14 Mountings with fixed oculars or oculars with restricted movement

A mounting incorporating an ocular which remains stationary during observation contributes somewhat to the comfort and convenience of the observer; and it should be remembered that the accuracy of his observations is intimately connected with his relaxation and general comfort whilst making them. Fixed-ocular telescopes also introduce the possibility of permanently mounted auxiliary instruments, and an enclosed and warmed observing-room. But even the elimination or reduction of the movement of the observing position, in the case of an instrument entirely (i.e. including the observing position) in the open air, is a convenience to the observer. For this reason, mountings which reduce the normal movement of the eyepiece, whilst not altogether eliminating it, are worthy of consideration.

The possibilities of design are summarised in the following Table. Mountings of type (a) (known as polar telescopes, for obvious reasons) provide a completely stationary observing position; so do telescopes fed by a siderostat or coelostat. Types (b) and (c) reduce the normal range of ocular displacement.

The two disadvantages to which these mountings as a class are subject (they are not individually subject to both, or even either, of them) are:

(a) difficulty in determining the orientation of the field;
(b) complexity and size, with resultant expense.

220

Axis of ocular	Observing position	Axis of ocular directed toward	Locus of ocular for all positions of the telescope
(a) in polar axis	fixed (ocular rotates)	N/S celestial pole	point
(b) parallel to polar axis	restricted	N/S celestial pole	circle centre: polar axis radius: distance of ocular axis from polar axis
(c) in Dec axis	restricted	polar axis	circle centre: polar axis radius: distance of ocular from polar axis
(d) telescope fed by siderostat	fixed (field rotates)	usually horizontal and in meridian	point
(e) telescope fed by coelostat	fixed (no field rotation)	usually horizontal, any azimuth, or in meridian	

14.15 The Grubb mounting

Structurally the simplest of the polar telescopes. Economically, however, it must be beyond the reach of most amateurs. Although some types of polar telescope, not adapted to the provision of an enclosed observing-room, are quite easily encompassed by the amateur pocket and home-making ingenuity, this is not true of the mountings which, like the Grubb, can conveniently isolate the observer from the instrument.

At the lower end of the Grubb telescope (Figure 109) is an equatorially mounted plane mirror, to whose English-type mounting the telescope is attached: PP' is the polar axis, DD' the Dec axis. Relative to a vertical line through the eyepiece, the field rotates; but if the ocular rotates to-gether with the rest of the instrument once in 24 hours there will be no relative motion between it and the field. The same result would be achieved by disconnecting the main telescope from the flat mounting, fixing it on a permanent mounting, and simply rotating the eyepiece at the same angular velocity as the mirror by means of a small driving clock (Figure 109(b)).

A typical existing example of this design of mounting is the 30-cm polar telescope at Harvard, designed in its details by Gerrish.

The specific weaknesses of the design are:

(a) the position of the objective near the ground makes it very sus-ceptible to convection currents, with deterioration of seeing;

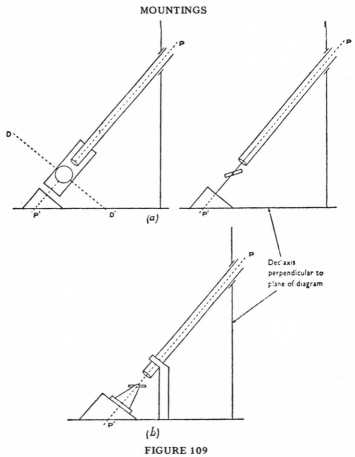

FIGURE 109

The Grubb polar mounting

(b) polar region inaccessible;

(c) low Decs are unobservable unless the flat is large compared with the objective.

For a description of an amateur-built polar telescope of this type, see B. 5.4, 10.22.

14.16 Cambridge (Sheepshanks) *coudé*

The main tube (Figure 110) forms the polar axis, PP'. A side tube, carrying the objective, is mounted so as to be able to move in N and S directions, i.e. it is mounted on a Dec axis. At the junction of the two axes is

FIGURE 110
The Cambridge *coudé* mounting

situated a rotatable flat which is connected to the Dec axis by a somewhat intricate linkage which ensures that as the latter is turned, the normal to the mirror always bisects the angle between the axes of the two sections of the tube.

As against the Grubb mounting, its objective is raised higher from the ground. On the other hand, the polar regions are still inaccessible, loss of aperture will occur at low S Decs unless the flat is considerably larger than the objective, and the image suffers a one-plane reversal (owing to the single reflection).

14.17 Paris (Loewy) *coudé*

The *coudé* (or 'elbowed') principle of telescope design was initiated by Loewy at the Paris Observatory in 1891. Figure 111 shows three variants, the third being for a reflector.

Its characteristic feature is the two large flats, permanently mounted at right angles to one another. The second reflection corrects the mirror-reversal of the first, and the introduction of a second flat makes all Decs accessible, though with an unavoidable small loss of light grasp. It shares the general disadvantages of polar telescopes: it is cumbersome, expensive, and outside the scope of most amateurs' constructional ability. Its specific disadvantage is that the orientation of its field varies with the Dec of the centre of the field. For a given Dec, however, the field remains motionless relative to the ocular. Reversal is generally necessary soon after the meridian passage of an object. See B. 10.11.

Figure 111

224

FIGURE 111 *(c)*

The Paris *coudé* mounting

14.18 The Ranyard mounting

Two variants of this type of mounting are shown in Figure 112. Others are described in B. 10.48. The cone of rays from the primary mirror is intercepted first by a small convex mirror, mounted axially near the focus, and then by a still smaller flat which reflects it up or down the hollow polar axis. The flat is geared to the Dec axis by a linkage which ensures that it rotates at half the angular velocity of the axis. An incidental advantage of this mounting is the large equivalent focal length provided.

14.19 Pasadena and Springfield mountings

The optical arrangement of these two polar mountings is the same (Figures 113, 114). In the latter—which was designed by Russell Porter, designer of the Mt Palomar 5-m—the Dec axis is reduced to a minimum, in the interests of stability. Both it and the polar axis are reduced to mere studs; these are attached to circular plates, again for stability. For further details of construction, see B. 7.48, 7.49.

The odd number of reflections introduces mirror-image reversal, and the light loss at three reflecting surfaces is far from negligible. A smoother

(b)

FIGURE 112
The Ranyard mounting

FIGURE 113
The Pasadena mounting

ray from
primary mirror

FIGURE 114
The Springfield mounting

drive would probably be obtained if the polar axis were prolonged. It has also been said that the downward-directed eye position causes trouble from motes, but of itself this can be of little weight compared with the mounting's advantages.

14.20 Manent's mounting

A 'semi-fixed' design providing restricted movement of the ocular during observation, the axis of the ocular being coincident with the Dec axis

(Figure 115). Thus, if the tube were made to sweep out all the parallels of Dec and hour circles the ocular would not sweep out a sphere, as with ordinary mountings, nor a point, as with polar telescopes, but a circle.

FIGURE 115
Manet's mounting

Additional advantages are its great compactness (allowing a long-focus optical system to be housed in a small tube) and the protection offered to the reflecting surfaces by their enclosed position—though this is not now such a telling advantage as it was in the days before aluminising.

Chief weaknesses of the design are: only one end of the Dec axis being available for the ocular, this assumes awkward positions when the tube is far from the meridian, and a diagonal has to be used; the loss of light at the two additional reflecting surfaces.

228

14.21 Pickering's mounting

In this design (Figure 116) the ocular may be mounted in either of the hollow Dec trunnions—the flat being rotatable about the optical axis of the primary mirror—which gives the choice of a less uncomfortable observing position when the tube lies far out of the meridian. The displacement of the ocular from the polar axis being so small, the observing position remains at virtually the same height above the ground at all times (cf. Newtonian). A further advantage of this mounting is the comparative ease

FIGURE 116
Pickering's mounting

with which it can be homemade. Being restricted to Declinations S of the zenith it is of value to the lunar or planetary, rather than the sidereal, observer.

Apart from the characteristic drawbacks of the fork type of mounting, Pickering's mounting involves the use of a heavy counterpoise, or alternatively a considerably longer tube than that of the equivalent Newtonian. A fuller description will be found in B. 10.31 or 10.32.

14.22 Siderostats

These are a development for astronomical purposes of the more primitive heliostat, which is chiefly used by surveyors. The heliostat—a two-mirror arrangement, one mirror being equatorially mounted—reflects the image towards the northern horizon (usually between NE and NW). Its disadvantage is that the field rotates in a non-uniform manner (the rate of rotation varying with hour angle), which makes its orientation impossible, or very difficult, to establish.

The field of a siderostat also rotates, but this rotation is uniform for a given Dec. The simplest form of the instrument is the polar siderostat, which is the basis of the Grubb mounting (Figure 109). The reflected ray can be directed towards either the N or S pole, the angular velocity of the mirror being equal to that of the Earth, i.e. the mirror's polar axis rotates once in 24 sidereal hours. It will be clear from the diagram that the normal to the mirror's surface intersects the star sphere at a point whose polar distance is half that of the centre of the field, and whose hour angle is the same as that of the centre of the field.

By employing a second plane mirror, mounting it either above or below the driven mirror and in line with its polar axis, the fixed-direction beam reflected by the driven mirror can be directed to any azimuth. This sort of two-mirror siderostat has the advantage over the heliostat that the orientation of its image rotates in a uniform manner, while, unlike the coelostat (section 14.23), different Decs can be viewed merely by altering the tilt of the driven mirror. It could, therefore, be used for planetary observation by feeding light into any fixed, orthodox telescope. Assuming the second mirror to be mounted above the driven one (the more convenient arrangement if fairly high Decs are to be observed), the smallest required diameter is obtained when the final beam is reflected approximately in the direction of the meridian.

The huge 1·5-m solar telescope at Kitt Peak National Observatory, Arizona, is fed by a single polar heliostat arrangement, the driven mirror reflecting light down an inclined tunnel set parallel with the Earth's axis.

Foucault's siderostat is adapted for use with a horizontal fixed telescope, which may or may not lie in the meridian. The principle of the mounting is shown in Figure 117. PP' is a polar axis fitted with a driving circle, A, at its upper end, and a fork at its lower. The rod C is carried on trunnions, B, in this fork; it is in two sections, the lower of which can rotate axially about the upper. At the lower end of C is a second fork, in which the sleeve E is mounted on trunnions at D. Sliding freely in E is the rod F, which is attached to the centre point of the back of the mirror cell; the mirror is mounted on horizontal trunnions at M. F and C always lie in a single plane, and $BD=BM$. The mirror can also be swung in a horizontal plane about G, which lies vertically below B.

If the telescope is mounted horizontally in the meridian, the mirror mounting is revolved about G until M is due south of B, i.e. BM lies in the same vertical plane as the polar axis. Since triangle *DBM* always remains isosceles, if *BD* is maintained parallel to *XY*, the mirror will orient itself in such a manner that the reflected ray remains parallel to *BM*. The polar axis is then driven to maintain the plane *PP'D* parallel to the hour circle of the object; angle $PP'D = \delta + 90°$. If, now, $\phi = $ latitude of the observing station, and $\theta = $ the polar distance of the centre of the field:

field is fixed when $\theta = \phi$,

field rotates clockwise when $\theta < \phi$,

field rotates anticlockwise when $\theta > \phi$.

FIGURE 117
Foucault's siderostat

If it is required to set up the telescope in some azimuth outside the meridian, *BM* is simply moved into the required position by revolving the mirror mounting about *G*. The pressure of *C* against *F* forces the mirror at the same time to turn about its two axes, the sleeve *E* both turning and sliding on *F*.

14.23 Coelostats

The siderostat, as its name implies, can make the image of a star stand still. But this is only so if the star lies at the centre of the field; the remainder of the field rotates about this fixed point. The coelostat, on the other hand, can make not only a star but the sky stand still—its entire field is fixed and non-rotating.

In its simplest form (the true coelostat) it consists of a single plane mirror, mounted on a polar axis and clock-driven at half the angular rate of the Earth—i.e. it rotates once in 48^h, the plane of the reflecting surface passing through the axis of rotation. A fixed telescope, normally horizontal, directed at this mirror will then always see the same field. The drawback of this scheme is, of course, that it makes no provision for the viewing of different Decs,* the reflected beams from objects in different Decs lying in different azimuths, only one of which contains the telescope. For any given position of the telescope, then, the Dec of the centre of the field reflected in the mirror is fixed; to observe other Decs, without using a second mirror, the telescope must be moved to a new position—in the case of rays reflected horizontally, to a new azimuth. The required azimuth, *a*, is given by

$$a = -\frac{\sin \delta}{\cos \phi}$$

where δ = Dec of centre of field,

ϕ = latitude of observing site.

Circumpolar objects can be reached only at the cost of (*a*) inclining the tube out of the horizontal, or (*b*) introducing a second mirror, which remains fixed throughout the observation. But there will necessarily still remain a restricted circumpolar region that is inaccessible, since the coelostat mirror has no adjustment in Dec.

The use of a second mirror is the easiest way of making a variety of Decs available to the objective—the only way, indeed, if the telescope is to remain in one position. It is then the position of this second mirror which allows a particular Dec to be fed into the fixed objective; in the

---

*This is no disadvantage in the observation of solar eclipses, where the single-mirror coelostat is perfectly effective.

case of the Sun it can be left for a week or so at a time (and is therefore often called the 'fixed' mirror—as in fact it always is through an observation) owing to the Sun's slow motion in Dec. In changing from one Dec to another, both the position and inclination of the fixed mirror must be altered.

The Sun presents the most favourable conditions for observation by means of a coelostat with a fixed secondary flat; not only is its rate of change of Dec slow, but it never leaves approximately equatorial regions. The coelostat can be arranged to deliver either a horizontal (e.g. the old Mount Wilson Snow telescope), or a vertical beam to the objective (e.g. the McMath-Hulbert telescope, Michigan).

When designing any instrument in which the objective is fed by one or more mirrors at some appreciable distance from its surface, care must be taken to ensure that they are sufficiently large for the purpose. If the beam reflected to the objective is of the same diameter as the aperture, then full illumination will be obtained only for images exactly on the optical axis, i.e. the fully-illuminated field of view will be zero. For full illumination across a field of view θ, a plane mirror set at a distance d in front of an objective of aperture D must reflect a beam whose diameter D' is given by

$$D' = D + 2d \tan \theta.$$

The true diameter must exceed this by a factor cosec i, where i is the maximum angle of incidence at which the mirror will be used. For example, a siderostat mirror set 15 cm in front of an OG of diameter 10 cm is required to reflect a fully-illuminated image of the Sun (diameter $\frac{1}{2}°$) at all seasons of the year. The maximum Dec of the Sun is $\pm 23\frac{1}{2}$; hence, at midsummer or midwinter (depending on whether the siderostat reflects the beam north or south) i will have a value of $(90° + 23\frac{1}{2}°)/2$, or about 75°. Then

$$D' = (40 \tan \tfrac{1}{2}°) \operatorname{cosec} 57°$$
$$= 18·8 \text{ cm}.$$

At the cost of some slight diametrical vignetting at the extreme solar Dec, a centimetre or so could be trimmed off this value.

It follows from this that the mirrors used in a coelostat, generally being set to smaller angles of incidence, can be smaller, for a given objective, than those used in the normal heliostat arrangement.

14.24 Setting circles

If the usefulness of a telescope is increased 100 per cent by mounting it equatorially, this is primarily because of the increased ease of following an object in its diurnal motion, increased accuracy in photographic work, and the possibility of using certain accessories which are impracticable with an altazimuth. The addition of circles to the equatorial increases the instrument's usefulness by perhaps a further 10 per cent. Circles are a convenience, but by no means a necessity. Given a good star map, all the preliminary work of identifying the neighbourhood of a faint object can be done during the day, and the amount of observing time wasted in locating it need be very small. Only for finding objects during the day-time do circles have an unquestionable advantage.

In any case, very finely divided circles are not only unnecessary but may be positive nuisance, owing to the time spent poring over the verniers and the probable destruction of the eye's dark adaptation by the use of a fairly bright light to make the graduations visible. Position circles are, after all, for finding and not for accurate position measures—for which the correct instrument is the micrometer—and this function they perform perfectly satisfactorily if they permit the identification, not of a point object, but of a LP field.

Finely divided circles justify the fitting of verniers, but it must be remembered that this is only true if the adjustment of the equatorial axis is of a comparable precision. Circles of sufficient precision to enable an object to be picked up with the finder or in a LP field can easily be home-made; they do not require reading verniers, a simple index mark being sufficient.

Two obviously faulty procedures should be avoided in the graduation of circles: (a) if the graduations are to be made direct on the rim of the circle with the aid of some form of protractor, the latter must be larger, not smaller, than the circle; otherwise errors in the graduation of the protractor will be increased, not reduced, in their projection on the circle; (b) if the graduations are made on a long strip of some material which is to be wrapped round and fixed to the rim of the circle, the two end points must be accurately fixed at a distance apart equal to the circumference of the circle, and the space between them filled in with the requisite number of equal graduations; to start at one end and work along to the other will inevitably introduce cumulative error which will be discovered when $360°$ is reached.

Method (b) is easier to carry out than (a). The strip to be graduated may be of thin brass, marked with a steel engraving tool; celluloid, painted black on one side, on which the graduations are cut with a sharp knife; paper, on which the graduations are marked with a fine-nibbed pen, which is fixed to the circle rim and protected from the weather with waterproof varnish.

Graduations can easily be made by any of these methods at 1-mm intervals; about 2-mm intervals, however, are generally to be preferred. Assuming that the Dec circle is to be graduated in degrees at 2-mm intervals, and the hour circle in 2-mm divisions of 6^m (preferable, for interpolation by eye, to 5^m intervals), the diameters of the two circles will be respectively 22·9 cm and 15·28 cm. Rather than decide on the exact linear value of one graduation and then calculate the diameter of the circle, however, it is better to decide on a convenient circle diameter which will allow the graduations to be about 2 mm, and then to sub-divide a distance equal to its circumference. Figure 118 illustrates, to scale, the appearance of the graduations of a Dec and hour circle, both 30 cm in diameter: they are easy to read at a glance, and with interpolation by eye can be read to 15′ and 2^m.

The method of marking the circles is as follows. Wrap the strip round the rim of the circle with an overlap of about an inch, and mark off the circumference on it. Pin the strip to a drawing-board and lay off a perpendicular to it through one of the two marks indicating the circle's circumference. On the edge of a separate sheet of paper divide any convenient length, greater than the circumference of the setting circle, into the required number of graduations (e.g. in the case of the 30-cm circle already mentioned, the graduations on the subsidiary scale could be

FIGURE 118
Circle graduations

5 mm apart). Now pin the scale to the drawing-board so that one end of it coincides with one end-mark of the strip, and the other coincides with the perpendicular through the other end-mark. With a set-square, drop a

perpendicular to the strip from every graduation of the scale. These are the required divisions of the circle.

Circles may be graduated in any of the ways illustrated in Figure 119. But bearing in mind that the Dec circle must read $0°$ and the hour circle 0^h when the telescope is lying in the meridian with the celestial equator passing through the centre of the field, and that, furthermore, reversal of the instrument involves a $180°$ rotation of the Dec axis and a 12^h rotation of the polar axis, it will be seen that method (a) (for both Dec and hour circles) is the most convenient to use.

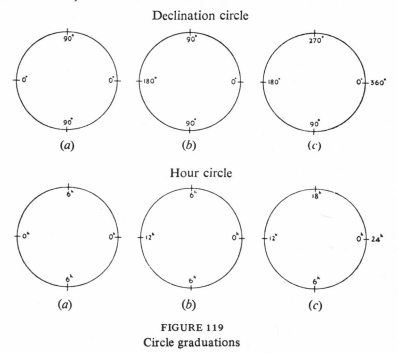

Declination circle

(a)   (b)   (c)

Hour circle

(a)   (b)   (c)

FIGURE 119
Circle graduations

With either the polar axis circle fixed to the axis and its index to the mounting, or vice versa (the circle then reading hour angle), the procedure for finding an object whose coordinates are known is as follows:

(i) Swing the telescope in Dec until it is pointing at the celestial equator ($0°$). Depress or elevate the telescope through an angle equal to δ, and clamp. The actual reading of the circle will depend on the way in which it is graduated: in case (a) (Figure 119) it will be δ; in case (b) it will only be δ in one position of the telescope; in case (c) it will only be δ in one position of the telescope and on one side (N or S) of the equator.

(*ii*) The RA of the meridian, Local Sidereal Time, is given by the observatory clock (T); the RA of the object is assumed known (a). Then:

If $a > T$, object is E of the meridian: swing telescope to W side of the pier.
If $T > a$, object is W of the meridian: swing telescope to E side of the pier.

Start the driving clock (if any). Set the telescope in the meridian (hour angle 0^h) and swing it eastwards through the angle $a-T$ or westwards through the angle $T-a$, as the case may be. Clamp. The object should then be in the field of the finder. Bring it to the intersection of the cross-wires with the slow motions. It will then be in the telescopic field.

In practice it is usual to add, say, 1^m to T, to cover the time taken in setting the telescope. Then the object will be in the field of the finder at time $T+1^m$.

To discover the approximate coordinates of an object at the centre of the field, the procedure is simply the reverse of the foregoing: δ is given by (or quickly calculated from) the Dec circle reading; $a = T \pm H$ (according to whether the object is E or W of the meridian), where T is given by the observatory clock and H by the hour circle.

With a fixed hour circle a separate calculation of $a-T$ or $T-a$ thus has to be made each time an object is found by means of its coordinates. To avoid this inconvenience a slipping polar axis circle with two verniers, or two reading indices, is usually employed. Off the single scale T is read by one index and RA by the other, eliminating the separate calculation of H each time. One index is fixed permanently to some immovable part of the mounting—generally, for convenience, central on the S side; the circle rotates freely on the polar axis but can be clamped in any position relative to this fixed index; a second circle, ungraduated but carrying an index, rotates with the polar axis; this index coincides with the fixed index when the telescope lies in the meridian. The procedure, having set the telescope to the correct Dec, is then as follows: set the telescope in the meridian (same circle reading with the two indices): slip the circle till the fixed index reads a; clamp the circle; rotate the polar axis till the second reads T. The object should then lie in the field of the finder.

14.25 Steadying rods and slow motions

Steadying rods are fitted to most altazimuths. They attach the eye end of the telescope to some point on the mounting—the pillar, the tripod, or the trunnion fork. Individual makers have their own particular designs, but the general principles are illustrated in Figure 94. When the locking screw *a* is slackened, the rod *b* can slide freely in the sleeve *c*, which is in turn connected with the sleeve *d*; *c* can rotate in a vertical plane and *d* in a horizontal plane, allowing the tube to be moved in altitude and azimuth.

With instruments over about 75 mm aperture, slow motions are so useful as to be virtually essential. Altazimuths require slow motions in both altitude and azimuth; equatorials require slow motion in RA to correct variations or inaccuracies in the drive, but a slow motion in Dec is less important providing the mounting itself in accurately adjusted. Whatever the type of mounting, the slow motions must be efficient: in particular, they must be slow enough, and there must be no looseness or backlash in their operation.

An altazimuth steadying rod is easily connected to a device for applying slow motion adjustments by the addition at *e* (convenient to the observer's hand) of a rack-and-pinion action, as in Figure 94. Alternatively, the end of the rod may be threaded and connected to the telescope by a long sleeve, threaded internally and milled externally. Rotation of this sleeve will then vary its position along the rod, applying slow motion in altitude (Figure 120). A combined slow motion and steadying rod for a reflector mounted on an altazimuth stand is illustrated in Figure 121. A threaded rod, perhaps 20 cm long, is pivoted on the underside of the tube. A large knurled collar is threaded on to this rod, the lower part of which passes loosely but without excessive play into a tube. This tube slides through a short sleeve, pivoted to an extension at the front of the azimuth axis, and can be set at any required telescope altitude by a thumbscrew. A long spring, or even stout rubber bands, tension the threaded rod inside the tube; fine adjustment in altitude is obtained by rotating the collar In instruments larger than about 20 cm in aperture, it may be found more satisfactory to have the threaded rod itself rotatable, screwing into the upper end of the adjustable tube; the telescope tube can be given a slight bias to take up any slack in the screw fitting.

Slow motion in azimuth (see Figure 94) is usually provided by means of an endless screw, actuated by a hand-rod *f*, to which it is attached by a universal joint (often called a Hooke's joint), and meshing with a toothed collar incorporated in the azimuth axis. The quick motion in azimuth is independent of the slow-motion screw. For many purposes, a simple tangent drive (a screwed rod fitted below the fork, bearing against a plate fixed to the stand and kept in tension by a spring) will prove adequate, affording perhaps an hour's controlled following before the rod has to be wound back and the telescope rotated bodily in azimuth.

Commercial equatorial heads are usually equipped with worm and wheel drives on both axes, the Dec adjustment being made by controlling

FIGURE 120
Steady rod and slow motion

Threaded
rod

Telescope
tube

Knurled
nut

Quick
release

Tensioning
spring

FIGURE 121
Steady rod and slow motion for a Newtonian reflector

239

the worm from the observing position via a flexible connection, Hooke's joint, or even a remotely-operated motor. Since large angular adjustments in Dec never have to be made once the object has been acquired, a more sensitive adjustment (and firmer locking of the tube) can be obtained by using a large-radius sector of, say, $20°$, instead of the much smaller complete wheel that the same money should buy. Preliminary setting is then made by freeing the sector to rotate around the Dec axis. It is not even necessary to have a worm and sector arrangement for so small an arc, a screwed-rod tangent drive (B. 10.25) being likely to prove more sensitive because of the smaller pitch of its thread.

A simple undriven equatorial head could also have a sector or tangent drive in RA to provide slow-motion adjustments; but the usual procedure is for the polar axis to carry a complete wheel, so that continuous driving, whether manual or automatic, is possible. The wheel is joined to the polar axis via a slipping clutch, so that the telescope can be moved through a large angle in RA without disengaging the worm; this clutch may be operated either by friction or by a collar with positive locking. If the instrument is motor-driven, there must be provision for over-riding the sidereal drive to obtain fine adjustment in RA. A purely mechanical arrangement could involve (a) direct control of the worm or other drive system feeding the polar axis; (b) separate drive wheel and polar axis, the two being linked either by a sector or tangent drive, or by a pair of 'sun-and-planet' pinions. Interference with the worm is objectionable because backlash in the driving chain between worm and motor will almost certainly result in lost motion; a slow-motion device employing two linked worm drives is described in B. 10.19.

The almost universal adoption of synchronous electric motors for telescope drives (see section 15.2) has made available a simple and convenient way of adjusting the telescope's position in RA by altering the speed of the driving motor. Since, for precise diurnal drive, the electric mains frequency must be stabilised by a variable-frequency oscillator, this oscillator can itself be used to speed up or slow down the drive and hence swing the telescope slightly in RA. This can be done by altering the output frequency of the unit, but a superior design was two frequency controls, one to set the basic rate and the other to superimpose a different rate temporarily for the purpose of fine adjustment. Assuming a standard rate of 50 Hz, the normal range of frequency to which a motor will respond is in the range 45 to 65 Hz. Variable-frequency control lacks the flexibiluty of a mechanically-operated slow motion, since the superimposed rate is severely limited unless a stepping motor (B. 10.27) is used. It has, however, been found perfectly satisfactory for adjustment of a telescope over small arcs of RA

240

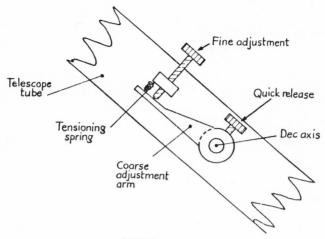

FIGURE 122
Tangent drive in Declination

14.26 The support for the head

An altazimuth head, a German equatorial head, or a fork mounting, if small, has to be supported at a convenient height above the ground.

Solidity and insusceptibility to vibration—whether from the wind, traffic, or the movements of the instrument itself—are all-important. A portable tripod, designed to be stood on a table, window-sill, or other makeshift foundation, is worse than useless. The best thing to do with such a mounting is to weld or bolt the toes of the 'claw' to three steel plates and mount these on 50-mm morticed hardwood stays arranged radially at 120° to each other and mounted firmly on the top of a single pier sunk in the ground.

Generally speaking, a single vertical pier is preferable to a tripod, since it renders the zenith accessible by elevating the telescope from any azimuth. A stout wooden tripod, firmly bedded in the ground, and preferably fitted with stretchers (see Figure 94), serves quite well with altazimuth or German heads carrying instruments of up to about 10 cm aperture. In the case of an equatorial it is preferable to fit the points of the tripod into sockets in the top of concrete columns sunk in the ground, so that the horizontality of the base of the head can be found again immediately, after the instrument has for any reason been moved; truly accurate adjustment of the polar axis cannot, however, be maintained in the way.

Vertical piers can be constructed of masonry, concrete cast in a mould (it should be reinforced), or a steel pipe or pillar embedded in concrete. An iron pillar is the most affected by traffic vibration; a wooden tripod, the least. Vibration of this sort consists of surface waves, and if instead of sinking a concrete pier 1·3 m in the ground it is sunk to a somewhat greater depth in the bottom of a 2·6 m hole (comfortably larger than the pier, and revetted), complete freedom from vibration can be secured, even in the vicinity of a railway or highway carrying heavy traffic. Since this involves an additional 2·6-3 m of concrete, it is a solution that will only be applied if the problem is acute.

Vibration caused by underground trains can only be circumvented by moving the observing site to another district, or by restricting observation to those few hours when the trains are not running—neither a very palatable solution.

If observation is carried out in a hut (see section 17) the pier must pass through a hole in the floor large enough to avoid all contact between the mounting and the floor-boards.

TELESCOPE DRIVES

15.1 Introduction

The drive is a mechanism that imparts to the polar axis of an equatorial a rotation whose angular velocity is such that the axis completes one turn in 24 sidereal hours; provision for varying the speed to the solar or lunar rate may also be incorporated.

Essentially it consists of (a) a source of power; (b) transmission connecting (a) with the polar axis; (c) some form of governor to control the drive at the required rate and to smooth out short-period variations in the output of (a).

Varying degrees of precision are demanded, according to the work that is going to be done with the telescope: (a) for visual work, it should at least be accurate enough to keep an object near the centre of a HP field for, say, half an hour; (b) photographic work with short-focus cameras demands greater precision, if guiding is not be become intolerably tedious; (c) photographic work with long-focus cameras demands the highest precision possible.

For (b) and (c) a drive may be considered essential—though surprising and perhaps rather pointless feats have already been achieved with hand guiding; for (a) it is not essential, but it is so easily effected and so convenient (especially when making drawings) that, once experienced, it will not subsequently be forgone.

15.2 Sources of power and methods of control

For approximately accurate driving, such as is an enormous boon to the visual worker, some most extraordinary mechanisms have from time to time been brought into successful service: the works of grandfather clocks, Meccano, aid-bladders with adjustable exhaust cocks, falling weights regulated by the flow of sand or mercury from a cylinder with an adjustable vent, and the like. With such drives, not suitable for large instruments, some form of belt transmission is adequate.

Omitting the more fantastic, methods in common use include:

- (a) clockwork, mechanical governors, and falling weights in a variety of combinations;
- (b) electric motors, controlled either by mechanical governor or pendulum;
- (c) synchronous motors, relying on a frequency-related grid supply or employing either some form of frequency regulation or a frequency-controlled private supply.

15.2.1 Clockwork and controlled falling weights: Ordinary clockwork (e.g. the works of an alarm clock) is neither accurate nor robust enough to be suitable as a source of power for telescopes. It can, however, be used with very light photographic apparatus with, at most, a small guide telescope (e.g. B. 10.4, 10.5). More importantly, clockwork can be used, not to drive the telescope, but to control the rate at which an independent source of power becomes available to turn the polar axis. At its simplest such an arrangement need consist of no more than a wire attached to the rim of a driving circle, thence passing over a pulley in the plane of the circle and the vertical plane through the polar axis, to a spindle mounted on the winding shaft of an alarm clock; it is given several turns round this spindle in the direction opposite to that in which the clock is wound up; a weight is attached to its free end. A slightly more elaborate arrangement, easily constructed at home, is described in B. 10.41.

Arrangements of this sort, while having the advantage that the weight and not the clock is driving the telescope (hence adequate power can be obtained merely by increasing the weight), are nevertheless suitable only for small instruments (10-cm or less); they are a blessing to the visual observer, but are not reliable enough for photographic work.

Clockwork (controlled by pendulum or escapement), whether used as a source of power or as a governor, introduces jerkiness into the drive, though this may be of negligible magnitude. An escapement beating four times per second, for example, advances the telescope in jerks whose angular value is $0.25 \times 15'' = 3''75$. Star images on the plate of a 450-mm focus camera would therefore be short lines of length

$$\frac{450}{206,265}.3.75 \cos \delta \text{ mm}$$

or 0.008 mm at the equator. This is quite negligible, and would be swamped by other errors in a drive of this sort, quite apart from atmospheric turbulence and emulsion diffusion. On the other hand, an electric motor or a falling weight will provide a perfectly continuous motion. Synchronous electric motors are now so widely used for telescope drives that mention of other systems, particularly the pendulum-controlled or governor-controlled non-isochronous electric motor drives, will almost certainly be of historical interest only.

15.2.2 Synchronous motors: These are simply electric motors whose running speed is, within certain limits, dependent solely upon the frequency of the current supplied to them. The speed of a synchronous motor is, in particular, unaffected by temperature variations over any normal range, or by small variations in its load; when a certain level of overloading is reached the motor will suddenly stall, and for this reason a slip clutch should be incorporated in the transmission.

The effectiveness of a synchronous motor for telescope drives therefore depends upon the precision of the frequency-regulation of the mains electricity supply. This frequency varies according to the load being put on the generators, which is usually at its highest during the evening, when the heaviest domestic consumption occurs; during this build-up, an electric clock can easily be gaining at the rate of 5 s to 10 s per hour (equivalent to an equatorial drift of $1\frac{1}{4}'$ to $2\frac{1}{2}'$ per hour). Even this is negligible for most visual purposes, and for short-exposure photography with small cameras. Moreover, variable-frequency control units, which generate their own AC, make mains instability unimportant, while a battery-powered unit can make the telescope completely portable. These can be purchased ready-made, or built up from the designs described in B. 10.23, 34, 38, 50.

Synchronous motors are available in a number of standard voltages and frequencies; in the UK, the mains supply is 240 volts at 50 Hz, while in the United States it is 110 volts at 60 Hz. Shaft speeds vary from 1/24 r.p.h. to several r.p.m., and power consumption ranges between about 2 and 12 watts. The performance of these small motors is surprising. In one case, two 2·5-cm/kg motors drove a 25-cm Cassegrain carrying a 10-cm refractor and photographic equipment weighing, in all, more than 450 kg.

The success of any telescope drive, particularly that of one in which the power source is relatively feeble, will depend in very great measure on the correct balancing of the instrument. For example, incorrect positioning of the counterweight in a German mounting may result in the motor stalling when the telescope is pointing near the meridian. The same thing can occur, at large values of H, if the telescope tube is not correctly balanced about the Dec axis. Even if the motor can handle considerable torque, severe imbalance can injure the worm and bearings. For a discussion on balancing an equatorial telescope, see B. 10.9.

15.3 Power transmission

The power from the motor can be transmitted from the initial driving shaft to the polar axis in a number of ways. The classic method is via a worm-and-wheel arrangement, but other cheaper, and not always less effective, methods are available.

15.3.1 Tangent drive: If the end of a screwed rod R is made to bear against a projection P fixed to the polar axis, rotation of the rod will cause the polar axis to turn (Figure 123). The drawback of this drive is that the rotary motion slows down as the angle Q departs more and more from a right angle. This is caused (a) by the component of motion of R decreasing (as cos Y), and (b) by effective length L of the projection increasing (as cosec Y). Causing the rod to bear at a fixed point on P eliminates effect (b) and reduces the loss (a) to a factor of cos Y/2. Assuming that R is driven so that the polar axis revolves at the correct rate when $Y=0°$ (i.e. the drive is tangential), the following Table shows the accumulated drift of an equatorial star from its initial position over given intervals of time:

Interval (min.)	Accumulated error ("arc)	
	L varying	L fixed
4	0·5	0·1
8	3·3	0·8
12	10·4	2·6
16	24	6·0
20	45	12
24	78	16
28	124	22
32	—	30
36	—	40
40	—	52

FIGURE 123

Tangent drive in RA

The accumulated error can be lessened considerably by employing a slightly faster drive rate than that appropriate for Y=0°. For example, if this rate is set 0·2 per cent fast, the accumulated error with variable L will come to $10''$ after 15 minutes, while with fixed L the drift after 40 minutes will be only about $15''$. Such precision is adequate for all normal visual work, and for unguided photographic work with lenses of fairly short focal length. Accepting a total drift in the focal plane of 0·025 mm as tolerable, the following Table indicates feasible exposure times for lenses of different focal length, photographing equatorial regions, with accelerated drive (the tolerance at other Decs increases as $\cos \delta$):

Focal length (mm)	Limit of acceptable exposure		
	Undriven	L varying	L fixed
50	6·9 sec.	26 min.	50 min.
100	3·5	21	40
200	1·7	16	30
500	0·7	12	19
1000	0·35	9	15

For a screw of pitch S, turning at x r.p.m., the length L is given by

$$L = \frac{228 \cdot 5 S}{x}.$$

B. 10.36 describes a telescope drive of this sort; see also B. 10.16 for the use of a compensating cam to produce a constant drive rate.

15.3.2 Friction drive: A large, plain wheel on the polar axis can be driven by, for example, a hard rubber pinion wheel, or by a belt drive. The former suffers from the possible disadvantage that local irregularities in the contact can affect the rate, but it has been found perfectly adequate for visual and photographic use. A belt drive, passing around a small pulley wheel, may be preferable (B. 10.30, 47).

15.3.3 Worm-and-wheel drive: In this standard arrangement, the drive is transferred to the polar axis via a worm wheel (the driving circle) and a tangential screw (worm). The worm should be single thread—i.e. in one rotation it turns a worm wheel of x teeth through $1/x$ of a revolution.

If the driving circle is reduced to a segment (as is often the case in English and fork mountings) whose angle is θ, the maximum length of uninterrupted run that is possible is $\frac{\theta}{15}$ hours. In such a case the worm has to be disengaged in order to reset the segment to the beginning of its run; in the case of a complete circle a friction clamp loosens the axis from the driving circle to allow hand setting in RA without interfering with the drive. Descriptions of home-made sector drives appear in B. 10.17, 40.

The worm and driving circle are the most important elements in the transmission from the viewpoint of accuracy of drive; the machining of the worm, in particular, should be of high precision. Within reason, the larger the worm wheel and the greater the number of its teeth, the better; in this way the effects of any inaccuracy in the cutting of the worm are reduced. 100 teeth should be regarded as the absolute minimum for a small driving circle; with larger instruments the number is more often 720 to 2880 (the worm turning at from $\frac{1}{2}$ r.p.m. to 2 r.p.m. respectively). Both the engaging and the idling faces of the teeth should be comfortably in contact with the worm, and backlash (end-to-end play in the worm itself) must be rigorously eliminated.

15.4 Transmission gearing

In the absence of some means of conveniently altering the drive rate at source, a conversion to sidereal time is necessary if the power source is regulated to mean time (e.g. synchronous motor with frequency control). This is in addition to the reduction of the motor spindle speed (if necessary) to the value required for transmitting rotation to the polar axis.

The tropical year consisting of 366·24219 sidereal days and 365·24219 mean solar days, a conversion of MT to ST involves a speeding up of $\frac{366 \cdot 24219}{365 \cdot 24219} = 1 \cdot 0027379$. Taking this as equivalent to $\frac{366}{365}$ introduces an error of roughly 0·0065 secs per hr, a negligible amount. The gear train must therefore include a 366/365 unit to effect the time conversion. Apart from that, the reduction is merely a question of slowing down the initial drive. If the input to the transmission is x r.p.m., the required reduction is (there being 1440 minutes in a day) $\frac{x}{1440}$, or, in the case $x=1$, 0·00069635. The actual gearing employed will be dictated by the spur gears available and the number of teeth in the worm wheel (driving circle). Six examples are given below, assuming an input from the motor of 1 r.p.m. (MT), and driving circles with from 96 (too few) to 504 teeth. In the last two the 366/365 unit is employed; in the first four the total reduction of 0·00069635 (or a close approximation to it) is obtained without treating the time conversion separately:

	A	B	C	D	E	F	G	Reduction
(a)	24	60	28	66	26	66	96	0·00069636
(b)	14	18	15	66	26	66	100	0·00069636
(c)	26	56	32	56	32	127	96	0·00069634
(d)	12	14	25	63	26	127	100	0·00069634
(e)	15	365	366	30	26	52	360	0·00069635
(f)	15	365	366	15	28	80	504	0·00069635

In practice, it will be found far more convenient to purchase a synchronous motor of approximately suitable speed to be connected to the driving worm or other component, and to make the final speed adjustment using a variable-frequency oscillator.

15.4.1 Solar-sidereal conversion: The true ratio between the mean solar and sidereal day may be taken, for practical purposes, as 1·00273791. Mechanical conversion ratios have been calculated to various degrees of accuracy, including the following:

Annual error

$$\frac{30 \times 61}{25 \times 73} = 1·002739726 \qquad 57·3 \text{ sec.}$$

$$\frac{32 \times 331}{247 \times 43} = 1·002737915 \qquad 0·2 \text{ sec.}$$

$$\frac{45 \times 71 \times 257}{29 \times 151 \times 187} = 1·002737909$$

15.5 Residual motion of the image

Possible causes of a slow drift of the image in the field are:

(a) maladjustment of rate: steady drift in RA;
(b) maladjustment of polar axis: drift in Dec;
(c) inaccuracies in transmission: periodic displacements in RA;
(d) atmospheric refraction: slow displacement in both RA and Dec. To compensate for differential refraction, the drive must run slow at the E horizon, accelerating to the true rate at the meridian, and decelerating again between the meridian and the W horizon.

SECTION 16

SETTING AN EQUATORIAL HEAD

16.1 Introduction

The successive stages in the setting up of an equatorial are:

(1) approximate orientation of the polar axis in altitude and azimuth;
(2) zero setting of the Dec circle index or vernier;
(3) final adjustment of the polar axis in altitude;
(4) final adjustment of the polar axis in azimuth;
(5) zero setting of the hour circle index or vernier;
(6) checking collimation: the perpendicularity of the optical and Dec axes;
(7) checking that the polar and Dec axes are mutually perpendicular.

These adjustments between them ensure that the polar axis is parallel to the Earth's axis of rotation, and that the Dec axis is perpendicular both to the polar axis and to the plane containing the optical axis. The order in which the operations are carried out is important: if the mounting is not fitted with circles, stages (2) and (5) are of course omitted; if, on the other hand, either of the circles is used in determining a particular error, its index error must first be determined; with one exception the methods described here for carrying out stages (3) and (4) require at most the Dec circle, while the hour circle is employed in stages (6) and (7).

Most small mountings make no provision for the adjustment of the optical axis relative to the Dec axis, or of the Dec axis relative to the polar axis, these being attended to by the makers. In the case of new instruments, at least, it is safe to assume that they are correct, and stages (6) and (7) may be omitted.

If any particular stage involves a large shift in the position of the head, it is likely that prior adjustments will have been affected, and these must therefore be repeated. A large change in the azimuth of the head in stage (4), for example, will also affect the adjustment made in stage (3); stages

(3) and (4) must therefore be repeated as fine adjustments after the first coarse adjustment of (4).

As a preliminary to the methods discussed in detail below, it may be worth mentioning the way in which errors of alignment of the polar axis manifest themselves.

(*a*) *Error in altitude:* The telescope's polar axis lying at an angle of, say, $\Delta l°$ above or below the true pole, its equatorial plane will then intersect the celestial equator at an angle of $\Delta l°$ at the E and W horizons; on the meridian, the two will be parallel to each other but separated by $\Delta l°$.

If a star were followed in its diurnal motion across a large arc of the southern sky, its drift in the field in a N-S direction would be most rapid at the E horizon, reducing to zero at meridian transit; it would then reverse direction and accelerate in amount towards the W horizon, the effect being symmetrical about the meridian.

(*b*) *Error in azimuth:* The azimuth of the telescope's polar axis differing from true north by an angle of, say, $\Delta z°$, its equatorial plane intersects the celestial equator practically at the meridian, but is N of the celestial equator at one horizon (either E or W) and S of it at the other, the amount varying as the cosine of the observer's latitude. Azimuth error causes the drift of a star in the field of view to be symmetrical not about the meridian, but about the E and W horizons; an equatorial star will continue to drift in the same direction throughout its passage across the southern sky.

It follows from these considerations that the most straightforward way of orienting the polar axis is (*a*) to manipulate the axis in azimuth until stars at equal and opposite sides of the meridian show equal and opposite drift (or until a star passing the meridian shows no drift); (*b*) to reduce this drift to zero by observing stars far from the meridian whilst adjusting the altitude of the polar axis.

A variety of methods are available to choose from at each stage; these vary among themselves in accuracy and in convenience and speed of operation. It is assumed in the following instructions that a HP and a LP ocular are available (if possible both, and certainly the former, provided with cross-webs), and that from the observing site there is a clear view of the pole and of the meridian point of the equator.

16.2 Approximate orientation of polar axis

(*i*) *Azimuth:*

(*a*) The equatorial head aligned on a meridian mark on the ground with the aid of a prismatic compass, the current magnetic variation being taken from the Ordnance Survey sheet of the district in which the observing site is situated.

(*b*) Meridian marked by the shadow of a plummet at true noon (i.e. equation of time applied to LMT).

(*c*) Meridian defined by two plummets aligned upon Polaris at upper or lower culmination.

(*d*) With hour circle reading 0^h 0^m 0^s—or, in the absence of circles, with the Dec axis made horizontal with a striding level—bring a star of known RA, α, to the centre of a LP field when LST$=\alpha$, by rotating the whole head in azimuth.

(*ii*) *Altitude:*

(*a*) The latitude of the observing site, correct to the smallest fraction of a degree to which the latitude scale on the mounting can be read, is obtained from a 2-cm O.S. map. The base of the head having been levelled, this angle is then set on the latitude scale.

FIGURE 124
Approximate elevation of polar axis

(*b*) If the mounting is not fitted with a latitude scale, employ a home-made clinometer as in Figure 124.

Note: Before making any adjustments of the head, the base supporting it must be shown to be at least approximately horizontal by levels placed on it in, and at right angles to, the meridian.

16.3 Index error of Declination circle

(*a*) Select any fairly bright star near to the meridian (whose altitude, therefore, will vary by a negligible amount during a few minutes). With the telescope E of the pier, bring the star to the intersection of the webs of the HP ocular; clamp, and read the Dec circle (δ'). Reverse the telescope, return the star to the cross-wires, and re-read the Dec circle (δ'').

Assuming that $\delta' \neq \delta''$, adjust the vernier or index to read $\dfrac{\delta' + \delta''}{2}$ when the star is at the intersection of the webs, or is bisected by the horizontal web if the webs have been oriented NS and EW. Repeat, until the readings E and W of the pier agree, or treat any residual error as a correction to be applied, with a change of sign, to all Dec circle readings. It is advisable, as a check, to repeat the procedure with a star of widely different Dec from the first. A distant and well-defined terrestrial object could equally well be used.

(b) The base supporting the head having been levelled both NS and EW, mark the head's position in azimuth. Then turn it through 180°, so that the upper end of the polar axis is lying due S of the lower. Bring Polaris, at its upper or lower culmination, to the intersection of the webs; clamp, and read the Dec vernier. Reverse the telescope and repeat. As in (a), if the two readings do not agree, set the vernier to their mean value when Polaris is at the centre of the field, and repeat.

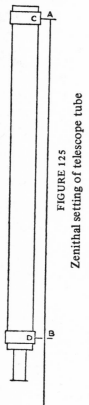

FIGURE 125

Zenithal setting of telescope tube

(c) A method that can be used by day, but is not otherwise convenient. Remove the OG cell and ocular; attach a plummet to the intersection of a wooden crossbar laid symmetrically over the top of the telescope. Adjust the tube so that the plummet falls centrally through the drawtube opening; the tube is then directed at the zenith. Set the Dec vernier to read the lattitude of the observing site; reverse the telescope; readjust its position relative to the plummet and re-read the vernier. If the two readings are different, set the vernier to their mean. Repeat until the circle reading is the same on either side of the pier, the plummet falling centrally through the ocular flange. Note the index error, if any, and apply it as a correction to all subsequent Dec circle readings. Alternatively, two brackets may be fixed to the tube by collars, as in Figure 125, CD being parallel to the optical axis; a plummet suspended from point A on one bracket will pass centrally through the hole B in the other, provided the telescope is vertical and that CA = DB.

Alternatively, again, the telescope may be pointed to the zenith by placing a spirit-level across the cell of the objective and centring the bubble in both E-W and N-S directions. Given an accurately constructed level the results are likely to be as accurate as those obtainable from the use of a plummet, which in practice is not easy to manipulate with precision.

16.4 Adjustment of the polar axis in altitude

(*a*) *Without circles:*

Polaris is situated (1980·0) at about 2^h 12^m in RA. At $\pm 6^h$ ST of this time, therefore (i.e. 8^h 12^m and 20^h 12^m), the altitudes of the NCP and Polaris are equal. As a guide, the NCP lies about $50'$ from Polaris in the direction of α Dra.

Set the Dec axis in the vertical plane through the polar axis, using a plummet, and clamp the polar axis. Adjust the elevation of the polar axis until Polaris can be brought to the intersection of the webs by rotation of the Dec axis only.

(*b*) *Without circles:*

A quick method of correcting both the elevation and the azimuth of the polar axis in a single operation.

Bring Polaris to the centre of the LP field and clamp the Dec axis. Rotate the polar axis through a wide angle, and note whether the image has moved relative to the webs. Reduce any such movement to a minimum by adjustment of the telescope in Dec; carry out the fine adjustment with the HP ocular. The tube and the polar axis are now virtually parallel.

Clamp both axes and adjust the head bodily (elevation and azimuth of the polar axis) so as to bring the NCP to the centre of the field. The adjustment can be made by eye estimation (the field diameter being known), with the aid of a micrometer, or by eye in conjunction with a map of the polar star field (see Figure 128).

(*c*) *Without circles:*

Orient the webs of the LP ocular NS, EW by trailing an equatorial star near the meridian (star over web, or web over star with Dec axis clamped).

Select a pair of stars in the NE or NW sky, differing in RA by several hours and in Dec by at most a few ' arc. Set the *f* member of the pair at the intersection of the webs and clamp the Dec axis. Rotate the polar axis to bring the *p* to the NS web. Knowing the difference in Dec of the two stars, estimate whether it is too far N or S in the field. Then:

p star too far N ⎫	both stars E ⎰	elevation of polar axis too great.
,, ,, ,, S ⎭	of meridian ⎱	,, ,, ,, small.
p star too far N ⎫	both stars W ⎰	elevation of polar axis too small.
,, ,, ,, S ⎭	of meridian ⎱	,, ,, ,, great.

Adjust the elevation of the axis by half the amount required to bring the *p* star to its estimated correct position on the NS web. Correct, if necessary, the orientation of the webs, and repeat the procedure first with the LP and finally with the HP ocular; in the latter case it may be necessary to use another pair of stars, of smaller $\Delta\delta$. The following pairs of bright stars may be found useful.

Stars	RA (1960) Dec		Mag
γ Cas	00h 54m3	+60° 30'	2·4
β Cam	04 59·9	60 23	4·0
δ Aur	05 56·3	54 17	3·7
ϕ UMa	09 49·4	54 16	4·6
α UMa	11 01·3	61 58	1·8
η Dra	16 23·4	61 36	2·7
ζ Dra	17 08·7	65 46	3·2
ι Cep	22 48·2	+65 59	3·5

(d) *Without circles:*

Observe a near-equatorial star when on the meridian and when at equal distances E and W. At the first (E) observation bring the star to the intersection of the webs, oriented NS and EW, and clamp the Dec axis.

The second and third observations consist in noting whether the star has shifted N or S from the EW web and, if so, estimating how much. Then:

If field position of star at meridian is midway between its positions when E and W of the meridian: elevation of polar axis is correct.

If position at meridian is N of the mean of its positions when E and W of the meridian: elevation of polar axis is too great.

If position at meridian is S of the mean of its positions when E and W of the meridian: elevation of the polar axis is too small.

Having reduced the error, and if necessary reoriented the EW web, repeat the observations at increased intervals before and after culmination.

(e) *Without circles:*

The most generally satisfactory method, and one capable of revealing very small errors in the elevation of the polar axis, is the following:

Select two stars about 6h E and W of the meridian respectively, and in about 45° N Dec. Orient the HP ocular with its webs NS and EW by trailing an equatorial star on the meridian.

Bring the E star to the intersection of the webs, clamp Dec and follow the star with the driving clock or RA slow motion; note whether it tends to drift N or S from the EW web. If the elevation of the axis is nearly correct this drift will be slow, and some time will pass before it becomes perceptible.

Repeat the observation with the W star. Then:

If E star drifts N
„ W „ „ S } elevation of polar axis is too great.

If E star drifts S
„ W „ „ N } elevation of polar axis is too small.

Correction of the elevation of the axis must be carried out by trial and error.

(f) *With circles:*

Select a star of known Dec, δ, and, to minimise the effects of refraction and of the still only approximate azimuth adjustment, small zenith distance.

Set δ on the circle, the index error having been eliminated or allowed for, and clamp the Dec axis.

Adjust the elevation of the polar axis so that the star, when on or very close to the meridian, can be brought to the intersection of the webs of the HP ocular by sweeping in RA only.

Reverse the telescope and repeat; finally check with another star.

(g) *With circles:*

With the set-up described in section 16.3, para (c), set the Dec axis horizontal by means of a striding level, and clamp the Dec axis.

Rotate the telescope about the Dec axis to make the plummet fall centrally through the hole in the lower bracket.

The Dec circle reading should now be the latitude of the observing site. If it is not, set this reading on the Dec circle and bring the plummet back to the centre of the lower bracket hole by adjusting the elevation of the polar axis.

Alternatively, a spirit-level may be used in the manner described in section 16.3, para. (c).

16.5 Adjustment of the polar axis in azimuth

(a) *Without circles:*

The same assumption is made as in section 16.4, para (a). The Dec axis is set horizontal with a striding level, and the polar axis clamped.

At or near the appropriate ST (or the upper or lower culmination of α Dra), the azimuth of the head is adjusted so that Polaris can be brought to the intersection of the webs by the Dec slow motion.

(b) *Without circles:*

See section 16.4, para (b).

(c) Without circles:

Orient the webs NS, EW on an equatorial star at the meridian. Select two near-equatorial stars of known α and δ, situated at equal distances (about 30ᵐ) E and W of the meridian, and differing in Dec by not more than a few ′ arc.

Bring the E star to the intersection of the webs and clamp the Dec axis. Swing the telescope westward in RA to bring the second star into the field. Knowing $\Delta\delta$, estimate whether it lies too far N or S in the field. Then:

> If W star is too far S: polar axis is lying NE-SW.
> ″ E ″   ″   ″   ″ N: ″   ″   ″   NW-SE.

Reduce the error by half, by means of azimuth adjustment of the polar axis. Repeat with a second pair of stars, differing in Dec from the first. Finally, repeat with pairs of stars of increasing $\Delta\alpha$ until no error is detectable over much larger arcs. The following Table lists some stars suitable for the purpose.

Stars	RA (1960)		Dec		Mag
μ Cet	02ʰ	42ᵐ8	+09°	57′	4·3
λ Ori	05	32·9	+09	54	3·4
α Tau	04	33·6	+16	26	0·9
γ Gem	06	35·4	+16	26	1·9
χ²Ori	06	01·6	+20	09	4·6
γ Leo	10	17·8	+20	03	2·0
ι Hya	09	37·8	−00	58	3·9
γ Vir	12	39·6	−01	14	2·7
ι Leo	11	21·8	+10	45	3·9
δ Ser	15	32·9	+10	39	3·8
α Her	17	12·8	+14	26	3·1
β Del	20	35·7	+14	28	3·6
δ Del	20	41·6	+14	56	4·4
α Peg	23	02·8	+14	59	2·5
γ Peg	00	11·2	+14	58	2·9
η Psc	01	29·3	+15	08	3·6

A single star, observed when E and again when W of the meridian, could of course be used instead of a pair of stars, but this procedure takes longer. In the early stages of correcting the azimuth it is sufficient to note the N or S drift of a single star (preferably near the zenith) when crossing the meridian. A northerly drift means that the axis is W of the pole, and vice versa.

(*d*) *Without circles:*

Employs the same principle as section 16.4, para (*c*); but the first setting of the telescope in Dec is made on a star at the meridian, and the comparisons are made with stars at equal distances E and W of the meridian. Then:

$$\left.\begin{array}{l}\text{If E star lies too far N} \\ \text{ ,, W ,, ,, ,, ,, S}\end{array}\right\} \text{polar axis is lying NE-SW.}$$

$$\left.\begin{array}{l}\text{If E star lies too far S} \\ \text{,, W ,, ,, ,, ,, N}\end{array}\right\} \text{polar axis is lying NW-SE.}$$

(*e*) *Without circles:*

Orient the webs NS, EW on an equatorial star at the meridian. Select two stars of known α, δ, and $\Delta\alpha$ of a few minutes, situated respectively in S Dec and high N Dec.

When they are in the neighbourhood of the meridian, clamp the telescope in RA slightly ahead of the *p* star, and time its transit at the NS web.

Rotate the telescope about the Dec axis to pick up the *f* star, and time its transit. Then:

If transit of *f* star occurs too early: polar axis is lying NE-SW.

If transit of *f* star occurs too late: polar axis is lying NW-SE.

(*f*) *With circles:*

Select a star of known α, δ, situated roughly midway between the meridian and the E or W horizon, and between the zenith and the horizon.

Set the Dec circle to δ, and clamp. Adjust the azimuth of the polar axis so that the star can be brought to the intersection of the webs by motion in RA alone.

Repeat with a second star, situated in approximately the same position across the meridian, making the final adjustments with the HP ocular.

(*g*) *With circles:*

A method of great precision (given finely and accurately divided circles), which requires the prior elimination, or determination, of the hour circle index error (see section 16.6). It involves the observation of a circumpolar star of known α, δ. A few minutes before upper or lower culmination note the time of its transit at the NS web of the HP ocular (T_1), and the hour circle reading (t_1). Reverse the telescope and note T_2, t_2.

Then
$$a = \frac{1}{15}\left(\alpha - \frac{T_1+T_2}{2} - \Delta T + \frac{t_1'+t_2'}{2}\right)\frac{\cos\delta}{\sin\phi\pm\delta}$$

where a = required azimuth correction in $''$ arc,
     t_1', t_2' = the values of t_1, t_2 corrected for index error,
     ΔT = clock correction,
     ϕ = latitude of the observing site.

The sign of δ in the denominator of the last term is + for lower culminations, − for upper. If the derived value of a is +, the polar axis is oriented NW-SE; if −, NE-SW.

Knowing the precise value of the required adjustment, and not merely its direction, involves a great saving of time, since the micrometer web can be laid off a'' from some reference point on the horizon, and the whole head rotated in azimuth until it is returned to the web, thus eliminating an often rather lengthy process of trial and error.

16.6 Index error of the hour circle

(a) Set the Dec axis horizontal by means of a striding level (bubble central in both positions of the level), and clamp the polar axis. Set the hour circle index or vernier to zero; if two verniers are fitted, one reading local ST and the other RA, their zero readings must coincide.

(b) Calculate the exact time of culmination of a star whose RA is known. Just before culmination, bring it to the intersection of the webs in a HP ocular, clamp both axes, and follow it with slow motion in RA. At the calculated instant of meridian transit stop following the star; set the hour circle vernier to zero, and the RA vernier, if present, to the RA of the star.

(c) Set-up as described in section 16.3, para (a). Note the hour circle reading, t. Reverse the telescope and reread the hour circle, t'. If $t \neq t'$, either 5, 6, or 7 (section 16.1) may require adjustment. Assuming that the two latter are in fact correct, the index error is

$$\frac{(t+t')}{2} - 24^{\text{h}}0^{\text{m}}0^{\text{s}}$$

which can either be removed by setting the circle to the reading $\dfrac{t+t'}{2}$ and adjusting the vernier to read $24^{\text{h}} 0^{\text{m}} 0^{\text{s}}$, or can be applied with a change of sign as a constant correction to all hour circle readings.

16.7 Collimation

With circles:

Select an equatorial star near the meridian; clamp the telescope ahead of it, and time its transit at the NS web (T_1). Read the hour circle (t_1).

Reverse the telescope; again clamp it ahead of the star, time its transit at the NS web (T_2) and read the hour circle (t_2).

Then if the collimation error is zero,

$$T_2 - T_1 = t_2 - t_1$$

If $T_2 - T_1$ is greater/less than $t_2 - t_1$, the angle between the OG end of the tube and the Dec axis is more than/less than a right angle. The collimation error

$$\frac{(T_2 - T_1) - (t_2 - t_1)}{2}$$

may be applied as a correction, or may be removed if the requisite adjustment is provided. With small instruments, errors of collimation can be corrected by inserting tinfoil or thin metal shims inside one end of the cradle, with frequent trials on a star. The procedure is bad for the temper, but practicable.

If the difference between $(T_2 - T_1)$ and $(t_2 - t_1)$ is expressed in seconds of time, and b=length of the cradle in cm, the distance that the end of the cradle must be displaced from its existing position is, in cm,

$$15b \left[\frac{(T_2 - T_1) - (t_2 - t_1)}{206,265} \right]$$

16.8 Declination and polar axes perpendicular

(a) *With circles:*

Set the Dec axis horizontal with a striding level, and read the hour circle (t_1). Reverse the telescope, level the axis, then reread the hour circle (t_2).

Then it should be that

$$t_1 = t_2 \qquad (12^{\mathrm{h}} \text{ graduation})$$
or
$$t_1 = t_2 \pm 12^{\mathrm{h}} \ (24^{\mathrm{h}} \text{ graduation})$$

Most small mountings make no provision for the correction of this error, and a faulty mounting should be returned to the makers or to any optical firm specialising in repair work.

(b) *With circles:*

This error may be distinguished from collimation error by observations of the type described in section 16.7. Both errors will have the effect of introducing inequality between $(T_2 - T_1)$ and $(t_2 - t_1)$. But if the cause of the discrepancy is lack of perpendicularity between the instrumental axes, the error will be a function of δ; more precisely, of $\tan \delta$. Therefore repeat the observations with a star of not less than 45° from the equator. The magnitude of error is

$$\frac{(T_2-T_1)-(t_2-t_1)}{2 \tan \delta}$$

16.9 Limits of accuracy required

The more accurate of the methods described above are capable of reducing the angle between the polar axis and the parallel to the Earth's axis passing through the mounting, to less than 1′. Providing this angle does not exceed 2′ or 3′ the setting will be perfectly good enough for ordinary visual observation under all circumstances.

Even for photographic work a divergence between the polar and terrestrial axes not exceeding a few ′ arc is permissible. The effect of such divergence is to make the photographic plate rotate slowly about the image of the guide star, so that the images of all other stars in the field will not be points but arcs; the lengths of these arcs are a function of their distance from the guide star, and of the guide star from the pole. From the curve in Figure 126 may be read the maximum permissible

FIGURE 126

Permissible error in polar axis alignment

error in the setting of the polar axis at various polar distances, assuming that a length of trail of 0·06 mm for an image situated 40 mm from the guide star in the focal plane, during an exposure of 2^h, is tolerable. Figure 127 reveals the approximate values of the permissible divergence of the polar axis, using the same criterion, for a variety of exposure times.

For micrometer work involving the measurement of position angles, as for photography, the tolerance for the orientation of the polar axis

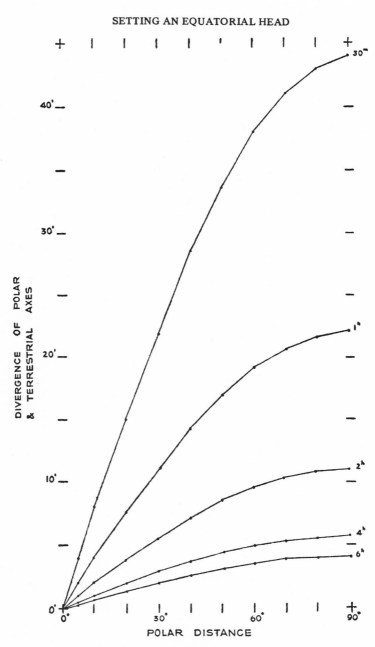

FIGURE 127
Permissible error in polar axis alignment

262

decreases rapidly with decreasing polar distance. Such work demands the greatest possible precision in the adjustment of the polar axis, which must be directed at the true pole—i.e. the effect of refraction must be taken into account (see section 16.10). Precision of this order is only obtainable if the instrument is fitted with finely divided circles, or by photographic methods based on star trails (see B. 10.12, 14.21).

16.10 Effect of refraction

The effect of atmospheric refraction is to increase the observed zenith distance of a star as it approaches its culmination, and to decrease it as it passes from culmination towards its setting. Hence a precisely adjusted polar axis would necessitate southward adjustments in Dec while following a star E of the meridian, and northward adjustments when the star is W of the meridian. At the same time the clock rate is falsified, since refraction keeps the star apparently above the horizon for a slightly longer period than it in fact is.

For photographic and ordinary visual observation it is convenient to effect a partial compensation of this by increasing the elevation of the polar axis slightly, so as to direct it at a fictitious pole as affected by refraction. The effects of refraction are virtually nil over an area of sky centred on the meridian in (for our latitudes) Dec about 20° N, whose radius is about 30°. Hence, if as a final adjustment the elevation of the polar axis is increased enough to keep, say, Arcturus, Aldebaran, or γ Leo at the intersection of the webs without use of the Dec slow motion— or, alternatively, to use one of these stars in the first place—then the effect of refraction will be largely eliminated over a wide area of the southern sky.

See also B. 10.49, 14.21.

16.11 Portable equatorials

Many users of small equatorials prefer to remove the head and telescope complete and keep it indoors when not in use, rather than go to the trouble and expense of making a shed which can be run off, folded back, or otherwise removed during observation.

The head can without difficulty be reset with sufficient accuracy (to within about 10′, say) for visual work—i.e. mechanical drive or slow-motion operation in RA will maintain an object near the centre of the field for considerable periods without any readjustment in Dec being required. The head is first set accurately, by any of the methods already described. It should be fitted with two adjustable levels, one in the NS plane containing the polar axis and the other at right angles to it. The bubble in each level is then centralised by adjustment of the level itself. When, subsequently, it is desired to set up the head again, it is placed on the mounting and the two levels are made horizontal. Correct orientation in azimuth can be achieved by viewing a distant prearranged mark through a sighting tube, or pre-focused simple telescope, fixed to the stand, or, at night, by observing the hour angle of a known star.

A method suitable for setting up a portable instrument at any site is to set the polar axis directly on the NCP, using a wide-field sighting telescope. For example, a German mounting could carry a small elbow telescope permanently fitted to the Dec plate where it joins the polar axis. The mounting having been accurately aligned, the polar axis is set to the hour angle of Polaris, which is then brought to the crosswires of the elbow telescope; resetting to the appropriate *HA* on any future occasion will restore the axis to the direction of the celestial pole. Alternatively, a fixed or removable telescope, with its optical axis accurately parallel with the polar axis, can be used; in this case, the intersection of the crosswires must be brought to that part of the sky near Polaris in which the NCP is located, offsetting from the field stars with a chart such

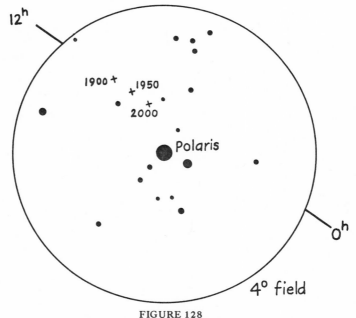

FIGURE 128
The position of the NCP, 1900-2000

264

as that in Figure 128. To check parallelism, bring Polaris to the intersection of the wires and rotate the polar axis through a complete circle; while this is being done, the star should remain stationary in the field.

Some convenient methods of aligning portable instruments equipped with setting circles are described in B. 10.24, while a more general discussion appears in B. 10.44.

The Mean Place of Polaris, 1979-2000						
Date	*RA*			*Dec*		
1979·0	02^h	10^m	53^{s}9	+89°	10′	08″
1980·0	02	11	47·6	89	10	24
1981·0	02	12	41·5	89	10	41
1982·0	02	13	36·1	89	10	58
1983·0	02	14	31·2	89	11	15
1984·0	02	15	27·0	89	11	31
1985·0	02	16	23·4	89	11	49
1986·0	02	17	20·4	89	12	05
1987·0	02	18	18·1	89	12	21
1988·0	02	19	16·4	89	12	37
1989·0	02	20	15·3	89	12	54
1990·0	02	21	15·0	89	13	10
1991·0	02	22	15·3	89	13	27
1992·0	02	23	16·3	89	13	43
1993·0	02	24	18·0	89	13	59
1994·0	02	25	20·4	89	14	15
1995·0	02	26	23·6	89	14	31
1996·0	02	27	27·4	89	14	47
1997·0	02	28	32·1	89	15	03
1998·0	02	29	37·4	89	15	19
1999·0	02	30	43·5	89	15	35
2000·0	02	31	50·4	89	15	51

(Reproduced by permission of the Director, Royal Greenwich Observatory.)

OBSERVATORIES AND SITES

17.1 Observing sites

If one were fortunate enough to have unrestricted choice of a telescope site, the following conditions could be satisfied in full:

- (*a*) remoteness from road and rail traffic (see section 14.25);
- (*b*) remoteness from any urban area, especially in the direction of the prevailing wind, and no inhabited buildings in the immediate vicinity;
- (*c*) absence of all artificial lights not controllable from the site;
- (*d*) clear view of the whole stellar hemisphere—at any rate of the whole meridian from the NCP to the S horizon, the greater part of the southern sky, and if possible the W and E horizons;
- (*e*) protection from the wind;
- (*f*) surroundings planted with low vegetation, or at least grassed.

The choice in practice usually being more or less restricted, precedence should be given to the satisfaction of (*c*) and (*d*) so far as possible. Condition (*e*) loses importance when the telescope is housed in a domed observing hut, and when the telescope is largely exposed during observation can often be partially satisfied by means of a movable canvas screen, mounted on a wood framework and secured by guy ropes.

Condition (*c*) is important in all types of work, but particularly so in photography and the observation of faint objects. If possible, any extraneous light should be screened from the observing site. Working lights must be dim, and controllable from the telescope. For a general light about the telescope, a low-wattage bulb may be used; it should be mounted at a sufficient distance to provide a distributed illumination at the telescope equal to about two Full Moons. For the illumination of charts, notebooks, etc, an ordinary torch can be used, the bulb being covered with enough thicknesses of red tissue paper to reduce the intensity to the required level. Alternatively, and more conveniently, a shaded bulb-holder may be mounted on the drawing-board, portable desk, or whatever the observer uses when at the instrument.

For observations requiring maximum dark adaptation it is a good plan to wear an eye-shade which can be flapped up against the forehead when observing, and down over the observing eye before the working light is switched on for consulting charts, making notes, etc. If the Sun and also faint objects, such as comets or variables, are habitually observed, it is advisable to use the right eye exclusively for the one type of work, and the left eye for the other.

17.2 Observatories

Small altazimuths can be moved indoors between observing spells, without trouble or waste of time; small, and not very precisely adjusted, equatorials can be set up in a matter of five minutes or so at the outset of each night's observations. Larger instruments cannot conveniently be removed from their mountings, and these must be protected from the weather when not in use. The minimum requirement is a small hut—or more accurately, perhaps, a large box—of just sufficient size to cover the telescope and its mounting, which can be removed when the instrument is wanted.

While considering the construction of a cover of this sort, it is worth remembering that protection is not only required by the instrument, and not only when it is not in use. In short:

(a) the instrument must be protected from the weather when it is not in use;

(b) the instrument must be shielded from the wind when in use;

(c) any form of protection tending to reduce dewing-up during observation is desirable;

(d) a comfortable observer tends also to be a better one, and even a simple and quite unpretentious observing hut vastly improves observing conditions, by giving protection from the wind, some protection from dew, and a permanent home for papers, notebooks, charts, etc.

In view of these considerations the construction of an observing hut—rather than a mere storage box—is strongly recommended wherever conditions allow.

A small observing hut is easily made at home by anyone whose carpentry is of the hammer-and-saw level of proficiency. The difficulties are likely to be economic and spatial, rather than constructional. Timber today is very expensive, while some types of hut (though not all) require two or three times the ground area covered by the instrument itself.

Recommended timber construction for small huts consists of: 100-mm × 75-mm bed-plates resting on a brick or concrete foundation, with damp-course between (it may be advisable to have the foundations put down by a professional bricklayer); 75 × 50 floor joists; air-space between floor and ground; 75 × 50 uprights; with tongued-and-grooved or weatherboard walls. Construction of the roof depends upon the method of opening to the sky that is employed; for sliding or folding roofs, white-painted marine plywood is the ideal material, being more permanent than canvas and much more durable than roofing felt. Domes may be constructed from panels of hardboard secured to a light timber framework, treated with sealant, and made waterproof with a covering of doped canvas; B. 11.8 describes a glass-fibre dome. Windows are not necessary, though one is certainly a convenience; the door must of course open outward; the pier supporting the telescope must be carried through the floor from a masonry foundation, and there must be a clearance of 10 mm or so between it and the floor so that no vibration is transmitted from the floor to the instrument, there being no physical contact between the telescope pillar and the hut structure.

17.2.1 Size: The minimum requisite size of an observing hut is determined by the focal length of the instrument, its optical construction (refracting, Newtonian, Cassegrain-Newtonian, etc), the type of mounting, and, in some cases where the instrument is extremely compact, the height of the observer. The walls are commonly made just low enough for the optical axis to clear them when the telescope is horizontal; but it is better to make them somewhat higher than this in the interests of wind protection, which is a more valuable advantage than the ability to observe objects within a degree or two of the horizon.

17.2.2 Types of observatory: A few basic types which have been proved successful are briefly described. Modifications to suit local conditions and personal requirements are possible in innumerable ways.

(*a*) Rectangular hut with roof, either pent or ridge, which slides off on horizontal rails supported by uprights mounted in concrete foundations. The roof may either move as a single unit or may be divided centrally, the two halves being removable in opposite directions. In the former case, rails should be provided, if possible, on both sides of the hut, so that the roof can be slid off either to eastward or westward; if rails can only be provided on one side of the hut this should be the N side. In the latter case the overlap at the roof joint needs careful weatherproofing, especially if it is central and therefore directly above the instrument when the hut is closed. The 100 mm × 75 mm rails should be surfaced with galvanised sheeting (26-gauge can be cut quite easily with tinsnips); side-plates and end-stops prevent the roof from jumping on the rails; roller-skate wheels are a convenient means of mounting the roof on the rails. The roof may be of marine plywood, but a great gain in lightness is obtained by stretching fabric directly on to a wood framework. Pins or U-clamps should be provided to lock the roof securely to the wall top-plates when it is in the closed position.

Advantages of this arrangement are the wide opening provided, resulting in infrequent adjustment of the roof during prolonged observations or photographic exposures, plenty of sky room for auxiliary instruments such as wide-angle cameras, and rapid establishment of thermal equilibrium between the interior of the hut and the night air. Its disadvantages are the comparatively large ground area required, and reduced protection from wind and dew.

(b) A rather more cumbersome arrangement is to mount the entire hut on rollers, and remove it bodily, a door at one end being large enough to clear the telescope and mounting. While in position over the telescope it is securely anchored by pins passing through the bed-plates and rails, and fitting into sockets in the foundations.

This type cannot truly be described as an observing hut at all, since it is not there during observation. It is, in fact, a mere storage box, and gives no protection to telescope or observer when the instrument is in use.

(c) Domed hut, the dome consisting of two quarter-spheres, both revolvable one inside the other on concentric rails. This arrangement has the advantage—shared by all domed, as against sliding-roof, huts—that the ground area required is no greater than that of the hut itself. It has the further advantage over the usual dome-and-shutter type that a greatly increased sky area is exposed.

Its main disadvantage is that frequent adjustments of the dome are required when observation is being carried out in the vicinity of the zenith, which is itself inaccessible.

(d) Dome-and-shutter type. The drawbacks of costliness and difficulty of construction have often been rather over-emphasised. Furthermore, by the right choice of materials domes can be made both light and absolutely weatherproof. They provide maximum protection from the wind and greater comfort for the observer, including reduced trouble from dew-formation. Their chief disadvantages are relatively frequent interruptions to observation—too wide a slit reduces the strength of the dome below the danger level—and increased trouble from thermal currents, particularly with small domes.

Figures 129 and 130 illustrate two designs of dome structure, the latter being preferable since its shutter opening passes the zenith. The members of the dome framework may most easily be built up of bent 30-40-mm struts, and stressed at the joins. A 4-m dome running on roller-skate wheels can quite easily be turned by hand. It is preferable to mount the rollers on the top-plate of the wall rather than one the base of the dome; with this arrangement, any tendency of the dome base-plate to warp or sag between its points of support can be corrected by occasionally changing its orientation when not in use. With sufficiently wide and numerous rollers it is not necessary to provide a metal rail for them to bear against—the wooden base-plate of the dome supplying a hard enough running

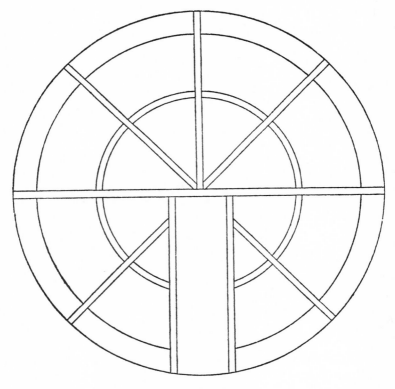

FIGURE 129
Plan view of dome structure

surface. Further rollers mounted vertically to bear against the side of the dome base-plate act as thrust bearings. The junction of the dome and the wall must be protected from the weather by a galvanised strip nailed round the bottom of the dome and overlapping the wall by 15 cm or so.

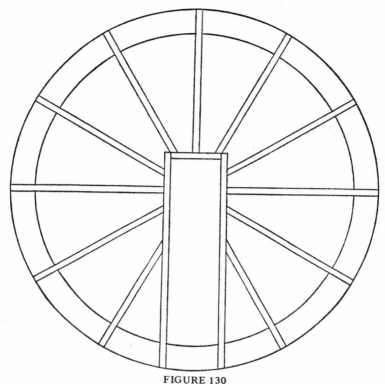

FIGURE 130
Plan view of improved dome structure

Of the many systems for opening and closing the shutter which are easy both to construct and to operate, two are illustrated in Figures 131 and 132. The lower section of the shutter, BC (Figure 131), merely hinges outward; it is retained in the closed position by means of a cord from B to a cleat on the inner side of the observatory wall; by means of this cord, also, it can be lowered gently into the open position. The counterpoise, D, is in the form of a roller, drilled centrally; the cord E is attached to a wire harness passing through D, so that as the shutter is raised until F (the other point of attachment of the cord) is directly below A, the counterpoise will have rolled down the full extent of its guide rails, G. As the opening of the shutter is continued, D will be pulled back into the position shown in the Figure. The shutter AB runs on castors or skate wheels in guide rails set on the inner side of the dome and continued down the sides of the slit. The shutter can be moved either by a pole with a hook at its end which engages a screw-eye near A or B, or by cords

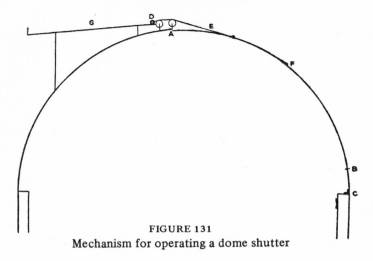

FIGURE 131
Mechanism for operating a dome shutter

FIGURE 132
Roller-blind type of dome shutter

attached in these positions and running over pulleys mounted on the inner side of the dome.

Figure 132 illustrates a roller-blind type of shutter. If the diameter of the dome is d, a strip of canvas of length $\frac{\pi d}{4}+30$ cm will be required; its width should be about 40 cm greater than that of the shutter opening. One end is tacked to the wooden roller A; the other is secured to the cross-member forming the upper end of the slit by a second cross-piece, B. Two cords, C, C, are secured to the projecting ends of the roller, and led over pulleys (fitted with guards to prevent the cords slipping, since their position is rather inaccessible) on the upper side of B, and thence to a cleat on the hut wall. When these cords are pulled together the canvas will roll up on A, which will travel towards the upper end of the slit; the cords are prevented from slipping off A by the discs D, D, screwed to the ends of the roller. The metal links E, E are attached to A so that they can swing freely, i.e. do not turn when A revolves. The cords, F, F attached to these links are for lowering the shutter and holding it in the closed position; they are attached to a cleat on the wall below the slit. A third pair of cords, G, is threaded through eyelets in the canvas, their ends being secured to a cleat at the bottom of the slit. These must be cast off before the shutter is opened; when secured, they hold the shutter tightly against the edges of the slit even in a high wind. H, H are battens tacked to the shutter at approximately 30-cm intervals to keep the canvas stretched across the slit without sagging.

(e) A cylindrical dome is considerably easier to construct than a hemispherical one, while retaining the particular advantages of domed as against sliding-roof huts. Both the horizontal and vertical sections of the shutter can be hinged, which again makes for ease of construction (Figure 133).

(f) The whole hut, rather than just the dome, may rotate; indeed, with this type of construction, there is no dome as such. The observing slit in the roof lies on the same diameter of the hut as the door, and is continuous with it—i.e. there is no horizontal jamb (or at any rate a removable one, giving strength to the hut when the telescope is not in use) between the slit and the door. Overall height of the hut can in this way be reduced. Rails can be avoided by setting vertically mounted roller bearings at intervals round a circular concrete foundation; these bear against the inner side of the wall base-plate. Small truck wheels mounted underneath the base-plate carry the weight of the hut and run directly on the concrete foundation.

It is probably true to say that no two identical observatories have ever

FIGURE 133

Twin-shutter arrangement on a cylindrical dome

been built, and reference to some of the many descriptions in the current literature is strongly advised before embarking on any construction project. Among the more recent references may be included the following:

Sliding-roof type: B. 11.4, 7, 10, 11, 14, 15
Wholly rotating type: B. 11.1, 5
Domed type: B. 11.8, 9

B. 11.6 describes a totally-enclosed, heatable observatory incorporating a 60-cm Newtonian reflector, while the use of transparent plastic to seal a dome slit, permitting observation to be carried on from the warmed interior, is described in B. 11.12.

17.2.3 Temperature control: The temperature conditions to aim for are: minimum excess, due to solar heating, of internal over external temperature during the day; speedy reduction of any such excess to zero after sunset.

Experiments have shown (B. 11.2, 3, 13) that as regards the first of these conditions, fresh white paint is markedly superior to aluminium paint—the only other competitor in the field. In urban atmospheres, however, this advantage is probably lost rather quickly, though no figures are available, and aluminium has the advantage of being more durable when applied to a rough and porous surface, such as hardboard.

To satisfy the second condition—minimum temperature lag between internal and external temperatures when the latter is falling—the hut should be opened to its fullest extent, including doors and windows, at about sundown; if it is of the domed type, the slit should be turned into the wind. In respect of this second condition, aluminium paint is superior to white.

Adequate ventilation of the hut when closed is important, since it tends not only to reduce the difference between the external and internal temperatures during the day, but also to decrease condensation and sweating.

The internal painting of the roof or dome, though it can influence the hut temperature to a small extent—by affecting the rate at which heat is radiated inward from the sun-warmed structure—is most commonly carried out in black, since a white or metallic-finished expanse tends to reduce the dark adaptation of the eye even when illuminated indirectly by a faint working light.

MICROMETERS

18.1 Principle of micrometer measures

Unlike the meridian instrument, which measures the absolute coordinates of celestial bodies, the micrometer determines the position of one object relative to another. For the observation to be of value, this second object (the comparison star or reference object) must be at least bright enough to be located with certainty on subsequent occasions; if possible, it should be brighter than the magnitude threshold of whatever catalogue is available. Its coordinates being known, and the micrometer having measured the differences $\Delta\alpha$, $\Delta\delta$ between its RA and Dec and those of the unknown object, the RA and Dec of the latter can be derived.

It is necessary to ensure that the values of $\Delta\alpha$ and $\Delta\delta$ are applied to the coordinates of the reference star which represent the apparent place of date. This is calculated from the mean place at the equinox of the catalogue (usually 1950·0) by the following process:

(1) Correcting the mean place at the catalogue equinox for precession and proper motion (this from the catalogue epoch) to Jan. 0 of the year of observation.

(2) Deriving the mean place of date from Jan. 0 of the current year by correcting for precession and proper motion.

(3) Deriving the apparent place of date from the mean place by correcting for aberration and nutation.

(4) Correcting the apparent place for refraction, where necessary.

The process of deriving apparent place is given in full in section 30.

Whereas all types of micrometer described below are capable of determining the rectangular coordinates (α and δ) of an object with more or less accuracy, the filar micrometer may in addition be used to obtain its polar coordinates (position angle and distance). The use of such coordinates is restricted to double stars; they are related to $\Delta\alpha$ and $\Delta\delta$ in the following manner:

If θ= position angle of unknown object with reference to the comparison star (i.e. the primary, in the case of a binary),

r= their angular separation,

α, δ= RA and Dec of the reference object,

α', δ'= RA and Dec of the unknown object,

$\Delta\alpha=\alpha-\alpha'$,

$\Delta\delta=\delta-\delta'$,

$\delta_0=\dfrac{\delta+\delta'}{2}$,

then

$$\left.\begin{aligned}\sin\frac{\Delta\alpha}{2}&=\sin\frac{r}{2}\sin\theta\sec\delta_0\\[4pt]\sin\frac{\Delta\delta}{2}&=\sin\frac{r}{2}\cos\theta\sec\frac{\Delta\alpha}{2}\end{aligned}\right\}\quad\cdots\cdots\cdots\cdots(a)$$

A simpler approximation, adequate when the two objects are very close together, or distant from the pole, or both, is given by

$$\left.\begin{aligned}\Delta\alpha&=r\sin\theta\sec\delta_0\\\Delta\delta&=r\cos\theta\end{aligned}\right\}\quad\cdots\cdots\cdots\cdots\cdots(b)$$

Of the types of micrometer described in this section, namely:

reticulated micrometer
ring micrometer
cross-wire micrometer
filar micrometer
double-image micrometer
comparison-image micrometer
objective-grating micrometer,

the first four employ graticules or webs located in the focal plane of the telescope; the last three employ other principles. Transit graticules and impersonal micrometers for use with meridian instruments are referred to in section 25.

18.2 Location of the graticule in the optical train

It is essential to the accurate use of the micrometer that the image of the star should lie precisely in the plane of the webs, i.e. the graticule must lie in the focal plane of the objective. If this condition is not satisfied parallactic displacement of the image will occur, its position relative to the webs depending to some extent on the position of the observer's eye. This parallax can be used to position the webs correctly: put an artificial star image on one of the vertical webs and move the eye laterally;

277

if the star moves off the web in the same direction as that in which the eye is displaced, the graticule is too near the eye; if the direction of its displacement is opposite to that of the eye, it is too near the objective.

Only positive oculars are suitable for use with a micrometer. If the graticule is placed in the objective's focal plane and observed with a negative ocular, it will be seen with the eye lens only, whose aberrations will therefore be unbalanced by those of the field lens. It will show strong colour, due to chromatic inequality of magnification, and, if a web is not diametrical, distortion. With a positive ocular, on the other hand, both the aerial image and the physical web are observed with the same optical system.

Fixed cross-webs defining the centre of the field, on the other hand, may be placed in the plane of the diaphragm of a negative ocular, for although they will be invested with spurious colour they will not be distorted.

18.3 Reticulated micrometer

The simplest and least accurate. A rectangular grid, with or without concentric circles, engraved on a glass diaphragm, is mounted in the focal plane of a positive ocular. The angular length of side of the divisions of the grid is determined by trailing a star of known Dec. The reticule can then be used to determine the approximate $\Delta\alpha$ and $\Delta\delta$ of any two objects which are in the field simultaneously:

(a) Rotate the eyepiece until the objects trail the webs. The number of squares and estimated fraction of a square, measured at right angles to the objects' diurnal motion, gives their $\Delta\delta$.

(b) The number of squares and estimated fraction of a square, measured in the direction of their diurnal motion, gives their $\Delta\alpha$.

For example (Figure 134): if object A is trailing the web XY, and the side of a grid square has been determined as $4'$, then

$$\Delta\delta = 18'$$
$$\Delta\alpha = 16'$$
$$= 1^m\,4^s \text{ (assuming field on equator).}$$

The main drawback of this device is its reduced accuracy when used with an altazimuth telescope.

FIGURE 134
Reticulated micrometer

18.4 Ring micrometer

A narrow, flat opaque ring mounted on a glass diaphragm at the focus of the objective. Its inner edge is accurately circular; better still, both edges are accurately circular and concentric.

Its advantages are: simplicity of construction and use; no field illumination is required; no particular orientation or type of telescope mounting is required. Its chief drawback is that the measures made with it require a relatively complicated process of reduction as compared with those made with other forms of micrometer.

18.4.1 Preliminary to use: Before the micrometer can be used for the measurement of differential positions, the diameter (internal and external, or internal only) of the ring must be determined with the greatest possible accuracy.

(*a*) *Simple method.* Trail a star of known Dec across a diameter (estimated) of the ring. Take the mean of a large number of timings and convert to ′ and ″ arc. Half this value is the required quantity, R, the radius of the ring. Since the exact position of the diameter can only be guessed at (the star is in fact probably trailing a chord), this method, even when a large number of measurements are averaged, gives a less accurate value for R than—

(*b*) *Accurate method.* Select two stars near the meridian (thus eliminating the effect of refraction on $\Delta\alpha$), of accurately determined Decs. The Pleiades (see Figure 135 and Table) will probably provide a choice of pairs whose $\Delta\delta$ is about three-quarters of the diameter of the ring, and whose $\Delta\alpha$ does not exceed about 15^s to 20^s. The effect of differential refraction must be allowed for, and may be taken as the difference of the refraction in Zenith Distance of the two stars, to be applied to $\Delta\delta$ (see section 26.3).

Suppose that:

Star $S\,(\alpha, \delta)$ enters and leaves the inner edge of the ring at t_1, t_2 (Sidereal Time) respectively,

Star $S'\,(\alpha', \delta')$ enters and leaves at t'_1, t'_2.

Then (Figure 136) the internal radius of the ring may be derived from:

$$R = \frac{\Delta\delta}{2\cos A \cos B}$$

where A and B are given by

$$\tan A = \frac{7{\cdot}5(t'_2 - t'_1)\cos\delta' + 7{\cdot}5(t_2 - t_1)\cos\delta}{\Delta\delta}$$

$$\tan B = \frac{7{\cdot}5(t'_2 - t'_1)\cos\delta' - 7{\cdot}5(t_2 - t_1)\cos\delta}{\Delta\delta}$$

The Pleiades

Star	Mag	α (1950)			δ (1950)		
16 Tau	5·43	3ʰ	41ᵐ	49ˢ5	+24°	8′	1″5
17 Tau	3·81	3	41	54·1	23	57	27·7
18 Tau	5·63	3	42	10·4	24	41	2·0
19 Tau	4·37	3	42	13·6	24	18	42·8
4498	8·2	3	42	39·0	24	10	50·7
20 Tau	4·02	3	42	50·8	24	12	46·9
21 Tau	5·85	3	42	55·4	24	23	59·9
4506	6·46	3	43	3·9	24	22	24·3
23 Tau	4·25	3	43	21·2	23	47	38·9
4531	6·68	3	44	0·3	24	22	0·3
4535	8·0	3	44	19·1	23	34	23·7
24 Tau	? 6·7–8·1	3	44	22·4	23	57	46·9
4538	6·94	3	44	25·9	23	45	42·3
η Tau	2·96	3	44	30·4	23	57	7·5
4542	6·81	3	44	30·6	24	8	7·6
4567	6·56	3	45	31·1	24	11	36·3
4582	6·57	3	45	58·4	23	42	21·0
27 Tau	3·8	3	46	11·0	23	54	7·5
28 Tau	5·18	3	46	12·4	23	59	7·5
4588	8·25	3	46	13·7	23	44	8·3
4591	6·63	3	46	22·6	24	13	46·8
4603	6·11	3	46	45·2	23	33	39·8
4609	7·34	3	46	57·5	24	11	54·0
4611	6·68	3	46	59·6	23	41	53·2
4620	7·26	3	47	28·8	24	20	42·0

The four-figure numbers in the first column are the designations in Boss's *General Catalogue* (B. 23.36).

Take the mean of not less than five sets of transits. If extreme accuracy is justified, the probable error can be calculated (see section 29) and the radius of the ring expressed in the form

$$R = x'' \pm y''$$

The same procedure is used to give the external radius of the ring, if the outer edge of the ring is also to be used in the measurements.

18.4.2 Measurement of differences of position: Direct the telescope so that the diurnal motion of the unknown and the reference object will carry them across the ring; they may transit the ring on the same side of its centre, or one to the N and one to the S, but both should be as far from the centre of the ring as possible. Clamp the instrument slightly ahead of the *p* object and note the times of the four transits, as before. Then time of transit of centre of ring by $S = \frac{1}{2}(t_1 + t_2)$,

and time of transit of centre of ring by $S' = \frac{1}{2}(t'_1 + t'_2)$.

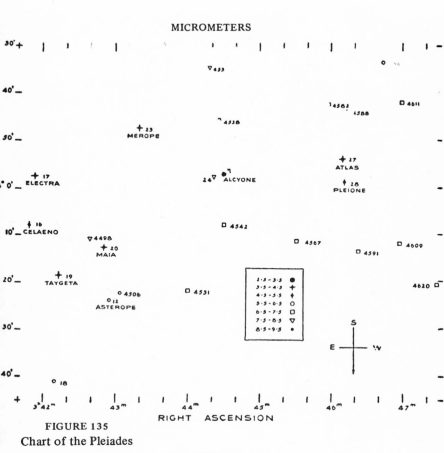

FIGURE 135
Chart of the Pleiades

FIGURE 136
Ring micrometer

Hence
$$\Delta\alpha = \frac{t_1' + t_2'}{2} - \frac{t_1 + t_2}{2}$$

Increased accuracy will be obtained if the times of disappearance and reappearance of each star at both edges of the ring are taken, the mean of four observations being taken, instead of only two.

The derivation of $\Delta\delta$ from the four (or, better, eight) observations depends upon the facts that (a) the intervals $t_2 - t_1$ and $t_2' - t_1'$ are determined by the Dec of the centre of the ring, and (b) diametrical transit of the ring in Dec δ takes $\frac{R}{15}$ sec δ seconds (R being expressed in $''$ arc).

The difference in Dec between the centre of the ring and each of the two objects, S and S', is given by
$$d = R \cos \gamma$$
$$d' = R \cos \gamma'$$
where γ and γ' are such that

$$\sin \gamma = \frac{t_2 - t_1}{\dfrac{R}{15} \sec \delta}$$

$$\sin \gamma' = \frac{t_2' - t_1'}{\dfrac{R}{15} \sec \delta'} \quad \cdots \cdots \cdots \cdots \cdots \quad (c)$$

Since the Dec (δ) of only one of the objects is known, δ' is taken as having the same value as δ, for a first approximation. Having derived d and d', the difference in Dec of the two objects is given by

$$\Delta\delta = d \pm d'$$

The sign will be $+$ when they are on opposite sides of the centre of the ring, $-$ when on the same side. δ and $\Delta\delta$ now being known, the second approximation to δ' is derived from $\delta' = \delta \pm \Delta\delta$. This value is then introduced in equation (c) and its accurate value derived.

In all micrometer measurements the signs of $\Delta\delta$ and $\Delta\alpha$ depend upon whether the unknown object is N or S, p or f the reference object.

At least four, and preferably eight, timings having to be made in fairly rapid succession, a number of stopwatches would be required. The most convenient method is to use a chronograph, or a tape-recorder in conjunction with radio time signals (see section 28.19). If MT is used, the measured time intervals must be converted to ST measure.

The simplicity of the ring micrometer favours both its use and its construction by amateurs. A large-scale ring (diameter about 15 cm) can be drawn with Indian ink and blacked in, then reduced photographically on to a thin glass slip; its reduced diameter will depend upon the focal length and field diameter of the ocular, but in most likely cases will be of the order of 10-20 mm. The slip is then mounted in the focal plane of a positive ocular. Alternatively, a thin metal ring may be accurately turned in a machine shop and cemented to a glass diaphragm with balsam.

18.5 Cross-wire micrometer

This consists of two wires or occulting bars in the eyepiece field. These wires need not cross at exactly 90°; in fact, they do not have to cross at all. Any two straight, non-parallel wires or occulting bars may be employed.*

In Figure 137, X marks the locus of the object whose RA and Dec are required, while A, B are the paths of comparison stars that have been

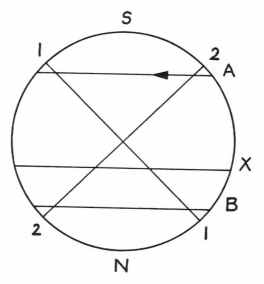

FIGURE 137
Cross-wire micrometer

identified. If the transit times of X, A and B on wires 1 and 2 are respectively X1, A1, B1 and X2, A2, B2, then the transit intervals are given by

*This account, a shortened version of the original article by G. E. Taylor (B. 12.19), is included by kind permission of the author.

$$a = (A2 - A1) \cos \delta_A$$
$$b = (B2 - B1) \cos \delta_B$$
$$x = (X2 - X1) \cos \delta_X$$

where δ_A, δ_B and δ_X are the respective declinations. It should be noted that, since A crosses the wire south of the intersection, a is negative.

By similar plane triangles,

$$\delta_X = \delta_A + \frac{(a - x)(\delta_B - \delta_A)}{(a - b)} \quad \ldots \ldots \ldots \ldots \quad (d)$$

$$\alpha_X = \alpha_A + (X2 - A2) + \frac{(a - x)}{(a - b)} \left[(A2 - B2) + (\alpha_B - \alpha_A)\right] \quad \ldots \quad (e)$$

The above equations assume that readings have been made in ST. If MT has been used, all intervals should be multiplied by 1·002738. Strictly speaking, the values of $(a - x)$ etc. in equations (d) and (e) should be divided by 15 cos δ, where δ represents the Dec midway between that of the two objects concerned. It is not advisable to use stars with very nearly the same Dec, because the denominator in these equations will then become very small.

Note that with nearer bodies, horizontal parallax would have to be taken into account. Differential refraction should also be considered and, where necessary, applied to the stars' positions before reduction of the observations; subsequently it should be removed from the derived position of the body in question.

It has been found that accuracies of the order of 0^s5 and $0'1$ are readily obtainable with this method. The average of several observations should be taken.

The use of a third reference star (C) will provide a valuable checking process, because it will enable three values of the position of the object X to be obtained—one each from A and B, A and C, and B and C. This is a powerful check. It should be noted that there is no need for all the objects to be visible in the field of view at the same time, there being no objection to a considerable spread in RA.

The cross-wire micrometer should not be confused with the older cross-bar micrometer. Use of the latter requires an accurately-orientated equatorial telescope, while the wires must cross at exactly 90°, with the cross bar at precisely 45° to the wires.

Advantages of the cross-wire micrometer include:

(*a*) Simplicity of construction and use.

(*b*) It may be used with an altazimuth telescope.

(*c*) If cross-bars rather than wires are used, no illumination is required.

284

18.6 Filar micrometer

A system of webs, one or more of which is adjustable, placed in the focal plane and observed with a positive ocular. The adjustable web is displaced by means of a calibrated head actuating a screw, the amount of the displacement being read from the calibrations.

In its usual form (Figure 138) it consists of webs fixed normally to one another, the transverse web, TT, the fixed web, FF, and an adjustable micrometer web, MM, which can travel along TT whilst remaining parallel to FF. For greater precision in determining $\Delta\alpha$ by the method of transits, the single transverse web may be replaced by a grid consisting of an odd number of mutually parallel transverse webs.

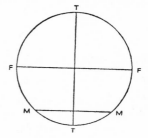

FIGURE 138
Filar micrometer

In addition, the entire micrometer is capable of rotation about the optical axis, its orientation being read off an engraved position circle with verniers. Any eyepiece containing a web can be used to determine approximate position angles by fitting a protractor to the drawtube flange and an index of some sort to the eyepiece itself.

If the filar micrometer is fitted with a second adjustable web, it moves along FF and MM whilst remaining parallel to TT. The single adjustable web MM is sometimes replaced by a pair of parallel webs.

The specific advantages of the filar micrometer are: accuracy; flexibility of use (in particular, it can be used to derive polar coordinates, by whose means the relative positions of the components of double stars are always expressed); it can be used (for deriving rectangular coordinates) with an altazimuth mounting.

Among the disadvantages of the filar micrometer, cost and non-availability tend to dominate. While one could be constructed by an individual with engineering skill (see, for example, B. 12.12, 22), a better choice would be the comparison-image micrometer (section 18.11). For precision work on double stars it suffers from the disadvantages common to all web micrometers: (*a*) the closeness of the pair may necessitate the use of a magnification which is too high for the state of the atmosphere, resulting in ill-defined and moving images, with reduced accuracy in the measures; (*b*) the illumination either of the webs or of the field will render very faint *comites* invisible.

The limit of accuracy of the micrometer can easily be determined. The smallest fraction of a turn that can be read is first determined: if the graduated head has x divisions, and the eye can estimate the reading to $\frac{1}{y}$ of a division, then it can be read to the nearest $\frac{1}{xy}$ of a turn. Usually there are 100 divisions, the minimum practicable reading being $0\cdot001R$ to $0\cdot002R$, where R is the value of one complete turn, according to the size of the drum. The pitch of the screw is then measured by dividing its length by the number of turns. Suppose it to be m millimetres. Then if the focal length is Fmm, one turn of the screw is equivalent to θ, where

$$\tan \theta = \frac{m}{F}$$

and the smallest movement of the web that can be read is $\frac{\theta}{xy}$. The magnification that can be used is dictated largely by the angular separation of the two objects, particularly their separation in Dec, since the run of the screw seldom exceeds $15'$ or $20'$. The accuracy of the measures falls off rapidly as magnification is reduced, though gross over-magnification is equally likely to introduce errors by accentuating atmospheric flaws.

18.6.1 Preliminary to use: (*a*) *Determination of the zero reading of the micrometer head.* Take the mean of 10 readings when *MM* and *FF* are judged to be exactly superimposed;* this reading is referred to below as R_0. The zero reading should be checked at regular intervals to guard against errors resulting from wear of the screw.

(*b*) *Determination of the zero reading of the position circle.* Allow an equatorial star near the meridian to trail across the whole field on *MM*, using a LP ocular (as low as × 10, if possible). Take the mean of 20 readings of the circle, each taken to the smallest fraction of a division practicable. If this mean is called P_1, the zero reading of the circle is given by

$$P_0 = P_1 + 90°$$

This value also should be checked at regular intervals.

(*c*) *Determination of the value of one turn of the micrometer head.* If the micrometer is in constant use, this should be carried out at approximately six-monthly intervals.

(*i*) By measurement of $\Delta\delta$ of two stars of known Declinations: The Pleiades fulfil the following requirements: accurately known positions; near the same hour circle; similar magnitudes; $\Delta\delta$ equal to from 50 to 100 turns of the screw, which, since this is more than a normal field diameter,

*Whenever adjusting the web, whether on to a star or on to another web, it is advisable to make the last movement of the head in the direction that increases the tension of the springs, thus taking up any slackness and avoiding a fruitful source of error.

necessitates the existence of intermediate stars to act as stepping-stones. The seeing should be average or better, and the observations made near culmination. The procedure is to measure $\Delta\delta$ of the selected stars, first from N to S and then back again, on seven separate nights; the mean of these readings of the micrometer head is then taken. $\Delta\delta$ must be derived from the known Decs of the two stars at a given epoch by the application of their proper motions during the interval between that epoch and the date of observation, precession, the reduction from mean to apparent places, and correction of the apparent places for refraction (see sections 30, 26.3). Comstock's formula is accurate enough for determining the refractions, the difference between which is added to the difference between the Decs of their apparent places.

(*ii*) By means of transits of a circumpolar star: With the position circle reading P_0, and using a LP ocular, observe Polaris just before upper or lower culmination. The telescope drive is stopped, with *MM* in the *f* half of the field just *p* the star; throughout the observations the telescope's position must not be altered. The time of transit of Polaris at the micrometer web is noted. The web is then advanced one turn, or convenient fraction of a turn, and the transit time at the web in this new position is noted. Observe in all 30 or 40 transits before, and an equal number after, culmination. Repeat on several nights, with the telescope on opposite sides of the pier on alternate nights. In the reduction of each night's observations, take the mean of transits 1 and 41 (assuming there to have been 40 on each side of the meridian), 2 and 42, 3 and 43, etc, and with these 40 mean values of *t* set up 40 equations of the form

$$(m'-m)R = \sin(t'-t_0)\frac{\cos\delta}{\sin 1''} - \sin(t-t_0)\frac{\cos\delta}{\sin 1''}$$

where $R=$ value of one turn of the micrometer head,

    $t_0=$ ST of transit of star in position S_0,

    $m=$ micrometer reading when star is in position S,

    $m'=$ micrometer reading when star is in position S', (positions $S_0, S,$ and S' being successively assumed),

    $t=$ ST when star is at S,

    $t'=$ ST when star is at S',

    $\delta=$ Dec of the star (for position of Polaris, see the Table on p. 265).

If a Mean Time clock is used, $(t'-t_0)$ and $(t-t_0)$ must be converted to ST intervals.

The value of R so derived, after the mean of the values obtained on the several nights is taken, must be corrected for refraction by means of

$$\Delta R = -R \tan 1'' \cot (\delta - \phi) r$$
$$\Delta R = -R \tan 1'' \cot [(180° - \delta) - \phi] r$$

for upper and lower culmination respectively, where ϕ = latitude of the telescope, r = mean refraction.

This is a special and more accurate case of the simpler method where the transit of any star is timed at MM, separated a large whole number of turns of the micrometer head from FF, and again at FF. Then

$$R = \frac{15(t_2 - t_1) \cos \delta}{n}$$

where R = the value of one turn of the head in $''$ arc,

n = the number of turns by which MM and FF are separated,

t = is expressed in seconds of time.

18.6.2 Methods of use: The filar micrometer may be used to measure either $\Delta\alpha$ and $\Delta\delta$, or (more accurately) the position angle θ, and angular separation r, of two objects. For the former, either altazimuth or equatorial mounting is suitable, and the position circle is not used; but for the measurement of polar coordinates a clock-driven equatorial is required.

It is convenient for S and S' to be reasonably close to one another—in the case of double stars this is always so. Except in the method of direct measurement (A (*ii*) below)—where the separation must be less than the field diameter in both RA and Dec—it is less important that $\Delta\alpha$ should be less than the field diameter than that $\Delta\delta$ should be thus restricted.

A. MEASUREMENT OF $\Delta\alpha$ AND $\Delta\delta$

(*i*) *Single adjustable web: any form of mounting*

1. Clamp the telescope with the micrometer so oriented that S trails MM. MM and FF are then coincident with parallels of Dec, and TT with an hour circle.

2. Clamp the telescope p the field containing S and S', adjusting it in Dec so that S trails FF.

3. Adjust MM by means of the micrometer head, so that S' trails it. If the micrometer reading is R_1,

$$\Delta\delta = R_1 - R_0$$

4. Note the times at which S and S' transit TT. If they are t and t', then

$$\Delta\alpha = t' - t$$

If a Mean Time clock is used, the usual correction must be applied to convert $(t'-t)$ to a ST interval. If the micrometer is fitted with a grid of transverse webs, take the means of the times of transit of S and S' at each.

5. Repeat the procedure several times (preferably not less than five) and mean the results.

(ii) Single adjustable web: clock-driven equatorial

1. Adjust FF parallel to the equator by trailing S along it.

2. Start the driving clock and measure $\Delta\delta$ by bisecting the images of S and S' by FF and MM respectively. Then

$$\Delta\delta = R_1 - R_0$$

3. Rotate the micrometer through 90° and again bisect the images of S and S' by the fixed and micrometer webs. Then, if R_2 is the new reading,

$$\Delta\alpha = (R_2 - R_0) \sec \frac{(\delta + \delta')}{2}$$

4. Rotate the micrometer through a further 90° and remeasure $\Delta\delta$ by bisecting the images of S and S' by MM and FF.

5. Take the mean of the two determinations of $\Delta\delta$.

(iii) Two adjustable webs: clock-driven equatorial

$\Delta\alpha$ and $\Delta\delta$ are measured as in *(i)* except that $\Delta\alpha$ as well as $\Delta\delta$ is measured by the displacement of MM from its zero position.

Although practice and care can achieve fairly reliable results with a hand-operated equatorial by this method—provided S and S' are not too remote in RA from each other, and that the instrument is supplied with good slow motions—a clock drive is nevertheless required for complete satisfaction with the measures to be assured.

B. MEASUREMENT OF θ AND r

Position circle, single adjustable web, clock-driven equatorial

1. (*a*) Rotate the micrometer so that S and S' are bisected by MM. Read the position circle (P_1). Take the mean of four independent readings, rotating the micrometer back and forth through roughly a right angle between each observation of the set. Then

$$\theta = P_1 - P_0$$

(*b*) An alternative method of measuring θ, not recommended for very close or unequal doubles, consists of separating MM from FF by a distance equal to roughly twice the diameter of the spurious disc of the primary, and then rotating the micrometer until it is judged that the line joining the centres of the discs is parallel to the webs lying on either side of them.

2. Rotate the micrometer through 90° from P_1, so that the line joining S and S' is now parallel to TT.

3. Bring S to the intersection of TT and FF by means of the slow motion in RA.

4. (a) Bisect S' with MM. Note reading of the micrometer head (R_1). Take the mean of several (three or four) independent measures of R_1. Then

$$r = R_1 - R_0$$

(b) An alternative, and more accurate, method of determining r is to measure the double distance: bring MM to the other side of FF and bisect the images of S and S' by MM and FF respectively. Note the micrometer reading (R_2). Take the mean of three or four measures of R_2. Then

$$r = \tfrac{1}{2}(R_2 + R_1)$$

The necessity of involving the zero reading in the expression for r is thus obviated; this is usually described as the method of crossed-zeros.

r and θ having been determined, $\Delta\alpha$ and $\Delta\delta$ could if required be derived from (a) or (b), p. 277.

18.7 Illumination of field or webs

The webs of a micrometer, or of a transit eyepiece, may be made visible at night by three means:

(a) silhouetting them against an illuminated field;
(b) illuminating the webs themselves, so that they stand out bright against a dark field;
(c) silhouetting them against an unilluminated field.

Method (a) is the most satisfactory with bright stars. It cannot be used when the observed star is fainter than, or only just as bright as, the minimum field intensity necessary to make the webs visible. It has been found that by using a red field illumination stars may be observed which would otherwise be invisible by this method. The illumination is provided by a light source in the telescope tube which either shines on an opaque white screen, whence light is diffused towards the eyepiece, or which is directed towards the OG, whence light is returned to the ocular by reflection from its air/glass surfaces. The light source must be fairly bright, and it is a great convenience if a variable resistance is included in the circuit, so that the strength of the field illumination can be adjusted as required.

Method (*b*) is applicable to fainter stars than method (*a*). The source or sources of illumination within the tube are so arranged that, while they illuminate the webs, no direct light can fall upon either the objective or the ocular; otherwise the field will also brighten. The webs may be illuminated from the side by a dim, screened bulb situated in their plane, facing which is a slip of plane mirror which ensures that they are equally illuminated on both sides, or they may be illuminated from the OG side by white opaque screens which are themselves illuminated by a bright screened source. A variable resistance overcomes the difficulty that if the webs are sensibly brighter or fainter than the star, the accuracy of the measures will be impaired. In the observation of very faint stars, it is helpful to try and align the image with a finite stretch of the web, rather than to restrict the attention to the actual point of the web with which it is in contact. When the highest degree of accuracy is sought, one source of systematic error can be removed by making the colour of the illumination the same as that of the star being measured, if the latter is at all marked.

Although an ordinary under-run low-voltage bulb, in combination with a filter, can be used, the modern light-emitting diode (LED) unit will give a satisfactory red illumination with much less current consumption, besides being far more compact (B. 12.4). The use of self-luminous beta-lights for the same purpose was described in B. 12.5.

Method (*c*) is confined to those cases where the object is so faint that it would be lost if either (*a*) or (*b*) were used. Spider web cannot be made visible at night if neither it nor the field is illuminated, and (*c*) relies upon the use of thick metal wires. It is the least accurate of the three methods.

For the measurement of the positions of points on a bright surface, no illumination is, of course, required. When the detail is faint, however, the close juxtaposition of a precisely defined web often makes it invisible, or so difficult to see that accurate measures are impossible.

18.8 Materials for the graticule

The webs must have the three characteristics of (*a*) fineness, (*b*) uniform diameter, (*c*) elasticity, which ensures straightness under tension and less trouble from breaks.

Of the various materials that have been tried, spider web is the most popular: (*a*) its diameter is only a few μ, (*b*) is extremely uniform, (*c*) adequately elastic. It is hygroscopic, but owing to the ease of replacement, this is not important.

Textile threads fail on (*a*) and (*b*), most metal and quartz threads on (*a*) and (*c*). Quartz nevertheless has the advantage of being unaffected by temperature, humidity, or the lapse of time. Tungsten wire is now being manufactured in very fine threads (about 0·005 mm thick), and is gradually replacing spider web because of its permanency and convenience; such unexotic materials as nylon string filaments and stands of polystyrene cement have, apparently, been found satisfactory for certain purposes (B. 12.10, 12).

Almost any species of spider will provide web that is satisfactory for micrometer graticules, though some are preferable to others in respect of (b). That most commonly used is the garden spider, *epeira diademata*. For collection of the web, prepare a two-pronged fork, the separation of whose prongs is greater than the diameter of the graticule mount. Stand on a chair and shake the spider out of the box in which it has been kept. Wind the web on the fork, taking care to keep each winding well spaced from its neighbours; the weight of the spider will keep the web sufficiently taut. It is useful to have several freshly varnished forks on hand, as the spider will often produce more web than can be wound on a single fork. After the web has been collected, the forks can be stored in a dustproof, and preferably airtight, box until required.

Certain types of reticulated ocular contain a glass diaphragm on which lines are engraved with a diamond. When such a diaphragm is illuminated, the light being excluded from direct entry of the ocular, the engraved lines show bright on a dark field.

18.9 Mounting web and wire graticules

(a) *Single cross-webs.* Lay the end of the web of the fork across the graticule with the web lying in the graved lines marking its correct position; seal each end with a spot of varnish, and when this is dry cut the web free from the fork with a razor blade. Repeat the procedure with the second web.

(b) *Graticule of parallel webs.* Here it is convenient to construct a four-sided metal frame, larger than the graticule mounting, and pivoted at its four corners. Lay the webs on the frame, as described above, the frame being locked rectangular. Unlock the frame and push it gently into the shape of a trapezium, thus bringing the webs closer together, while maintaining their parallelism. Clamp the frame at the required degree of lopsidedness. Varnish the graticule mount and lay the webs across it. When dry, cut the projecting ends of the webs with a razor blade and remove the mounting frame.

(c) *For coarse graticules*—employing fine fuse-wire, for example—it is convenient to use an accurately machined mounting table, such as any machine-shop will make to specification quite cheaply. This consists (Figure 139) of a steel plate, drilled and plugged at four points (*A–D*),

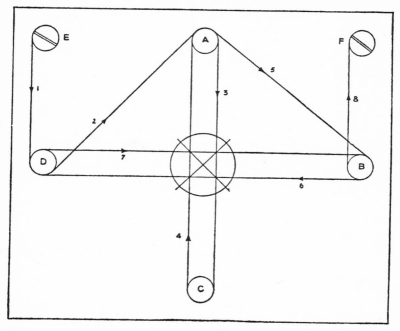

FIGURE 139
Mounting table for winding a graticule

drilled and threaded at two points (*E, F*), and engraved with a circle to mark the position of the graticule mount, as well as with two diameters to mark its centre and to act as a check on the accuracy with which the wires are placed. The direction of winding is indicated by arrows and consecutive numerals. The engraved diameters should be exact diagonals of the central square formed by the cross-wires.

The same mounting table can be used for a single cross-wire by choosing a different winding sequence (Figure 140).

The wire having been wound, tautened, and secured by the screws *E* and *F*, it is fixed to the ring by drops of solder, and the ring cut free.

18.10 Double-image micrometer

The last four micrometers to be described have been designed to avoid the disadvantages which are inherent in web micrometers generally. These

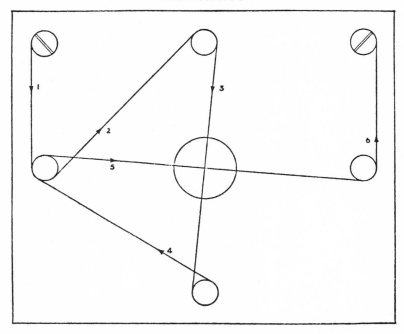

FIGURE 140
Mounting table—alternative winding

disadvantages, as already emphasised, are attendant upon the necessity of illuminating either the field or the webs, and also upon the difficulty of bisecting accurately a possibly ill-defined stellar or nebulous image by a well-defined web.

In the double-image micrometer the image of each star of the pair to be measured is divided by one of several available methods—a divided OG or Barlow lens, or a polarising prism mounted on the optical axis between the objective and the ocular. By separating the components of a divided lens, or by moving the prism along the axis, the two images of each star may be separated or closed; furthermore, the direction of their motion depends upon the orientation of the double-image device.

To measure θ and r of a binary, the micrometer is rotated until the direction of separation of the secondary images coincides with the line joining the primary images. The orientation of the divided lens or prism, read from a position circle, then gives θ (Figure 141(b)). The separation of each pair of images is increased until S_2 and S_1' coincide (Figure 141(c)). The angular separation of the primary images, r, is then a function of the separation of the components of the divided lens or of the displacement of the Wollaston prism along the optical axis.

The double-image micrometer can be used on nights of such inferior seeing that a filar micrometer would be useless. For whereas the operation of the latter relies upon the simultaneous bisection of the two stellar

FIGURE 141
Double-image micrometer settings

images—both suffering independent atmospheric distortion and movement—by clearly defined and motionless webs, the double-image micrometer involves the juxtaposition of two images, S_2 and S'_1, whose 'boiling' and distortion are identical.

The decreased brightness of the secondary images amounts to very little in practice, since it is compensated by the fact that no web or field illumination is required. The main source of inaccuracy lies in the difficulty of exactly superimposing, with precise concentricity, two stellar images of different apparent sizes.

An additional advantage of the polarising-prism type of double-image micrometer is its adaptability to the measurement of the difference in brightness of the components of double stars. This is achieved by placing a rotatable analyser between the prism and the eyepiece, mounted so that the orientation may be measured. Since the two rays from the Wollaston prism are polarised at right angles, the difference in apparent brightness between the two stars may be varied by rotating the analyser.

See further: B. 12.6, 12.8, 12.9.

18.11 Comparison-image micrometer

This device is similar to the double-image micrometer in that a polarising prism is used to produce a secondary image. In this case, however, the second pair of stars is produced from a single pinhole, which is converted by the prism into two polarised images; these are projected into the eyepiece, alongside the real double star, as shown in Figure 142. By altering the pinhole-prism distance, the separation of the artificial stars can be increased or decreased. The prism itself is mounted in a rotatable

FIGURE 142

Comparison-image micrometer

housing to allow the artificial pair to assume any p.a., the value of which can be read from a scale located on the prism mounting. Again, because a polarising element is used, the intensity of each of the artificial stars can be controlled by using a system of crossed Nicol prisms. This system is used with the 71-cm refractor at Greenwich; it is described fully in B. 12.3.

In order to make the device as sensitive as possible, it is necessary to keep the distance between prism and eyepiece large. In the case of the 71-cm refractor, the images are projected on to an elliptical flat mirror fixed to the telescope tube, which in turn reflects them into a prism near the eyepiece. The achromatic focusing lens has a focal length of 30 cm, and can be adjusted by using a pair of focusing rings; a 1-mm movement in the prism produces a change of $0''0794$ in the angular separation of the two stars.

The principle of the comparison-image micrometer was first described by F. J. Hargreaves (B. 12.8, 9) for use with a reflecting telescope. An alternative system for a reflector is described in B. 12.7.

Advantages:

(*a*) No field illumination required.

296

(*b*) No webs required, leading to easier measurement of double stars in poor seeing.

(*c*) Very little maintenance is required.

(*d*) The micrometer constant does not depend upon the telescope's *F*.

(*e*) The brightness of the artificial pair can be increased or decreased individually, leading to reduced systematic error when measuring pairs with large differences of magnitude.

Disadvantages:

(*a*) Range of measurement is limited and depends on the focal length of the projection system.

(*b*) Engineering ability is required for its construction.

(*c*) A Wollaston prism is likely to prove an expensive item.

18.12 Objective-grating micrometers

18.12.1 Theory: When a screen containing two narrow slits, one near each end of a diameter of the objective and perpendicular to that diameter, is placed between the objective and a star, interference between the two halves of the divided beam changes the observed diffraction pattern of the stellar image: instead of being a single spurious disc it is spread out into a short series of fringes—alternate bright segments and dark interspaces—the axis of this elongation of the image being parallel to the line joining the slits.

If the two slits are multiplied to form a grating of parallel bars whose interspaces all act as individual slits the same phenomenon occurs, though the images are brighter than before. To account for the formation of the fringes it is simpler to consider the case of a pair of slits at opposite ends of a diameter of the objective.

A parallel beam from a point source off the axis is represented (Figure 143) as falling from the objective, the whole of which is diaphragmed except for two widely spaced slits. After refraction the two rays converge to a point *X* in the focal plane *FG*. The linear distance of *X* from the axis is *d*, the angular distance θ. θ is also a measure of the distance, *BC*, between the slits measured in a plane perpendicular to the wavefront *AB*.

When $BC = \lambda(n + \frac{1}{2})$, where *n* is an integer and λ the wavelength of the radiation, crests and troughs of the two converging wave systems will combine at *X*. Destructive interference will occur, with the formation of a dark interspace. It can be seen that there will be a series of positions for *X* along *FG*, each corresponding to a different value of *n*. At all intermediate positions, however, *Y*, the distance between the slits (measured normally to the wavefront), will be $n\lambda$, a whole number of

FIGURE 143
Objective grating micrometer—
action of two slits

wavelengths; the two systems will reinforce each other at all positions Y, with the production of a bright band at each.

The spacing of the fringes is a function of the wavelength, λ, and the linear separation of the slits (which, in Figure 143, may be taken as D):

$$\tan Z = \frac{d}{F} = \frac{\frac{1}{2}\lambda}{D}$$

or

$$d = \frac{F\lambda}{2D}.$$

Consider now a coarse grating (Figure 144) consisting of a series of slits of width a, distance between them b. In this case,

$$\tan Z = \frac{d}{F} = \frac{n\lambda}{(a+b)}$$

for constructive interference. When $n=0$ (zero order), a bright image of the star is formed at X. Similarly, constructive interference occurs at integer values of n, a pair of images corresponding to each value of n being formed at either side of the zero-order image. For images formed near the optical axis, the angle Z is small and tan Z can be taken as equal to Z (in radians). Thus, converting Z into " arc,

$$Z = \frac{206265\lambda}{(a+b)} \text{ (when } n=1).$$

In the above relation, λ represents the effective wavelength of the star being observed. For a star of spectral type F, this is, typically, 550 nm, depending upon the telescope-and-eye combination used. If $a=b=1$ cm, for example, with $\lambda = 5\cdot5 \times 10^{-5}$ cm, then the value of Z is 5″72. Hence, the zero-order image is accompanied on either side by a first-order image 5″72 distant, with corresponding images further away, all of

FIGURE 144
Objective grating micrometer—
action of multiple slits

which lie on a line perpendicular to the orientation of the slits. These higher-order images are progressively more rectangular than circular, since they have been dispersed by the grating and are, in reality, short spectra. The brightness of the subsidiary images depends mainly on the width of the grating bar and slit, but is also reduced by the increasing dispersion.

18.12.2 Method of use: When a double star is viewed through the grating, each component is accompanied by its own series of subsidiary images. When the grating is rotated about the optical axis, these images will be seen to rotate about the zero-order images of the two main components. In the case of a grating where $a=b$, it will be found that, except for very bright stars, the second and higher-order images are faint in relation to the zero- and first-order images.

Figure 145 shows the situation when a double star is viewed through a coarse grating at two different orientations, the separation P of the pair being similar to the value of Z. The orientation of the grating in (*i*) is $\alpha_1°$; in (*ii*), $\alpha_2°$. In both vases, the grating has been rotated until the distance AA'' appears equal to BB'', so that $AA''B$ form an isosceles triangle (the other images, A' and B', can also be used). If the grating has been calibrated so that the orientation reads 0° when the line of images is parallel with the celestial equator, then θ, the p.a. of the pair, is given simply by

$$\theta = \frac{(\alpha_2 - \alpha_1)°}{2}.$$

The separation, P, is given by

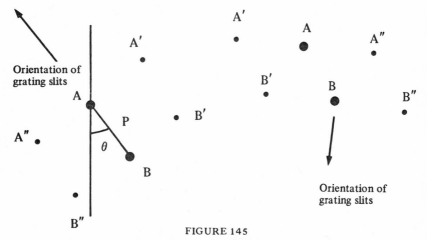

Orientation of
grating slits

Orientation of
grating slits

Orientation of
grating slits

FIGURE 145

Appearance of double star with different grating orientation

$$P = 2Z \sin \frac{(\alpha_2 - \alpha_1)}{2} \text{ '' arc.}$$

Thus, knowing Z and the two grating angles α_1 and α_1, it is possible to measure the position angle and separation of the double star.

FIGURE 146

Deformation method of
measuring a position angle

An alternative method of measuring p.a., communicated by M. V. Duruy, consists of detecting the point at which a right angle becomes slightly deformed, as shown in Figure 146. The 'deformation', subtending an angle ϵ, is observed on both sides of the true right angle. The mean of the two readings thus gives the p.a. directly. (For other methods, see B. 12.2).

However, the measurement of position angle and separation requires an accurate value of Z, and this, in turn, requires an accurate value of the effective wavelength, λ. There are two main methods of determining λ:

(a) *Drift method:* Time the passage of Polaris and its system of images across a graticule in the eyepiece which is perpendicular to the apparent drift of the star, the line of the images being along a parallel of Dec (Figure 147).

300

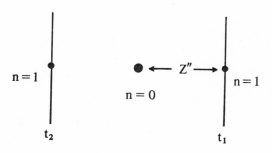

FIGURE 147
Determination of λ by observing Polaris

The times at which the first-order images transit the fixed wire being t_1 and t_2,

$$\lambda = \frac{(a+b)}{206,265} \times \frac{(t_2-t_1)\ 15 \cos \delta}{2}$$

δ being, in this case, the Dec of Polaris. The timings t_1 and t_2, noted in seconds of time, should be repeated many times in order to improve the accuracy of the value of λ.

(b) *By measurement:* Measurement of bright binaries of known position angle and separation can lead, by reverse calculation, to a value of λ. Orbit predictions can be found in the articles by Meeus in B. 28.20, 21.

Unfortunately, λ is not a constant quantity, depending as it does upon the spectral types of the stars being observed. Richardson, using an 11-cm OG, found a variation of from 561·6 nm for a star of type $B0$ to 578·3 nm for one of type $M2$. He also showed that the relationship between spectral type and λ is not linear (B. 12.15). It is fortunate that most common double stars have components in the range F-G, so that a value of λ obtained from transits of Polaris (spectral type F8) may be used over this range with reasonable accuracy. Anyone wishing to use this device to the limits of its accuracy will, however, have to make a thorough investigation of the variation of λ with spectral type. The accompanying Table gives the spectral type of a number of bright, wide pairs, along with the position angles and separations of fixed or slowly-moving pairs. It should be noted that with this micrometer it is necessary to know the spectral type of only one component.

The most suitable grating is one with slit separation equal to slit width.

(i) Northern hemisphere stars

Star	Spectral type†		Mag vis.°		PA	Sep
	A	B	A	B	°	"
λ Ori	08	0e5	3·7	5·7	43·0	4·4
1 Cam	B0		5·4	6·5	308·1	10·3
ζ CrB	B7	B6	5·1	6·0	305·5	6·3
ζ UMa	A2	Am	2·4	4·0	151·7	14·4
95 Her	A7	G5	5·1	5·2	257	6·3
σ CrB	F6	G1	5·7	6·7	(Meeus)*	
η Cas	G0	M0	3·5	7·1	(Meeus)**	
ι Tri	G5		5·1	6·5	71·0	3·9
γ Del	K1	F7	4·3	5·1	267·3	9·6
γ¹ And	K3	A0p	2·3	5·1	63·3	9·7
α Her	M5	Comp.	3·1	5·4	106·5	4·6

(ii) Southern hemisphere stars

Star	A	B	A	B	PA	Sep
α Cru	B1	B3n	1·6	2·1	114	4·2
h Cen	B5		4·7	7·9	185	14·9
Q Cen	B8		5·7	7·1	163	5·3
θ Eri	A3	A2	3·4	4·4	89	8·2
ν Car	A9	F0	3·2	6·0	127	5·0
κ Tuc	F6		4·9	7·1	32·8	5·6
α Cen	G2	K1	0·3	1·7	(Meeus)*	
γ Vol	G8	F4	3·9	5·8	300	13·6
36 Oph	K0	K1	5·3	5·3	(Meeus)*	
h5003 Sgr	M2	G8	5·3	7·0	105	5·5

† B. 25.10   * B. 28.21
° B. 28.17   ** B. 28.20

The equation below is from Wesselink (B. 12.20):

$$\frac{I_n}{I_0} = \left(\frac{\sin x}{x}\right)^2$$

where $x = n\rho\pi$ and $\rho = a/(a+b)$. I_n is the intensity of the nth order image, while I_0 is that of the zero-order image. If I is the intensity of the image without the grating, then

$$\frac{I_n}{I} = \left(\frac{\sin x}{n\pi}\right)^2$$

The maximum intensity of the first-order image occurs when $\rho = \frac{1}{2}$, or when $a=b$. (In fact, for odd values of n, all n images have the maximum possible brightness, whilst for even values of n, all n images have zero intensity.) Putting $\rho=\frac{1}{2}$ gives

$$\frac{I_1}{I}=0\cdot101 \quad \text{and} \quad \frac{I_1}{I_0}=0\cdot405,$$

corresponding to a zero-order image which is 1·50 mags fainter, and a first-order image 2·48 mags fainter, than the star image seen in the telescope without a grating.

Advantages:

(*a*) Quick, easy and cheap to construct.

(*b*) With care, it can give accurate results.

(*c*) Once Z is known, determination of position angle and separation requires only the measurement of two orientations of the grating.

(*d*) The spectral type of only one component need be known.

Disadvantages:

(*a*) Up to 50 per cent loss of incident light.

(*b*) Several gratings are required to provide a reasonable range of Z.

(*c*) The micrometer is awkward to use with refractors, a linkage system between the grating and the eyepiece end being required.

The drawback noted in (*b*) can be obviated by tilting the grating. This is the basis of the device used by Richardson in the 1920s (B. 12.13, 14, 15, 16). Richardson's micrometer is a simplified version of the instrument used by Schwartzchild (B. 12.17); a gauge, mounted on the grating, measures directly the degree of foreshortening of the slits and bars, and hence the reduced effective values of a and b. The six images (three from each star) are brought into line, and the grating is tilted until the dark spaces between the images of one star coincide with the images of the other star. If ϕ is the angle through which the grating must be tilted, then

$$P=\frac{206{,}265}{2\,(a+b)\sin\phi}\;'' \text{ arc.}$$

The exact point at which the six images mesh together being difficult to judge, Richardson also made a setting of the grating where the two sets of images are in close juxtaposition.

The diameters of celestial objects, particularly planets, can be measured by tilting the grating until the zero-order image and the two first-order images just appear to touch. The diameter, d, in $''$ arc, is given by

$$d = \frac{206,265}{(a+b)\sin\phi}.$$

The grating for this type of micrometer necessarily has to be larger than the aperture, and the range of measurement depends upon how large a grating the observer is prepared to make. It is necessary to bevel the edges of each grating bar, so that the thickness of the bar does not interrupt the incident light at large angles of tilt.

18.13 Binocular micrometer

A variation of the comparison-image micrometer. It dispenses with the polarising prism, projecting the artificial pair directly into the second eye of the observer (Figure 148).

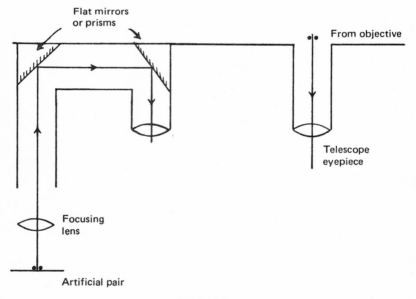

FIGURE 148
Binocular micrometer

To measure a double star, the separation and position angle of the artificial pair are altered until it appears to match the real pair. According to M. V. Duruy, who originated the device, in France, in 1936, a feeling of rest is then experienced. Results have shown that the performance is not markedly inferior to that of the filar micrometer. For a fuller account, see B. 12.2.

19.1 Prismatic spectra

A ray of polychromatic light in traversing a prism undergoes two modifications, deviation and dispersion. A monochromatic ray, on the other hand, suffers deviation only. Figure 149 represents a section through a prism taken in a plane perpendicular to its base; AB and AC are the refracting surfaces, and A the refracting angle. The following relations are worth noting:

refracting angle: $\qquad A = r + r'$

total deviation: $\qquad \omega = (i + i') - (r + r')$

$$A + \omega = i + i'$$

and $\qquad \omega = (i + i') - A$

whence is seen that ω is minimal when $i = i'$. This is the position of minimum deviation, in which the ray passes through the prism symmetrically:

$$AB = AC$$

$$i = i' = \frac{\omega + A}{2}$$

$$r = r' = \frac{A}{2}$$

$$\sin i = \mu \sin \frac{A}{2}$$

$$\mu = \frac{\sin \dfrac{A + \omega}{2}}{\sin \dfrac{A}{2}}$$

The refractive index of a material $\left(\dfrac{\sin i}{\sin r} \right.$ when one of the media is air) is inversely proportional to the wavelength of the radiation with which it is measured. Hence ω is a function of λ, and the measure of the differential deviation of radiations of different wavelengths is the dispersive power, or relative dispersion, of the prism. Using the suffixes F, C, and D

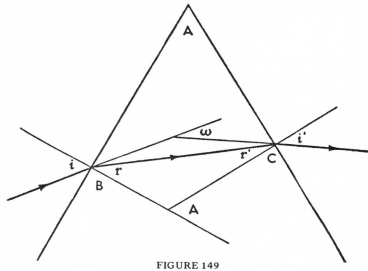

FIGURE 149

Refraction of monochromatic light by a prism

to denote quantities measured in the wavelengths of the F and C lines of hydrogen and the D line of sodium (roughly, the limits and centre of the visible range) respectively:

angular length of spectrum (when A is small):

$$\omega_F - \omega_C = (\mu_F - \mu_C)A$$

relative dispersion for F and C:

$$\frac{\omega_F - \omega_C}{\omega_D} = \frac{\mu_F - \mu_C}{\mu_D - 1}$$

and

$$\omega_F - \omega_C = \frac{\omega_D}{\nu}$$

where ν is termed the constringence of the prism (see section 6.1).

The position of minimum deviation is one of great importance, having the following characteristics:

(*i*) maximum purity of spectrum is obtained for a given slit width;

(*ii*) maximum light transmission;

(*iii*) minimum contribution of astigmatism by the prism;

306

(*iv*) minimum displacement of the spectrum results from any accidental displacement of the prism.

19.2 Diffraction spectra

A diffraction grating may be considered as a system of parallel, equidistant, opaque strips separated by transparent interspaces of the same width. An idealised section through such a system is represented in Figure 150, where *a* and *b* are two adjacent interspaces.

Each interspace acts as a secondary source, emitting radiation in all directions. Diffraction within each of these secondary wave systems is

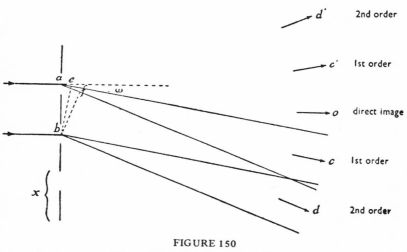

FIGURE 150

The action of a diffraction grating

such that, in the case of incident light normal to the plane of the grating, reinforcement will occur when

$$x \sin \omega = n\lambda \ldots \ldots \ldots \ldots \ldots \ldots \ldots (a)$$

where *n* is an integer, and maximum destructive interference when $n+\frac{1}{2}$ is an integer. Where the incident light is not normal to the grating, but inclined to the normal at an angle ϕ, the grating equation takes the form

$$x(\sin \omega - \sin \phi) = n\lambda$$

In the direction O will be formed, not a spectrum, but an integrated image of the slit, since all points in the wavefronts advancing in this direction are in phase on a plane surface. In the directions c and c' spectra of the 1st order will be produced, providing these directions are such that $ae=\lambda$, $(n=1)$, i.e. the wavefronts advancing towards c and c' are in phase on a plane surface, but the fronts from adjacent secondary sources are one complete wavelength behind one another. Similarly, the 2nd-order spectra will be formed in the directions d and d', provided $af=2\lambda$, $(n=2)$, i.e. wavefronts from adjacent interspaces are two complete wavelengths out of step with one another. And so on for spectra of higher orders. At positions intermediate between those just considered no image whatever will be observed, since for such positions n is not an integer, and destructive interference occurs. It will be noticed that the short-wave end of each spectrum is that which is nearest the direct image.

In practice, of course, the different orders are brought under observation most conveniently, not by shifting the eye or the camera, but by rotating the grating, if it be of the reflecting type.

By substituting the actual values of x and λ in equation (a), and taking successive integer values for n, the location of the various spectra for normal incident light can be calculated: taking $\lambda=570$ nm (yellow), and $x=1\cdot81.10^{-4}$ cms (5,500 rulings to the cm), we derive:

n	ω	
0	$0°$	direct image
±1	$\pm18°\ 21'$	1st-order spectra
±2	$\pm39°\ 2'$	2nd-order spectra
±3	$\pm70°\ 50'$	3rd-order spectra

Spectra in successive orders are increasingly deviated, dispersed, and faint.

Figure 150, though adequate for the explanation of principle, is misleading as to the spatial disposition of the different orders of spectra. They are distributed in a succession of planes, each one normal to the line joining its centre to the centre of the grating. Nor are the different orders separated as suggested by the Figure. Equation (a), p. 307, shows that the position of a wavelength λ in any order is given by

$$\sin \omega = \frac{n\lambda}{x}$$

Hence the wavelengths λ_1, λ_2, λ_3 will coincide in the spectra of the three orders n, $2n$, and $3n$, if

$$n\lambda_1 = 2n\lambda_2 = 3n\lambda_3$$

The result of this, so far as it affects spectra of the first four orders, is shown in Figure 151, where it can be seen that the overlapping of spectra becomes increasingly acute as n increases.

It is thus desirable to use spectra of low orders so as to obviate the necessity of colour filters or plates of restricted colour sensitivity. On the other hand, if n is small, high dispersion (see section 19.3.1) can only be obtained by reducing x, i.e. increasing m, the number of rulings per cm.

The surplus field light which is encountered with a grating, and which reduces contrast in the 1st and 2nd-order spectra, can be eliminated by

Each spectrum extends from 400 nm to 800 nm. Wavelengths are
shown in units of 100 nm.
FIGURE 151
Overlapping orders of spectra covering 400-800 nm;
wavelengths are shown in units of 100 nm

means of an ocular mask covering the two segments of the field above and below the image itself.

19.3 Efficiency of a spectroscope

Three characteristics of the performance of a spectroscope are relevant to a consideration of its efficiency:

its dispersion (analogous to the magnification of a telescope):
its resolving power (analogous to that of a telescope);
the brightness of its spectrum (analogous to the light grasp of a telescope).

19.3.1 Dispersion: The dispersion of a spectrum may be defined as the rate of change of deviation with wavelength. If we write ω and ω' for the deviations of two rays, and λ, λ' for their respective wavelengths,

then $\omega - \omega' = \Delta\omega$, $\lambda' - \lambda = \Delta\lambda$, $\dfrac{\lambda' + \lambda}{2} = \bar{\lambda}$, and we have for the wavelength λ

$$\text{angular dispersion} = \frac{\Delta\omega}{\Delta\lambda}$$

expressed in, e.g., degrees or radians per Ångström. The linear dispersion in the focal plane of the spectroscope depends upon the focal length, f, of the camera lens forming the image (or the magnification of the view telescope):

$$\frac{\Delta l}{\Delta\lambda} = f \cdot \frac{\Delta\omega}{\Delta\lambda}$$

whence $\qquad \Delta l = f . \Delta\omega$

309

expressed in, for example, mm per Ångström.

If the refractive indices of the prism for λ and λ' are N and N', and we write, as before, $N-N'=\Delta N$, and $\bar{N}=\dfrac{N+N'}{2}$, then when the prism is set at minimum deviation we have:

$$\Delta\omega=\Delta N\,\frac{2\sin\dfrac{A}{2}}{\left(1-N^2\sin^2\dfrac{A}{2}\right)^{\frac{1}{2}}}$$

and

$$\frac{\Delta\omega}{\Delta\lambda}=\frac{2\dfrac{\Delta N}{\Delta\lambda}\sin\dfrac{A}{2}}{\left(1-N^2\sin^2\dfrac{A}{2}\right)^{\frac{1}{2}}}$$

which is the expression for the angular dispersion of a prism of refracting angle A, for a wavelength $\bar{\lambda}$ whose refractive index is $\bar{N}$. Angular dispersion at a given wavelength is thus a function of the refractive index, position (minimum deviation, giving minimum dispersion, is here assumed), and refracting angle.

If we put $A=60°$, the last equation reduces to

$$\frac{\Delta\omega}{\Delta\lambda}=\frac{\Delta N}{\Delta\lambda}\cdot\frac{1}{\left(1-\dfrac{N^2}{4}\right)^{\frac{1}{2}}}$$

To compare the average angular dispersion of 60° prisms of typical crown and typical flint glass, substitute the values of N for the wavelengths of the F and C lines as given in section 6.1:

		Crown	Flint
	ΔN	0·008	0·018
	$\Delta\lambda$	\multicolumn{2}{c}{$1702\text{Å}=1702\times10^{-8}$ cm}
$\therefore$	$\dfrac{\Delta N}{\Delta\lambda}$	$\dfrac{0\cdot008}{1702\times10^{-8}}=470$	$\dfrac{0\cdot018}{1702\times10^{-8}}=1058$
$\therefore$	$\dfrac{\Delta\omega}{\Delta\lambda}$	$\dfrac{470}{\left(1-\dfrac{1\cdot519^2}{4}\right)^{\frac{1}{2}}}=724$	$\dfrac{1058}{\left(1-\dfrac{1\cdot625^2}{4}\right)^{\frac{1}{2}}}=1814$

Hence the average overall dispersion of the flint is roughly two and a half times that of the crown. In both cases the dispersion is about seven times as great in the ultraviolet as in the far red.

The angular dispersion provided by a grating is independent of N, the total number of rulings illuminated, being dependent solely on the separation of the rulings and the order of the spectrum:

$$\frac{\Delta\omega}{\Delta\lambda} = \frac{mn}{\cos\omega} \quad \dots\dots\dots\dots\dots\dots (b)$$

where m = number of rulings per cm,
$\quad\;\; n$ = order of the spectrum.

Since $m = \frac{1}{x}$ (where x is the distance separating the centres of adjacent rulings) this can be rewritten

$$\frac{\Delta\omega}{\Delta\lambda} = \frac{n}{x\cos\omega}$$

whence it can be more readily seen that the dispersion of a grating depends, ultimately, on nothing but ω—i.e. on the position of the spectrum—since an increase in x is automatically compensated by a larger value of n of the spectrum that is formed there.

If f is the focal length of the view-telescope objective or camera lens forming the image, the linear dispersion is given by

$$\frac{\Delta l}{\Delta\lambda} = f \cdot \frac{n}{x\cos\omega}$$

Once can deduce from expression (b) that a diffraction spectrum, unlike a prismatic spectrum, is normal, since $\cos\omega \doteqdot k$, and therefore $\omega \propto \Delta\lambda$. Also that the dispersion of high-order spectra is greater than that of low-order spectra—whence, also, they are fainter.

19.3.2 Resolving power: The resolving power of a prism is its ability to resolve into visually perceptible intensity-maxima the different wavelengths of which the incident radiation is composed. If, with the slit at minimum width, λ and $\lambda + \Delta\lambda$ are just resolved, then the prism's resolving power in that region of the spectrum is given by

$$R = \frac{\lambda}{\Delta\lambda}$$

R is a function of the refractive index and of the difference between the longest and shortest light paths through the prism, and is independent of the prism's refracting angle. If the whole face of the prism is illuminated, the difference in light path is equal to b, the length of the prism's base, and

311

$$\frac{\lambda}{\Delta\lambda}=b\cdot\frac{\Delta N}{\Delta\lambda}$$

Thus b is related to prismatic resolution in a manner analogous to that of D with telescopic resolution.

For average optical glass, $\frac{\Delta N}{\Delta\lambda}\eqsim 10^3$, where λ is expressed in cm. Hence, for a prism,

$$R\eqsim 10^3 b$$

when b is also measured in cm.

Analogously to the magnification and resolving power of a telescope, an increase of prismatic dispersion leads to increased separation of the spectral lines, but since it involves a corresponding increase in their width, resolution is unaffected.

Thus, determining the resolving power of a prism, we have:

(*i*) Slit width: maximum theoretical resolution being obtained with an infinitely narrow slit. Hence the purity and brightness of a continuous spectrum are inversely related to one another, in so far as they are dependent upon slit width. For a given value of f_{col}, spectral purity will vary inversely with slit width; for a given slit width it will vary directly with f.

(*ii*) Effective base length of prism; or, in the case of a prism train, the sum of the bases; or, in the case of reflection back through the same prism, the base length multiplied by the number of such reflections. To resolve the 0·6 nm-apart D lines, a base length of not less than 1 cm of dense flint is required.

(*iii*) Refractive index: hence a flint prism has a greater resolving power than a crown prism, other factors being the same.

(*iv*) Wavelength: as with dispersion, resolution with a prismatic spectroscope is seven or eight times as great in the ultraviolet as in the extreme red.

The resolution of a normal spectrum, produced by a grating, is similarly measured in terms of λ and $\lambda+\Delta\lambda$, two wavelengths which are just resolved:

$$R=\frac{\lambda}{\Delta\lambda}=nN$$

where $N=$ total number of rulings illuminated

$n=$ order of the spectrum.

nN then expresses the wavelength at any point in the spectrum divided by the least difference of wavelength that is perceptible there.

Hence with a given grating, the greater the area that is illuminated, the greater will be the resolution. To avoid impossibly large gratings, the finer the engraving the better; the commonly employed scale is about 6000 per cm. Gratings thus possess a signal advantage over prisms in this respect: the resolving power of the latter can only be increased by increasing the thickness of glass traversed—a procedure which is limited by the absorption becoming prohibitive (apertures of 75 mm are near the maximum that can be usefully employed)—whereas that of a grating can be increased by increasing the number of rulings per cm, or the number of cm illuminated, or both. Since about 1000 rulings will resolve the D lines in the 1st-order spectra, a 100-mm grating, normally engraved with 6000 per cm and fully illuminated, will have a 1st-order resolution equivalent to that of a dense flint prism train of base about 60 cm.

As in the case of a telescope, the visibility of resolved detail assumes the use of adequate magnification ($\geqslant M_r$). This condition may, and should, be fulfilled with small spectroscopes; but in the case of high-dispersion spectrographs, to increase the linear dispersion to an extent that would separate the images of adjacent resolved wavelengths would so dilute the intensity of the image that exposures would have to be impracticably long.

It should be stressed again that incomplete illumination of the prism or grating will result in wasted resolving power. Hence the collimator lens must be large enough, and must be filled with light from the slit.

19.3.3 Brightness of the spectrum: The brightness of the spectrum of a source of unit intensity is dependent on:

(*a*) *Size of objective*

(*i*) Point objects: Owing to the increased size of stellar images with increased D, owing again to their increased susceptibility to atmospheric turbulence, the brightness of the spectrum of a given spectroscope varies more nearly with D than with D^2, as would be expected from theory. Increasing the aperture of the collimator (and therefore its focal length, to preserve the relation $D/F=d/f$) will increase the purity but not the brightness of the spectrum, which in the case of stellar images is determined solely by D.*

(*ii*) Extended objects larger than the slit width, but smaller than the angular aperture of the telescope: Here the brightness of the spectrum is determined by the effective aperture of the collimator, irrespective of the fact that the intensity of the focal plane image of such an object varies inversely as F^2 for unit telescopic aperture. Provided only that

---

*The increase in F necessitated (with a given spectroscope) by an increase in D will also widen the spectrum of an extended source, and the dispersion; this springs necessarily from the fact that F and M are directly proportional.

equality is maintained between the focal ratios of the telescope and collimator, variation of F has no effect upon the brightness of the beam emergent from the collimator, and therefore none upon that of the spectrum. This can easily be demonstrated:

Putting F and D for the focal length and aperture of the telescope,
$\quad\quad$ f and d for the focal length and aperture of the collimator,
$\quad\quad$ x and a for the diameter and area of the focal image, dependent upon F,
$\quad\quad$ b for its brightness,

then for maximum efficiency,

$$\frac{D}{F}=\frac{d}{f}$$

If, now, F is halved, so that

$$\frac{D}{F'}=\frac{2d}{f}$$

only one-quarter of the radiation from the slit will be intercepted by the collimator, and $b'=\frac{b}{4}$. But at the same time $x'=\frac{x}{2}$, hence $a'=\frac{a}{4}$ and $b'=4b$. By the combination of these two factors, $b'=b$, and the brightness of the image will remain unchanged.

If f is also halved, to reinstate the relation

$$\frac{D}{F'}=\frac{d}{f'}$$

all the radiation from the slit will indeed be intercepted by the collimator, but the effective width of the slit (its angular width at distance f) will have been doubled, and to achieve the same resolution as formerly its linear width will have to be halved, so that b' will again be reduced to b.

(*iii*) Extended objects whose angular size exceeds the angular aperture of the telescope: These are best observed by doing without the telescope altogether, when the brightness of the spectrum will be directly proportional to d.

(*b*) *Width of slit*

The significant measure of slit width is its angular width as seen from the centre of the collimating lens, i.e. from a distance f. Effective slit width can therefore be reduced by a linear reduction or by an increase of f.

With normal slit widths and average seeing, from 60 to 90 per cent of the light from a point source transmitted by the objective passes the slit; in the case of a large instrument, this figure is often reduced to 50 per cent or less owing to the expansion of the theoretical spurious disc by atmospheric turbulence.

The brightness of a bright-line spectrum is independent of slit width, and dependent on (*a*) and (*c*) solely: opening the slit merely widens the lines.

(*c*) *Transmission factor of the optical train*

Light loss in the optical train of a high-dispersion spectrograph may be as high as 45 to 48 per cent. Assuming that 50 per cent of the incident light has passed the slit, this gives a transmission factor of about 27 per cent. With smaller instruments of lower dispersions, it is very much higher.

In addition to the inevitable light loss by absorption in, and reflection at, prisms and lenses, the transmission factor of the instrument will be needlessly reduced if:

> The focal ratios of objective and collimating lens are not the same, in which case a peripheral ring of one or the other will be out of commission;
>
> any part of the optical train is not large enough to pass the whole of the beam delivered to it;
>
> it is, on the other hand, needlessly large;
>
> the prism or prisms are not in the position of minimum deviation; even at minimum deviation from 10 to 20 per cent of the light incident upon a 60° prism is lost by reflection: the denser the glass the greater this loss.

Loss by absorption is also greater in flint than in crown; whilst the light loss from this cause is increased by either increasing the density of the glass or the length of the light path through it, this additional loss is not borne equally well by all wavelengths, but more heavily by the short than by the longer wavelengths.

(*d*) *Dispersion*

Doubling the dispersion while retaining the original width of a continuous spectrum reduces its brightness to one-half its former value. But if the doubling of the dispersion is achieved by an overall magnification of the spectrum, the brightness will be reduced to one-quarter. The brightness of an emission spectrum would be unaffected, since increasing the dispersion merely increases the separation of the monochromatic images.

(*e*) *Grating spectra*

The brightness of a normal spectrum is, within limits, a function of the closeness of the rulings (brightness increasing as the number of rulings per cm increase) and of the size of the grating, while being inversely proportional to the order of the spectrum.

19.4 Spectroscope construction: optical requirements

1. Slit in focal plane of telescope objective.

2. Slit in focal plane of collimator (which then transmits a parallel pencil to the prism or grating).

3. Slit parallel to the prism edge or the grating rulings.

4. Focal ratios of telescope objective, view-telescope objective and collimating lens the same ($F/D = f/d$), to utilise their full aperture.

5. Prism just large enough to accept the whole of the parallel beam from the collimator. (If too small, light grasp is being wasted; if much too large, unnecessary light loss through absorption.)

6. Prism in position of minimum deviation for the central wavelength.

7. Camera lens or view-telescope objective to reconverge the beam from the prism or grating (unless the latter is concave) to a focus at the photographic plate or the eyepiece.

19.5 Diffraction gratings

Reflecting gratings are engraved with a diamond on aluminised glass. Transmission gratings are engraved on clear glass. Both are expensive.

Replica gratings are collodion impressions of the original, and are comparatively cheap—comparable with prisms of equal aperture, and about one-tenth as costly as an original. Replicas are mounted on plane or concave glass, silvered or unsilvered, and may be used as either transmission or reflecting gratings. Their definition is inferior to that of original gratings when the slit is 'closed'; but when widened—for the observation of prominences, for example—there is nothing to choose between them.

Concave reflecting gratings have the advantage that they both form and focus the spectrum, thus eliminating the view telescope or camera objective with its unavoidable additional light loss.

Most modern gratings have about 6000 rulings per cm. For maximum efficiency the separation of the rulings should be comparable with the wavelength of light, i.e. of the order of 0·0005 mm. Concentration of most of the light into a particular spectrum (usually of the 1st order) is determined by the form of the lines ruled with the diamond.

The outstanding advantages offered by gratings are (*i*) normal spectra i.e. dispersion uniform throughout the spectra of a given order; (*ii*) choice of dispersions, that of the 2nd order being approximately twice that of the first, of the 3rd three times that of the 1st, etc.

The greatest enemy of gratings is dust and, in the case of reflecting gratings, tarnishing and corrosion of the reflecting surface. They must always be kept under airtight cover when not in use; dusting, as infrequent as possible, should be done with a sable-hair brush, the strokes being very light and made in a direction parallel with the rulings.

19.6 Direct-vision prisms and gratings

By suitable choice of refracting angles and refractive indices, it is possible to design two-prism combinations which (*a*) deviate an incident polychromatic ray without dispersing it, thus giving an achromatic prism,* or (*b*) disperse it without deviating it, thus giving a direct-vision prism. Such prisms may be regarded as a halfway stage from the plane parallel plate (which displaces the ray laterally but neither deviates not disperses it) to the ordinary pprism.

In the case of (*b*), nil residual deviation is arranged for a mean wavelength (usually in the region of the *D* lines of sodium), other wavelengths diverging on either side of the selected undeviated wavelength.

By means of two prisms (Figure 153), nil deviation can only be achieved with lateral displacement. To make the emergent *D* ray not

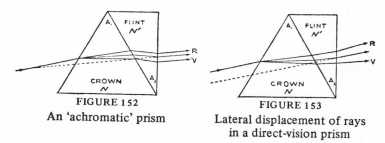

FIGURE 152
An 'achromatic' prism

FIGURE 153
Lateral displacement of rays in a direct-vision prism

simply parallel to, but coincident with, the incident ray, a similar but reversed unit must be added, producing a three-prism combination (Figure 154) known as an Amici prism. Any odd number of components may be used, each additional pair increasing the dispersion without introducing deviation. Three- or five-prism combinations are the usual d/v spectroscope arrangement, the first and last components being always of crown glass. Total internal reflection is avoided by cementing the components. The convenience of d/v for the centre of the spectrum is to some extent offset by the heavy absorption in such combinations, though this is of little account in solar work.

*Achromatism, in this sense, indicates that the emergent rays of different wavelengths, though separated, are mutually parallel (Figure 152).

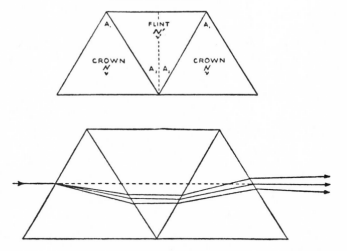

FIGURE 154
The Amici prism

The refracting angles, refractive indices, and constringences of the two (crown and flint) components are related to the dispersion of the combination as follows:

Given that $\omega_D = 0$

then $\dfrac{A_1}{A_2} = \dfrac{N_D'-1}{N_D-1}$

and $\omega_C - \omega_F = (N_D-1)\left(\dfrac{1}{v'}-\dfrac{1}{v}\right)$

where the constringence, v, is given by

$$v = \frac{N_D-1}{N_F-N_C}$$

Two types of d/v combination using one liquid component are illustrated in Figures 155 and 156. The first is known as a Zenger prism: it consists of two right-angle prisms, cemented at their hypotenuse faces and set with the base of the first component normal to the incident beam. If the material of which each prism is made has the same

FIGURE 155
The Zenger prism

318

value of N_D but different dispersions, λ_D will be transmitted undeviated, while other wavelengths will be deviated and dispersed. The second prism

GLASS   GLASS   R

LIQUID

V

FIGURE 156

Alternative form of glass-liquid prism

usually contains liquid of suitable refractive index. Not only is d/v secured, but the light loss from reflection from the prism faces is much reduced owing to the normal incidence.

The Wernicke prism is in effect a combination of two Zenger prisms. Apart from its high transmission factor, it is characterised by abnormally large dispersion—of the order of six times that of a 60° prism of dense flint.

Direct vision with a grating spectroscope is obtained by mounting a replica grating on the hypotenuse face of a prism of suitable A and N

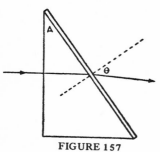

FIGURE 157

Diffraction
grating-prism
combination

(Figure 157). The value of A to give zero deviation for the centre of the 1st-order spectrum depends upon the number of the grating's rulings per cm, and upon the refractive index of the prism. The angle θ, for any wavelength λ, is given by

$$\lambda = x(N \sin A - \sin \theta)$$

where x = separation of adjacent rulings; and if λ is to be undeviated, we have

$$\lambda = x(N-1) \sin A$$

Substituting the values, $\lambda = 570\text{nm}, N = 1\cdot6$, $x = 1\cdot81.10^{-4}$ cm, we have approximately

$$A = 32°$$

In practice a 30°/60° prism is usually employed in small spectroscopes of the type sold for the observation of prominences.

D/v grating spectroscopes have the double advantage that their spectrum is normal, while zero deviation for any desired wavelength can be obtained merely by juggling with N and x.

A Thorp replica grating mounted in this way gives sufficient dispersion for the D lines of the solar spectrum, or the lines in the spectra of bright A-type stars such as Sirius, to be separated without further magnification.

19.7 Direct-vision and slitless spectroscopes

Although there is not, in the UK, any current manufacture of spectroscopes designed expressly for astronomical purposes, some of the following types may from time to time appear on the surplus market, or in collections, and the descriptions may be of use to the amateur wishing to construct his own equipment.

The slit is often omitted from small d/v spectroscopes. These commonly go by the name of 'star spectroscopes' or 'eyepiece spectroscopes'; the McClean and Zollner d/v spectroscopes are examples. They are employed either in place of the telescope's ocular, or as an attachment to be screwed on to the ocular.

Having no slit they can be used only on point sources if the spectrum is not to be hopelessly impure. The stellar image is in effect a very short length of a slit whose width is equal to the diameter of the spurious disc. For this reason good seeing is important, or an impure spectrum will result from the enlargement of the spurious disc.

Since absorption lines are impossible to see in a spectrum which, whatever its dispersion, is a mere line of colour, the spectrum must be widened in some way.* This is usually effected by means of a cylindrical lens (or a cylindrical curve may be incorporated in the prism itself) which draws out the spurious disc into a linear streak in a direction perpendicular to that of the dispersion, the amount of widening achieved in this way being inversely proportional to the radius of the curve. Effectively, therefore, such spectroscopes incorporate a slit whose width is equal to the diameter of the spurious disc of the star under observation.

The McClean spectroscope employs a concave cylindrical lens mounted close to the prism on the side furthest from the eye, and is screwed into the drawtube in place of the ocular. The Zollner employs a convex cylindrical lens mounted between the prism and the eye, the spectroscope being screwed on to an already focused LP ocular ($M=8D$ is recommended) in the manner of a suncap. Espin's star spectroscope consists of a d/v prism without either a slit or a cylindrical lens. It is mounted on the objective side of an ordinary ocular, the dispersion varying with the distance of the ocular from the prisms.

*Over-widening of the spectrum, however, will reduce the visibility of the details again, owing to excessive dilution.

Horne and Thornthwaite used to manufacture a useful 'universal' eyepiece spectroscope, with d/v prism, cylindrical lens, slit, and achromatic viewing lens. For work on the Sun, the cylindrical lens is omitted; for the telescopic observation of stellar spectra, the slit is omitted; and for meteors, aurorae, the Zodiacal Light and the like, the 'star' spectroscope is used without the telescope.

Inevitably both dispersion and spectrum width are small, and spectroscopes of this type are not suitable for photography, though in conjunction with quite small telescopes they will show the main spectroscopic type characteristics of stars down to about mag 3 or 4. A Zollner used with a 75- or 100-mm refractor, for example, shows the prominent hydrogen emissions of the B-type stars, the fluted spectra of the brighter M-type, the major A-type lines, and some lines can be glimpsed in the spectra of bright M-type stars when the seeing is good.

More powerful spectroscopes can also be used without slits in stellar work, and for the photography of stellar spectra often are. The usual practice in this case is to place a negative collimating lens a short distance on the objective side of the focal plane, so that a parallel pencil is delivered to the prism train, and thence to the camera objective in the usual way.

On the other hand, not all 'eyepiece' spectroscopes are slitless. The function of the slit is to increase the purity (and therefore resolution) of the spectrum, since it can be made narrower than the spurious disc, and also because wanderings of the primary image due to turbulence have no effect upon the definition of the photographic image. Spectrum width can be obtained, in photographic work, by trailing the image along the slit during exposure: the slit is oriented EW, and the driving clock set to run a little slow or fast. For visual work a cylindrical lens must still be resorted to: set in front of the slit it draws out the spurious disc into a line which must be made coincident with the slit. Maintaining this coincidence requires a very accurate drive and smoothly working slow motion in RA, and with small-dispersion 'eyepiece' spectroscopes little is to be gained by incorporating a slit.

19.8 Short-focus spectroscopes

Occasionally, spectroscopes of either the prismatic or diffraction type are encountered which are more powerful than the d/v 'eyepiece' spectroscopes, but are still compact and light enough to be used with telescopes of not more than moderate aperture.

The type often described as a 'student's' spectroscope consists of a single, usually 60° prism, with collimator and view telescope. A more elaborate form, giving greater dispersion, employs a train of three or more prisms, and should incorporate a pivoting arrangement whereby any required wavelength may be refracted at minimum deviation and the view telescope made parallel to its emergent ray when it is in this position.

In the type known as the constant-deviation spectroscope the latter provision is not required, the collimator and view telescope being set permanently at 90° to one another. Different wavelengths are brought to the axis of the view telescope by means of a constant-deviation prism (Figure 158); this may be either a single block of glass of the shape shown,

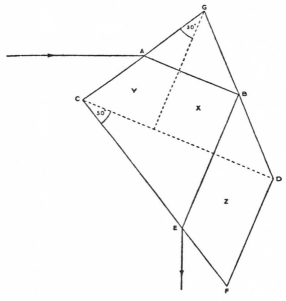

FIGURE 158
Constant-deviation prism

or may be made up of a right-angled prism, X, and two 30° prisms, Y and Z. As such a prism is rotated, successive wavelengths become incident in turn at 45° to the face GD, and are consequently deviated through a total of 90°. It is to the action of the 30° prisms that the dispersion of the combination is due, and the latter is therefore equivalent to a single 60° prism set at minimum deviation.

Typical layout of a small reflecting grating spectroscope is shown in Figure 159. Diffraction by a transmission grating, cemented to the hypotenuse of a 30° prism (giving nearly d/v, collimator and view telescope being inclined at only 11°), is made use of in Sellers' simple but extremely

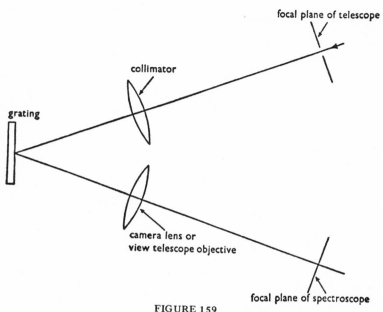

FIGURE 159
Optical layout of a grating spectroscope

effective prominence spectroscope. The optical layout is shown dia-grammatically in Figure 160. It was designed for ease of construction at home, and for use with small telescopes in the 50- to 150-mm range; for constructional details, the reader should refer to B. 13.5. The rotation of the spectroscope about the optical axis of the telescope, allowing the offset slit to travel tangentially round the limb of the solar image, is effected by rotating the drawtube to which it is attached; correct focus is maintained during this operation by means of a sleeve round the drawtube, whose length is such that one end of it bears against the end of the tele-scope tube and the other against the flange at the ocular end of the draw-tube when the focus is correct.

By modifying this design to allow the view telescope to pivot about an axis passing through the centre of the grating, a useful all-purpose solar or stellar spectroscope is obtained.

An even simpler arrangement for the observation of the prominences, though less powerful than Sellers', consists merely of a d/v prism with slit, offset in the drawtube by an amount equal to the semidiameter of the solar image in the focal plane.

Still another type of prominence and general solar spectroscope was

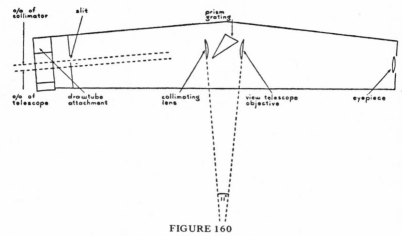

FIGURE 160

Optical layout of Sellers' prominence spectroscope

that designed by Evershed for Messrs Hilger; it was light and compact, but of good dispersion and extremely convenient to use. Its basic elements were two d/v prisms and a plane mirror (Figure 161). The prisms

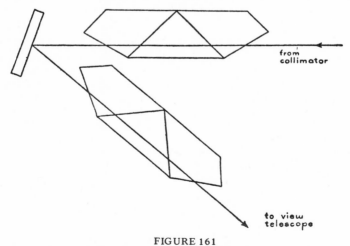

FIGURE 161

Combination of two d/v prisms in a small spectroscope

were fixed, different wavelengths being brought to the centre of the field by rotating the mirror in the plane of dispersion. Means of rotating the whole spectroscope were provided for prominence work.

The more powerful spectroscopes described in this section are capable of showing considerable detail in the solar spectrum, although anyone wishing to make a serious study of the Sun would probably envisage a more powerful long-focus instrument of the type described in section 19.10. The fairly recent availability of extremely narrow-wavelength filters has completely replaced the old type of 'prominence spectroscope' in which individual prominences could be observed through a widened slit set at the appropriate wavelength. A description of a prominence telescope suitable for amateur construction will be found in section 22.5.

19.9 Home-made short-focus spectroscopes

The basic specifications for any spectroscope are listed in section 19.4. Gratings are, on the whole, preferable to prisms, since they are lighter and easier to mount; their superior dispersion in the long wavelengths is also an important advantage.

The only part of the spectroscope that is liable to cause practical difficulty is the slit, and the amateur building his first spectroscope should not be surprised by several failures, and should be prepared to scrap these as being valuable experience which will, at the third or fourth attempt, enable him to produce a satisfactory slit.

The vital part of the slit is the small section of the jaws which defines the pencil passing to the collimator: the spectrum itself is composed of innumerable images of this aperture, each emission or absorption line being one such image. The following are the essential points:

(*i*) The edges of the two jaws must be absolutely straight and absolutely parallel.

(*ii*) The edges of the jaws must be sharp, to prevent light being reflected from them into the collimator; they must be bevelled almost to a knife-edge, the bevelled side being that facing the collimating lens.

(*iii*) The two jaws must lie in the same plane, so that when closed they come precisely together and do not overlap.

(*iv*) If the slit is adjustable, the parallelism of the jaws and their location in a single plane must be maintained at all slit widths.

Slit jaws may be made from 1·5-mm brass or aluminium sheet, which can be cut with scissors. Each edge that is to form the jaws is first worked on a fine-toothed file till it is absolutely straight, perpendicular to the sides of the jaw, and, in section, accurately square (Figure 162(*a*)). This work is finished off on an oil-stone, and examined with a magnifying glass; any serrations or other imperfections must be ground down. Then, starting with the file and finishing on the stone, one edge is worked down to

325

FIGURE 162
Form of slit jaw
for spectroscope

an increasingly deep bevel, this operation being stopped when it is just short of becoming a knife-edge, i.e. of interfering with the edge already perfected (Figure 162(b)). The work is examined with a powerful magnifying glass; the bevel on the two jaws should be approximately equal.

For a fixed slit (Figure 163) cut two 25-mm diameter semicircles of the metal sheet and work the diameters to the bevelled raw edges. Cut two rings of fine-grain 3-mm cork, outer and inner diameters 50 mm and 3 mm. The central hole is the stop limiting the height of the spectrum*; it should be bevelled on the side facing away from the slit, and finished very carefully with fine-grade glasspaper. One cork ring is coated with carpenter's glue and one of the slit jaws laid on it, with the edge that is to form the slit edge lying diametrically across the central aperture. The second jaw is then laid approximately in position and moved up to the first with a needle or sharp pencil point; with the slit supported over a bright light on a sheet of glass, and examined with a magnifying glass or ocular, the position of the second jaw is adjusted until a very fine and parallel-sided streak of light is visible. A 50-mm/28-mm ring of the same sheet of metal from which the jaws were cut is them glued to the cork outside the jaws; when it and the jaws are

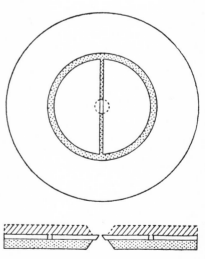

FIGURE 163
Construction of a fixed slit

*The diameter of this aperture should be about one-third of the diameter of the focal image of the Sun.

set firm, a second cork ring, of the same dimensions as the first, is glued over the jaws and the metal ring.

A simple arrangement for an adjustable slit is described by Sellers (B. 13.5). Another is illustrated in Figure 164. The slit plate, B, fixed

FIGURE 164
Construction of an adjustable slit

jaw, C, adjustable jaw, D, slides for the adjustable jaw, E, and bases for the slides, F, must be cut from the same sheet. Clearance of a few thousandths between E and D can be obtained by packing tin foil between E and F. All screws must be countersunk.

Not only should HP examination against a bright light show the slit to be accurately rectangular at all widths, but when closed it must exclude

327

light completely along the whole of its length. Slits must always be kept scrupulously clean, and be treated with great care. Specks of dust between the jaws will produce dark lines running the length of the spectrum, i.e. at right angles to the absorption lines.

The slit is positioned at the focus of one of a pair of achromatic OGs (whose focal lengths need not be equal, but usually are for convenience) of diameter at least equal to the minimum width of the grating or prism as presented to it. For optimum efficiency, the focal ratio of the collimator should match that of the telescope with which the instrument is to be used (see section 19.3.3), but for solar work a collimator of shorter focal ratio makes little difference; using a standard replica of, say, 25 mm width, a focal length of 25 cm to 30 cm would be convenient when used with a standard refracting telescope. The collimator and view telescope are then arranged as in Figure 159. Normally, for solar work, the second-order spectrum would be utilised, the angle between the two optical trains being about 37°. Rotation of the grating, whose centre must lie at the intersection of the two optical axes, brings the various wavelengths to the axis of the viewing telescope. Particular attention must be paid to the following details:

(a) View telescope and collimator axes must lie in the same plane, which must itself be perpendicular to that of the grating.

(b) The direction of the slit must be parallel to the direction of the grating rulings.

(c) The slit must be at the focus of the collimator lens. This is conveniently achieved by focusing the view telescope for infinity, and then viewing the slit, through the telescope, via the collimator OG, adjusting the slit-OG distance until the former appears well-defined. Secondary spectrum causes the focal length of a normal OG to be increased in blue light, and slight refocusing of the view telescope's ocular will be necessary when the short-wave end of the spectrum is being examined.

See, further, B. 13.1, 3.

19.10 Long-focus spectroscopes

The advantages of long focal lengths in solar observation—whether visual, photographic, or spectroscopic—are emphasised in *O.A.A.*, section 1.15, which is concerned with the layout of a specially designed horizontal telescope whose fixed components are fed by a coelostat. The incorporation of a long-focus spectroscope or spectrograph in a layout of this sort is a comparatively simple matter, and one such is described.

The Littrow, or autocollimating, spectroscope is another design that the arrangement with a coelostat-fixed image renders possible. The essential feature of the autocollimating spectroscope is the use of a single

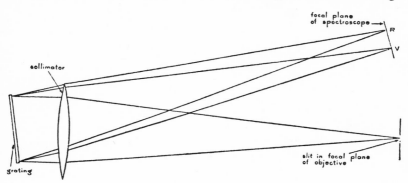

FIGURE 165

Autocollimating spectroscope

convex or plano-convex lens as both collimator and camera lens or view-telescope objective (Figure 165). The collimator delivers a parallel beam to the grating which returns part of the spectrum to the collimator; this, now in its capacity of view-telescope objective, forms an image of the spectrum at the same distance from itself as the slit, but, owing to the grating being slightly tilted, at a convenient distance above or below the slit. The collimating lens should be provided with a few inches of horizontal adjustment for focusing; the grating can be mounted on an alta-zimuth fork, altitude adjustment varying the position of the image at the photographic plate or eyepiece, and azimuth adjustment bringing spectra of different orders, or different regions of a given spectrum, to the obser-ving position. There should be provision for these adjustments to be made from the observing position.

The 45-m Sun telescope at Mt Wilson incorporates a 22·5-m Littrow spectrograph (30-cm OG of 60 m F, 25-cm collimating lens of 30 m F, 20-cm grating) yielding a dispersion of 7 mm per Å in the 4th-order spectra: a linear dispersion from the B to K lines of about 20 m. But even the 6-m, 12·5 cm objective suggested in $O.A.A.$, used with a 2·5-m, 50-mm collimating lens, will provide a solar spectrum which is a revelation to anyone whose experience has been confined to telescope attachments.

Another way to secure a longer focal length than is possible with an eye-end telescopic attachment is in effect to turn the whole telescope into a spectroscope, just as in prime-focus photography the telescope is turned into a camera. The use of an objective prism achieves this. Another

329

method is Butterworth's use of a concave grating with a Herschelian reflector (Figure 166). Much greater convenience and adaptability are, however, offered by a fixed-image arrangement, as in the Sun telescope, allowing the fixed mounting of the spectroscopic apparatus.

FIGURE 166

Reflecting telescope incorporating a concave grating

19.11 Objective prisms

See *O.A.A.*, section 15.

SECTION 20

PHOTOGRAPHY

20.1 Introduction

The power of photography as a weapon in the astronomer's hand springs from:

(a) its precision and impersonality, which adapt it to positional work, in which it is comparable with visual methods;

(b) the cumulative effect of light upon the photographic emulsion, adapting it to the investigation of very faint objects, in which it is superior to visual methods;

(c) its wholesale character, which, together with (b), adapts it to the discovery of new objects, in which it is superior to visual methods;

(d) the permanence of its records.

Its outstanding weaknesses are its greater dependence upon atmospheric turbulence and the comparatively poor resolution of the photographic emulsion, which make it much inferior to visual observation in those fields where the resolution of fine detail is the paramount requisite.

The applications of photography in astronomy usually divide into three main categories:

(a) photographing the Sun, Moon and planets;

(b) photographing faint extended objects, e.g. nebulae, comets, and star fields;

(c) meteor photography.

20.2 Image scale

In each of the categories mentioned above, different techniques or apparatus are used, one significant factor being image scale.

The scale of the image in the focal plane of a lens, whether camera or telescope objective, is given by equation (a), p. 29:

$$h = \frac{F\theta}{57 \cdot 3}$$

where θ is measured in degrees, and F and h in the same linear unit, whence the primary image scale in, e.g., degrees/cm is

$$\theta = \frac{57 \cdot 3}{F} \cdot$$

To fix ideas, the diameter of the primary focal image of the Sun or Moon $\simeq 0 \cdot 009F$.

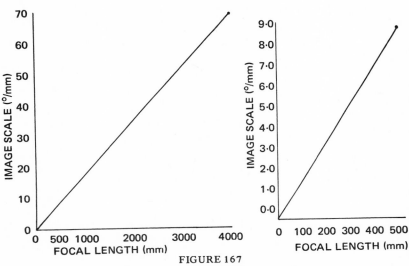

FIGURE 167

Image scale at various focal lengths

The above graphs show, for a range of F from 50 to 3,000 mm, the angular values of various values of h, from which it can be seen, for example, that a $10 \times 12 \cdot 5$-cm film used in a camera of 20 cm focal length will cover about $28° \times 36°$. A focal length of about 19 cm would give a plate scale very nearly equal to that of Norton's maps.

The field of acceptable definition will be less than this by an amount depending upon the quality of the lens, the focal ratio at which it is used, and the purpose for which the photograph is taken; it must be determined by trial, comparing the contents of the acceptable field at different relative apertures with a star chart or atlas.

The linear size of the field of tolerable definition is primarily a function of focal ratio, and within a certain range is largely independent of focal length. Thus the linear area of acceptable definition is approximately the same with a 50-cm $f/4 \cdot 5$ lens as with a 25-cm $f/4 \cdot 5$ lens, and not four times as great. As the focal length is increased the linear usable field does indeed increase also, but very much more slowly. This relative independence of linear field and focal length breaks down when things are carried to extremes: for instance, the $f/4\cdot5$ 25-cm lens could cover a quarter-plate satisfactorily, but a $f/4\cdot5$ 25-mm certainly could not.

If any material increase in the linear size of the tolerable field is to be achieved at a given focal ratio without enormously increasing F, the design of the lens must be modified; this will involve increased thickness and density of glass, which in turn involves increased absorption of the short wavelengths, and therefore increased exposures.

20.3 Image measurement

The only normal occasion on which the amateur wishes to measure positions from a photographic plate is as an alternative to the micro-metrical determination of the position of a comet or asteroid. With care, and provided that, in the former case, a well-defined central condensation is present, the photographic method can produce results of useful accuracy. It was formerly thought that the rigid glass backing of a plate is necessary if sufficiently precise measurements are to be made from a photograph, but recent experience has proved sheet film to be suitable for amateur purposes, over at least a short term. With sufficient magnification, and a micrometer screw, position of a stellar image can be determined to within about 0·005 mm without difficulty, being equivalent to about $2''\!\!\cdot\!5$ on an emulsion exposed in a camera of 500 mm F; with a standard Newtonian, this precision will be increased in proportion to F, assuming, of course, satisfactory guiding. For example, Manning has photographed asteroids down to mag 16, using a 26·5-cm $f/8$ Newtonian and small squares of sheet film, from which measures of position to within about $1''$ have been made. Depending on the speed of movement of the object, the time of beginning and ending of the exposure must be noted with appropriate precision; professional positional measurements normally record the time of mid-exposure to 0·0001 day (about $8\frac{1}{2}$ sec.).

For greatest possible accuracy, a plate-measuring machine is essential. For lesser precision, however, one may be improvised, or projection used instead. In principle, any device incorporating controlled motion in two mutually perpendicular directions—such as, e.g., a microscope mechanical stage, or a double-slide plate carrier—could be adapted to plate measurement. With such makeshifts, however, error is liable to be introduced by the fact that insufficient attention was devoted in their construction to the precision of the mutual perpendicularity of the two motions. A more satisfactory plan is to use a filar micrometer for its magnification and its micrometer screw. With a single-web micrometer, it or the plate must be mounted in a frame which can be rotated to allow the second coordinate to be measured, and the machining of the frame must be of a high degree of precision to ensure that the two positions of the web relative to the plate are accurately perpendicular to one another. Measurements with such adaptations are liable to be rather laborious, and the field is small; but for the measurement of cometary positions—of its nature, not a very frequent operation—with short F cameras these disadvantages are not serious.

Alternatively, plates may be measured by projection on to a rectilinear grid (1-mm squares). It is important that the planes of the grid, of the plate, and of the projecting lens should be accurately parallel. This can be checked by projecting a photographicaly reduced circle or square on to a screen on which a similar figure has been drawn; the grid is then, after the necessary adjustments have been made, placed on this screen. It is also wise to confine measures to the central region of the field. (See B. 14.19).

Reductions can be conveniently made by Comrie's method of dependences using three comparison stars (B. 14.4); see also B. 14.23 for Merton's two-star method, a modification of Kaiser's interpolation method, and discussion of these papers, B. 14.5. Schlesinger describes a much shortened method in B. 14.37. See also B. 14.45, 46.

20.4 The photographic emulsion

The light-sensitive emulsion consists of a crystalline silver halide (commonly bromide), further sensitised by an organic dye, which is suspended in gelatin on a base of film or glass. The photographic speed, or sensitivity, of the emulsion is related to the size of the silver salt particles: the grain of modern emulsion ranges for the most part from $0·3\mu$ to $1·5\mu$. The graininess of the final image is also influenced to some extent by the development.

The precise mechanism whereby the latent image is formed by the action of light on the silver crystals is still imperfectly understood. However that may be, the particles of the latent image are reduced to dark metallic silver by the action of the developer. Those halide crystals which were not exposed to light are then removed by the fixing solution, in its simplest form sodium thiosulphate or hyposulphite ('hypo').

The term optical density, or just density, referred to developed photographic emulsions, is defined as follows:

If the intensity of light falling on a plate or film is I_0 and the intensity of light transmitted through it is I, then the density of the exposed area

(denoted by D) is

$$D = \log \frac{I_0}{I}$$

Thus for a totally clear area, $D = 0$
for an area transmitting 0·1 of the incident light, $D = 1$
for an area transmitting 0·01 of the incident light, $D = 2$, etc.

If a graph of emulsion density produced by a given exposure (in logarithmic units) is plotted, the characteristic D-log E curve results, sometimes called the H and D curve after the pioneers Hurter and Driffield (Figure 168). This curve indicates that the photographic emulsion is far from being the perfect light detector. Ideally, the curve would be a straight line, and density would be directly proportional to exposure. In fact a maximum density, D_{max}, is reached, usually with a value of 3 or 4. At the other end of the scale, even the emulsion which has no exposure at all will have a certain density, the fog level, notwithstanding any density due to the emulsion support, film or plate.

FIGURE 168
The characteristic curve of a
photographic emulsion

The transition region between the fog level and the straight line portion of the curve is called the 'toe'. The slope of the straight line is the contrast, usually referred to as gamma (γ), given by tan θ. The term 'contrast index' is becoming used—this is measured taking into account the effects of the toe. In order to differentiate faint details, emulsions with high contrast should be used. Such emulsions are generally slow, and have a desirably fine grain.

An ideal emulsion would obey what is known as the 'reciprocity law', which states that if exposure is defined to be the product of the intensity of light and the duration of exposure to it, then equal exposures will give rise to equal image density regardless of the actual intensity and duration values. In practice it is found that exposure of an emulsion to light of high intensity for a short time will in general produce an image of greater density than exposure to a low intensity for a long period, even though the exposures be calculated to be equal. This effect is a failure of the reciprocity law as applied to photographic emulsions, and is commonly known as 'reciprocity failure'. For a photographic grain to be made developable, it must absorb a certain number of light 'quanta'. However, if the light intensity is low, it becomes probable that the thermal motions of the molecules within the grain will cancel the effect of the first quanta to arrive, and it may be that the required number to sensitise the grain may never be achieved. It is possible to reduce the effect of reciprocity failure by cooling the emulsion with dry-ice or liquid nitrogen during exposure, thereby reducing the thermal agitation of the molecules in each grain. Some so-called 'cold cameras' are described in B. 14.11, 12, 18, 44. Ordinary films, exhibiting high reciprocity failure, can gain 2 mags or more over their sensitivity at ambient temperature; astronomical emulsions, of low reciprocity failure, do not show the same improvement.

A less effective way of overcoming reciprocity failure is to give the emulsion an overall pre-exposure before use, in order to produce a certain amount of fog. This will reduce the number of additional photons required to make the grains developable. This has the effect of rendering visible the fainter images which might not otherwise have been developed. It has most effect upon the lower density part of the D-log E curve.

A chemical method of achieving higher emulsion speed is hypersensitization, in which pre-soaking the emulsion in liquid ammonia, or more simply but less effectively, in distilled water, has the effect of removing anti-fog chemicals which suppress some of the sensitivity but produce more acceptable pictorial results. The emulsion must be dried quickly and used as soon as possible. This method is most frequently used with infrared sensitive plates, which are normally particularly slow. (B. 14.36).

The technique of baking the emulsion at $50°C$-$70°C$ for 12-48 hours is also widely used professionally, and creates more sites within the emulsion in which photons can be trapped. (B.14.16).

When an emulsion leaves the manufacturer, it is not necessarily at its highest sensitivity—if it were, it would be foggy and unpredictable. Treatment to bring the emulsion to its optimum performance is left to the experienced user, and no guidance is generally given by the manufacturer. Some of the methods described above could be used in combination, but in general only one is used at a time.

20.5 Magnitude thresholds and exposure limits

The effectiveness of an emulsion is influenced by so many variables that it is not feasible to give exact values for the faintest magnitude detectable with a given aperture. It is better to detail the various factors and allow the astronomer to experiment on the basis of these.

If the stars were seen against a perfectly black ground, and if a perfect detector were used, there would be no limit to the magnitude reached with any aperture other than the exposure time. In practice the limits are set by the inefficiency of the emulsion, and by the general brightness of the sky background which tends to drown out the fainter stars' images. The problems connected with the emulsion itself are summarised at section 20.4. Exposures cannot be prolonged indefinitely because fogging will occur and will eventually reduce the ratio between the image and background densities (the signal to noise ratio) to a level at which the image becomes indistinguishable. There are two kinds of fog present: sky fog, which is entirely due to the exposure, and emulsion fog which is in turn made up of chemical fog and any pre-exposure fog. The emulsion fog normally has little effect on the detectability of the image, but merely raises the base level above which the density of subsequent exposures is measured.

We have seen in section 1.7 that whereas the image intensity of an extended object is proportional to $(D/F)^2$, that of a point object is proportional to D^2. We can apply this property to the sky background itself, to show that the larger the value of the focal ratio of the objective, the smaller will be the amount of sky fog recorded. In practice the star images themselves are extended to a degree depending on the seeing conditions, which will slightly reduce the image contrast. Thus an instrument working at $f/8$ would in general have a fainter limiting stellar magnitude than one of $f/3$ having the same aperture. It should, however, be remembered that the faster instrument would perform better in the recording of extended objects such as nebulae.

A variety of factors influence the degree to which sky fog will be present on a given occasion, among them:

(a) Diffuse auroral light and sunlight reflected from interplanetary matter, which largely accounts fro the brightness of the moonless night sky, usually taken to be equivalent to about 13·5 mag per square minute of arc, though this varies rather widely from night to night, from one part of the sky to another, and with the geographical position of the observing site.

(b) Distance of Sun below horizon. If less than about 10° fogging of the plate is rapid. Severe trouble is encountered by the cometary photographer, for instance, when his object is near perihelion.

(c) Distance and phase of the Moon.

(d) Azimuth distance of the field from the Sun.

(e) Altitude of the field.

(f) Clarity of the atmosphere.

337

(*g*) Presence of artificial lights. These, aggravated by the products of combustion suspended in the atmosphere, were responsible for the retreat of the Royal Observatory from Greenwich to Herstmonceux. Again, diffusion of the lights of Los Angeles from the 'smog' layer which normally overlies that city reduces the maximum exposure time of the $f/5$ instruments at Mt Wilson from 80 mins to about 30 mins, as was demonstrated during the war when Los Angeles was put under a partial blackout.

Under the best conditions (complete absence of moonlight and artificial light), the effect of (*a*) limits the effective exposure of a film of low reciprocity failure, such as Kodak 103a-0, to about $1\cdot6$ x (focal ratio)^2 minutes. For a given value of F, therefore, the permissible exposure complements the area of the objective, and hence the practical limiting magnitude is independent of D. However, this argument applies only to exposures which are prolonged sufficiently to bring up sky fog, and with F sufficiently short to produce practically a point image, unaffected by seeing motion, on the film. (See B. 14.14.)

Moonlight being polarised, up to a maximum value of about 60 per cent at $90°$ from the lunar disc , the use of a polaroid filter placed in the light beam will have good effect on background fogging if the field being photographed is at the optimum angular distance from the Moon. However, the filter factor (typically about 3) will involve at least trebling the exposure for the advantage to be fully realised.

Although in practice the threshold magnitude appropriate to the equipment and materials to hand may be discovered only by experiment, it can be said that using a 35-mm camera with the standard lens of 45-mm or 50-mm focal length at the common widest aperture of $f/2\cdot0$, in conjunction with a fast film (speed of 400 ASA or more) it is possible to record images of stars fainter than the eye can see using exposures of only a few seconds duration. It should, however, be noted that these lenses suffer from considerable vignetting; light loss at the edge of the longer dimension of a 35-mm frame can exceed $0\cdot5$ mag, probably reaching more than 1 mag in the extreme corner. An analysis of vignetting in two typical lenses, a Sekor 50-mm $f/2$ and a Takumar 55-mm $f/1\cdot8$, showed the following effective light loss, through vignetting, at various distances from the centre of the frame:

Distance from centre (mm)	Light loss (mag)
$5\cdot0$	$\cdot03$
$8\cdot9$	$\cdot14$
$12\cdot5$	$\cdot30$
$15\cdot8$	$\cdot48$
$18\cdot6$	$\cdot70$

The limiting magnitude attainable with a given astronomical telescope used photographically is subject to the many factors mentioned above. Using ordinary fast film (e.g. Kodak Tri-X), exposed at the prime focus of an orthodox 30-cm Newtonian, a limiting magnitude of about 15 can be expected with an exposure of 10 minutes under rural conditions.

Methods of calibrating star magnitudes by measuring the diameter of their photographic images are discussed in section 21.5.

20.6 Equivalent exposures

The intensity of a true point image is, for a given exposure, directly proportional to D^2. Thus the exposure E' with an aperture D' which is equivalent to the exposure E with an aperture D is given by

$$E' = \frac{D^2}{(D')^2} \cdot E$$

A 4-sec exposure with an aperture of 15 cm is equivalent to a 1-sec exposure with an aperture of 30 cm, irrespective of F in the two cases. By the same token, a 4-sec exposure with a 15-cm lens is equivalent to a 4-sec exposure with any other 15-cm, irrespective of their focal ratios.

In the case of an extended image the photographic speed (the degree of darkening of the emulsion with unit exposure) is again a function of D^2, but in addition is inversely proportional to F^2 since the area of the image increases as F^2. Photographic speed in this case is therefore a function of $\left(\frac{D}{F}\right)^2$, and equivalent exposures of $\left(\frac{F}{D}\right)^2$. The exposure E' with a lens D', F', which is equivalent to an exposure E with another lens D, F, is given by

$$E' = \left(\frac{F'}{D'}\right)^2 \left(\frac{D}{F}\right)^2 \cdot E \quad \dots \dots \dots \dots \quad (a)$$

When equivalent exposures with a single lens used at different focal ratios are under discussion, $F = F'$, and equation (a), above, simplifies to

$$E' = \left(\frac{D}{D'}\right)^2 E$$

F being the same in each case, D and D' are given by the respective $f/$ numbers of the two stops. Thus if a film has been exposed for 1 hr with the full aperture of a $f/5\cdot6$ lens, the equivalent exposure with the lens stopped down to $f/8$ is given by

$$E' = \left(\frac{8}{5\cdot6}\right)^2 \times 60 \text{ mins}$$
$$= 122 \text{ mins}$$

If the focal length is doubled, therefore, the length of the equivalent exposure is quadrupled. Similarly, doubling the exposure is equivalent to decreasing the focal ratio by $\sqrt{2}$; or in the case of a series of stops to be used with a particular lens (F therefore constant) to increasing D by $\sqrt{2}$. The following exposures at their respective focal ratios are therefore a measure of equivalent light quanta reaching the photographic emulsion:

$f/1$	1^s		$f/8$	1^m 4^s
$f/1\cdot4$	2		$f/11$	2 8
$f/2$	4		$f/16$	4 16
$f/2\cdot8$	8			
$f/4$	16			
$f/5\cdot6$	32			

Due to the effect of reciprocity failure in all but specialist emulsions, this table has no practical meaning when converted into a photographic guide. Taking HP5 as a typical example, the graph in Figure 169 shows that the theoretical exposure at $f/4$ (16 seconds) would have to be extended to almost 50 seconds for an equal negative density to be obtained.

FIGURE 169

Reciprocity failure curve for a typical fast commercial emulsion
(*reproduced by courtesy of Ilford Ltd*)

340

It will be clear from equation (*a*), that, for example, a 50-mm lens working at *f*/2·8 is faster than the 5-m at *f*/3·3. Where a large lens or mirror gains over a smaller one—even though the latter's focal ratio, and therefore photographic speed, may be the same—is, of course, in its greater resolution and (because of correspondingly larger *F*) image scale. Thus a *f*/15 50-mm and a *f*/15 25-cm will provide equally dense images of the Moon with exposures of the same length, but one will be 33-mm in diameter while the other will be less than 7·5 mm.

20.7 Photographic resolution

The resolving power of the photographic emulsion is much inferior to that of the eye, and this is not connected (as is often supposed) with the size of the grain.

Grain diameter, depending upon the speed of the emulsion and the developer, is of the order of $0·3\mu$ to $1·5\mu$. But owing to diffusion in the emulsion the diameter of the image of a point source may be anything from about 20μ (0·002 mm) if under-exposed, to 1000μ (0·1 mm) if grossly over-exposed. Taking 50μ as a rough figure for the diameter of a 'normally' exposed stellar image, two such images will be resolved on the plate if their separation is not less than about 25μ. Substituting 0·025 mm in equation (*a*), p. 29, we have, for the threshold of photographic resolution,

$$R_p = \frac{516}{F} \text{" arc}$$

(*F* being measured in cm), and also, equation (*g*), p. 22,

$$R_v = \frac{11·58''}{D}$$

(*D* also expressed in cm). Now $\frac{620}{F} < \frac{11·58}{D}$ if $\frac{F}{D} > 54$, approximately, which is to say that at all normally employed focal ratios photographic resolution is inferior to visual. For the photographic plate to compete with the eye in this respect, focal ratios larger than about *f*/50 must be used. In practice, lunar and planetary photographers employ effective focal ratios of up to *f*/200, either by the use of orthodox instruments with an enlarging eyepiece to project the image on to the film, or (preferably, because it preserves the perfect achromatism of a reflecting telescope) by using a high-amplification Cassegrain system. See, further, section 20.10.

Successful resolution of limiting detail therefore depends, inter alia, on (*a*) separation of such detail above the resolving limit of the emulsion;

(b) fixity of this detail on the emulsion during the exposure. It is (b) which concerns the high-resolution photographer, the almost inevitable turbulence in the atmosphere smudging out finer detail while the exposure is being made. Increasing the film speed, while reducing the necessary exposure time and hence the effects of turbulence, also decreases the resolving power of the emulsion. The radical solution is to move the telescope to a better site; which is why some of the world's finest lunar and planetary photographs have been taken, using relatively modest apertures—e.g. the 60-cm reflector at the Pic du Midi—from observatories enjoying particularly favourable seeing conditions.

For lunar and planetary photography, it is, however, possible to achieve some improvement by the technique of superimposing several negatives each deliberately underexposed to give a lower density than normal. Resolution of the emulsion varies directly with contrast and inversely with the square root of the density, so the optimum is with high contrast and low densities, which have associated low granularities. Thus the exposure should be made to give a density on the upper part of the toe of the D-log E curve, just before the straight line region, with a density of about 0·2. This is also the region of maximum light detecting efficiency of the film. Tolerating a less dense exposure means that a shorter exposure time (where seeing is troublesome) or slower, finer-grain film may be used.

Superposing a number of such exposures will give a printable density which is less grainy and shows more detail than a single exposure. The effectiveness of the technique increases only as the square root of the number of exposures used, the optimum being about six. It should be remembered that this technique depends for its success on the accurate registration together of the multiple negatives. More details of this and other printing techniques are given in B. 14.41.

20.8 Emulsions and developers

The choice of the best emulsion and its development for a specific job are matters in which experience alone brings confidence. The popularity of the 35-mm film size camera means that this is now the most widely used material, whereas twenty years ago the glass plate was nearly always used for serious work even from the amateur point of view. The currently available materials can be divided into four main categories:

(a) 35-mm-width film with sprocket holes for miniature cameras, supplied in cassettes for easy loading.

(b) Roll film, most commonly 120 size, for larger cameras. This is backed with a detachable paper roll with numbers on the reverse side so that the film can be wound on an appropriate distance.

(c) Glass plates, usually about 2-mm thick, in recognised sizes of which those most often used are 5 x 4 ins, $4\frac{3}{4}$ x $6\frac{1}{2}$ ins (half plate), and 10 x 8 ins. The astronomer may also find the $3\frac{1}{4}$ x $3\frac{1}{4}$ in size useful.

(d) Sheet film (sometimes called cut film or flat film), available in the same size as glass plates.

Of the above, glass plates are the most expensive and, for amateur purposes, the least widely used. The advantages of plates are their flatness and dimensional stability: sheet film can easily curl if not held properly, although most plate holders can be modified to hold sheet film as flat as will be required for purposes other than astrography. In professional work, however, dimensional stability is often of paramount importance since the image positions will be measured accurately, possibly over a span of many decades, and the glass plate will reign supreme in such work for years to come in the absence of an alternative flat and stable film base.

The Kodak Spectroscopic series of plates are almost exclusively used in professional astronomical photography. These plates have specific colour sensitivity ranges and are thus most suitable for photometry and the examination of features in various wavelength ranges. Some emulsions are manufactured so as to have little reciprocity failure, and are thus suitable for recording faint sources. The cost of these plates, however, generally prohibits their use by amateurs.

The spectral sensitisation of the emulsion is of great significance when a selection is to be made for a specific purpose. An unsensitised emulsion is responsive to blue and ultra-violet light only, an orthochromatic one is sensitive to green light as well, and a panchromatic emulsion is sensitive over most of the visible spectrum in addition to the ultra-violet. Emulsions with infra-red sensitivity in addition to panchromatic sensitivity are also available.

An exhausive list of the properties and availabilities of emulsions suitable for astronomical work is not feasible, but the following is offered as a guide. Spectroscopic emulsions, with very low reciprocity failure, are not generally available in the UK and so are not included. Emulsions are listed in approximate order of normal ('instantaneous') speed.

(a) 35-mm films:
   Kodak 2475 Recording Film
   Kodak Tri-X
   Ilford H.P.4
   Kodak High Speed Ektachrome (colour reversal)

Kodak Plus-X Pan
Ilford F.P.4
Kodak Infrared Film
Kodachrome 64 (colour reversal)
Ilford Pan F
Kodak Panatomic-X
Kodachrome 25 (colour reversal)

All the above emulsions are panchromatic, with the obvious exception of the colour films.

(b) Roll films:
Kodak Royal-X Pan
Kodak Tri-X
Ilford H.P.4
Ilford Selochrome
Kodak Verichrome Pan
Ilford F.P.4
Kodak Panatonic-X

All of the above emulsions are panchromatic.

(c) Sheet films:
Kodak Royal-X Pan
Kodak Panchro-Royal
Ilford H.P.4
Kodak Tri-X Pan
Kodak Tri-X Ortho
Kodak Plus-X Pan
Ilford F.P.4
Ilford Commercial Ortho
Kodak Commercial Ortho

Except for the three orthochromatic emulsions, the above are panchromatic.

The characteristics of fast and slow emulsions are, in summary, as follows.

Fast films are more sensitive, permitting shorter exposures with minimum loss of definition through atmospheric turbulence; they fog more rapidly, produce lower contrast, and have a coarser grain which may limit possible degree of enlargement.

Slow films are less sensitive and require longer exposures. They run a lower risk of fogging, have higher contrast, with better consequent discernability of detail, and have a finer grain allowing greater enlargement without loss of detail.

FIGURE 170

Sensitivity curves for different emulsions compared with the average eye
(*reproduced by courtesy of Kodak Ltd*)

Processing of emulsions is an art in itself, and largely empirical. Plates and sheet films are processed in dishes which should be considerably larger than the plate or film itself, and contain an adequate depth of the solution. Roll and miniature films are processed in spiral holders in small tanks; once the film is loaded, the rest of the process may be carried out in the light. Very rarely is it possible to inspect a film with a safelight, so all processing should be done by the time-temperature method. Particularly when developing, continuous agitation of the dish or tank is advisable.

All solutions should be at the same temperature. The penalty for ignoring this rule is reticulation, which is often mistaken for graininess. Although raising the processing temperature accelerates development, the development becomes less controllable. A temperature of 20°C (68°F) is a good standard.

Fine-grain developers suffer from the disadvantage that they reduce the speed of the emulsion and increase the effective exposure necessary to give a satisfactory density. Medium fine-grain developers, with a high sodium sulphite content and less alkaline than the normal fast hydroquinone types of developer, are generally preferable except where the highest speed possible is required, and if much subsequent enlargement is envisaged they are essential. The detailed technique of development depends on the type of work being undertaken, and there is always considerable room for experimentation. The development times recommended by manufacturers are generally intended to give the most pleasing pictorial results, with a minimum of fog and fine grain. It is usually possible to use longer times with increased film speed resulting from the consequent full development, and this is a normal technique when the aim is to achieve the faintest limiting magnitude on a photograph. Kodak D-76 or, for high contrast, D-19, are two widely-used developers.

Polaroid film is now quite popular, and offers the astronomical photographer some advantages. The process is physically slightly different from conventional photography in that printing paper is squeezed into contact with the exposed negative material and the chemicals, and the image is transferred onto the paper by chemical diffusion. As only a thin emulsion layer is needed to produce a normally dense image on paper, a markedly slower, less grainy emulsion can be used. Polaroid film is widely available with a speed rating of 3000 ASA, and the process takes only a few seconds. The technique is comparatively new and cannot yet yield the same high quality that conventional photography has achieved, but it offers many interesting possibilities.

It is not necessary to use a Polaroid camera for this work; indeed, these have such small apertures and large focal ratios that they are most unsuitable. A device for holding the film and carrying out the processing can be bought or made.

20.9 Focusing and squaring-on

The surface of the emulsion must lie accurately at the photographic focus of the objective, and at the same time be accurately perpendicular to the optical axis. Insufficient attention given to these adjustments is one of the common causes of disappointing results.

What the correct focus is depends upon the equipment, and can be determined by various methods. A reflector being perfectly achromatic,* the focal plane is in the same position for all colours, and the camera can

*Though unequally efficient in different wavelengths; see section 6.

can be focused by either visual or photographic methods. This is virtually true, also, of a photovisual refractor. The correct photographic focus of a visually corrected objective (see Figure 21) can also be determined visually, assuming that a panchromatic emulsion is being employed. However, the out-of-focus blue and violet light, which contributes little to the visual image but is recorded strongly by the photographic emulsion, must be eliminated by interposing a yellow (minus-blue) filter, such as the Wratten No. 8.

20.9.1 Visual determination of focus: (*a*) *Reflector or photovisual.* Traditional methods of focusing the image, whether prime-focus or amplified, involved observing the image through the back of a focusing screen loaded into the camera and making adjustments until the surface of the screen, and the image, appeared sharp simultaneously. This is not a particularly sensitive method unless the depth of focus of the telescope image is large, while the general adoption of 35-mm film mitigates against the use of a focusing screen.

If, for any reason, a plate camera and screen are being used, the latter should consist of a piece of clear glass just frosted with fine emery to give it no more than an opalescent appearance. The aim is to have sufficient surface pits to allow a magnifier to focus on the plane of the emulsion, without obscuring the astronomical image being viewed. An ocular of 9 to 12 mm focal length will be found suitable for use with ordinary objectives. With telescopes working at pre-amplified focal ratios, or with Cassegrains, a focusing eyepiece of longer focal length, say 25 mm, will give a cleaner image.

The use of 35-mm format cameras has led to alternative focusing methods being employed. For lunar, planetary, and high-resolution work generally, the most convenient solution is a special camera unit incorporating its own focusing system. Rackham (B. 14.34, 35) and Dall (B. 14.8) have described their own variations: Rackham uses a HP ocular fixed into the back of the camera, focused through a hole punched through the film leader, with an orthodox Newtonian; Dall's design, for use with a Cassegrain, incorporates a diagonal and eyepiece unit which is withdrawn from the light path before the exposure is made. Both systems have proved satisfactory.

Another method, commonly employed when working with short and medium focal ratios, is so-called 'knife-edge' focusing. A bevelled metal straight-edge is laid across the film guides so that its sharp edge is in the plane to be occupied by the emulsion. The image of a star is brought to the knife edge so formed, and the focus is adjusted until the fully-illuminated mirror, viewed with the eye placed just behind the knife-edge, darkens evenly all over as the star image traverses the edge. If the mirror begins to darken at the side opposite the knife edge, the latter is outside the focal position; and vice-versa. B. 14.10 describes a convenient, detachable knife-edge focusing device.

(*b*) *Visually-corrected objective.* Any of the above procedures can be used, with the addition of the appropriate yellow filter. Standard gelatin photographic filters are too thin to have any noticeable effect on the image quality, but, in the interest of economy, any filter should be as small as possible, which means that it should be placed in or near the narrowest part of the light beam. If an ocular is used to enlarge the image, the filter should be placed near the exit pupil (not near the primary image formed by the objective, when any marks on its surface would be imaged near the plane of the emulsion); if prime-focus photography is being undertaken, it can be positioned a few centimetres in front of the film. If a filter is being incorporated permanently in a photographic lens, the ideal position is next to the aperture stop.

20.9.2 Photographic determination of focus: (*a*) *Star trails.* Expose five trails of a mag 2 star situated near the equator and the meridian. These trails should be distributed around the visually determined focus; in the case of a visually corrected objective a yellow filter must be used (see section 20.9). It is essential that the adjustable element (lens or film-holder of a camera, drawtube of a telescope) should be fitted with a rigid and accurately divided scale, so that its position corresponding to any particular trail can be repeated as required. The trails can either be spaced diametrically across the field (the suitable exposure time depending upon the image-scale and the Dec of the star; the first trail, on the *f* side of the field, should be given double the exposure of the others for identification purposes), or, preferably—since it keeps the trails more nearly central in the field—the instrument can be raised or lowered in Dec and moved back in RA between each exposure; identification of the trails in this case is facilitated if each is exposed for a constant interval longer than its predecessor. The two procedures would produce plates represented schematically below:

The trails are examined on the negative with a HP magnifying glass. If one trail is narrowest, and that on either side of it is equally wide, then it was made at the focus. If one trail is narrowest, but the trails on either side of it are unequally wide, then the focus lies between that trail and the narrower of its neighbours but near the former. If two adjacent trails are equally narrow the focus lies between them. If there is a progressive narrowing of the trails from one end of the series to the other, the focus may lie beyond the series of film positions used, and a further series of exposures must be made. If the objective-film distance was varied between exposures by more than about 0·25 mm (as certainly it would have been unless the position of the focus was already known fairly precisely) a second series of exposures must be made, at about this value of Δf, centred upon the best position as shown by the trails of the preliminary series.

(b) *Star images.* The general procedure is the same as in (a), except that the images are not trailed. Equal exposures are given, and the telescope readjusted in Dec between each exposure; for identification purposes, this displacement is doubled between the first and second exposures. The weakness of the method is that any errors of guiding will enlarge the images and thus confuse the issue. On the other hand, if the instrument is to be used for photography at all it ought to be safe to assume that it has an efficient drive and slow motions—otherwise its unsuitability for the job renders the whole procedure of accurately determining the focus rather beside the point. The outstanding advantage of the method is the detailed information that the appearance of the images in different parts of the field provides regarding both focus and squaring-on. This is discussed fully in B. 14.15, to which the reader is referred for further treatment of the subject.

In the case of a portrait lens it will probably be found that there is a certain amount of latitude in the focusing, there being no material change of focus over a small range of movement of the plate or objective along the optical axis. In these circumstances it is best to keep the distance objective-film to a minimum, but if this is exaggerated it will be found that the focus at the centre of the field becomes inferior to that in an annular zone surrounding the central region.

20.9.3 Hartmann method: An application of the principle of the Hartmann test provides a simple photographic method of determining the focus of a lens. The objective is covered by a mask in which two circular holes are punched; these should be equidistant from the centre and should lie on the same diameter; their diameters should not be less than about 1 mm. Two photographs of a mag 1 star are then taken, one well inside the focus and the other outside. Then if s_1 and s_2 are the separations of the two images in the two photographs, and f_1 and f_2 the corresponding lens-plate distances, the focal length is given by

$$F = f_1 + \frac{s_1}{s_1 + s_2}(f_2 - f_1)$$

20.10 Photography at the telescopic focus; image amplification; exposure

The obvious way of employing a telescope photographically is to position the emulsion in the focal plane of the objective: the prime focus. In the case of older British-made instruments, which had a $1\frac{1}{4}$-inch (31·75-mm) screw-in eyepiece fitting, and modern American telescopes, which have a push-in fitting of the same diameter, a coupling unit to link the camera to the drawtube can be bought or made. However, many modern British and continental instruments use a 24·5-mm push-in eyepiece fitting, giving a seriously restricted field across a 35-mm frame. To overcome this, it will be necessary to construct a new focusing unit. In the interests of gaining the largest possible fully illuminated field, the camera in a Newtonian system should be situated as close as possible to the side of the tube, and hence to the diagonal (see section 9.3).

Prime-focus photography is usually aimed at the recording of faint objects. Hence it involves relatively long exposures, and accurate guiding becomes of primary importance (section 20.12).

Without amplification images at the prime focus are too small for effective examination unless telescopes of much larger F than are normally found in amateur hands are used; the primary image of the Moon, for example, does not attain to a diameter of 50 mm until F reaches 5·5 m. The image can either be amplified before it falls on the plate, or an enlargement can be made from the unamplified negative later. Generally the former is preferable. If the required detail is below the limit of resolution of the emulsion no amount of subsequent enlargement from the negative will reveal it: such a procedure is exactly analogous to trying to reach detail or separate points which are beyond the resolving power of an objective by increasing the magnification.

Taking the resolving power of the normal eye as $2'$ to $3'$ (section 24.5), which approximately equals 0·25 mm at a distance of 25 cm, and taking the minimum photographic 'point' image as 0·02 mm (section 20.7), an enlargement of about 12 times is the maximum feasible; and since 20μ is an optimistic figure for the photographic resolution, it may be expected to be smaller. Thus at a focal length of 2·4 m the image of Jupiter at opposition is 0·5 mm in diameter. All detail smaller than 1/50 of the diameter of the disc will be submerged by emulsion diffusion. For subsequent enlargement to reveal smaller detail than this, the image must be amplified before it reaches the emulsion. As stated in section 20.7, effective focal ratios of between 50 and 200 have been employed by lunar and planetary photographers. Taking $f/80$ as an average figure, the following diameters (in millimetres) will represent the various solar system objects using instruments of different D:

D (cm)	Moon	Mercury	Venus	Mars	Jupiter	Saturn
10	68-78	·18-0·50	·38-2·5	·14-0·97	1·2-1·9	·57-0·80
15	103-117	·27-0·75	·58-3·7	·20-1·5	1·8-2·9	·86-1·2
20	137-156	·36-1·0	·77-5·0	·27-1·9	2·4-3·9	1·10-1·6
30	205-234	·55-1·5	1·20-7·4	·41-2·9	3·5-5·8	1·70-2·4
40	274-312	·73-2·0	1·50-9·9	·54-3·9	4·7-7·7	2·30-3·2
50	342-390	·91-2·5	1·90-12·4	·68-4·9	5·9-9·7	2·90-4·0

In the case of Jupiter, which is, after the Moon, the most popular subject for planetary photographers, an easily remembered rule is that the image diameter in an $f/70$ system will be about $D/70$..

In theory, at least, the exposure to photograph a point image of given brightness is a function of D^2 only (neglecting reciprocity failure); in practice, since the image is never quite point-like, the exposure is weakly proportional to F as well. The brightness of extended images, as explained in section 20.6, is inversely proportional to the focal ratio of the system, regardless of D. The following exposures, using medium-speed FP4 or Plus-X film, have been found appropriate for $f/80$ instruments:

Moon	0·2-2 sec (according to phase and distance of photographed region from terminator)
Mercury	1 sec
Venus	0·1 sec
Mars	0·5 sec
Jupiter	1 sec
Saturn	4 sec

See, further, the article by Dall in B. 14.8.

The amplifying lens to be used, assuming that a sufficiently large effective focal ratio cannot be obtained by using a Cassegrain system, requires some consideration. A Barlow lens is satisfactory for low amplification (x2 to x3), and, being placed inside the focal point of the primary, involves a shorter extension than would a positive eyepiece. However, in cases where amplifications of x5 to x10 are required, a HP ocular of the orthoscopic type may prove more satisfactory, its aberrations having been corrected for very unequal conjugates. The amplification, A, afforded by an ocular of focal length f used at a distance d from the final image plane, can be derived from

$$A = \frac{d}{f} - 1.$$

20.11 Aberrational tolerance of a photographic reflector

We have seen (section 20.7) that the diameter of a stellar image due to photographic diffusion varies between wide limits. For a well-exposed image 60μ may be taken as a rough figure, and aberrations may be disregarded so long as they do not enlarge the image beyond about this size. That is to say, coma and astigmatism can be ignored if their combined effect is to produce a star image whose longest dimension ξ is less than 60μ. Consider the case of an $f/5$ 30-cm reflector. From equation (a), p. 29, the plate-scale is $7\cdot4\mu$ per $1''$. Hence the maximum tolerable value of ξ is $8''1$. From Figure 16 it can be seen that $\xi = 8''1$ occurs at a distance of about $17'$ from the axis. Therefore the usable field, limited by coma and astigmatism, is about $34'$ in diameter; and correspondingly larger or smaller for over- or under-exposed images respectively.

A grossly over-exposed image, however, even though the aberration may be masked by the photographic diffusion, cannot be accurately measured. A fainter or less exposed star, lying outside the region of tolerable definition, will be asymmetrically comatic, and similarly not susceptible of accurate measurement. A still fainter or more under-exposed star may have only the brighter part of the image recorded, in which case its position can be measured with reasonable precision. Finally, a star which with a given exposure would just be recorded at the centre of the plate will be lost altogether towards the edge of the field, owing to the dilution of its light throughout the enlarged aberrational image.

Similarly with other focal ratios and linear dimensions—see Table and Figure 171 below. It should be re-emphasised that these figures are necessarily very approximate, owing to the assumptions involved in their derivation. They are indicative of the order of magnitude, however, and furthermore show very clearly how the reduction of F with D constant, or the increasing of D with F constant, results in the rapid reduction of the size of the usable field.

Diameters of fields ('arc) limited to photographic tolerance by coma and astigmatism.

		D (cm)			
		15	22·5	30	37·5
f/ratio	$f/3$	41	28	21	17
	$f/5$	63	44	34	27
	$f/7$	82	58	44	36
	$f/9$	110	80	64	52

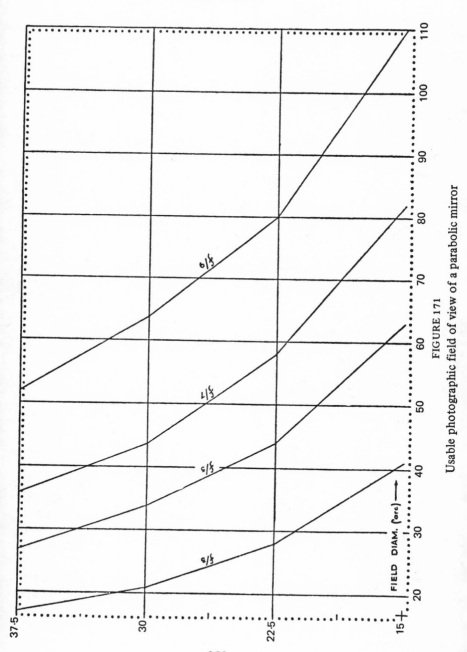

FIGURE 171

Usable photographic field of view of a parabolic mirror

20.12 Guiding

Given a bright enough object and a fast enough lens, exposures may be so short that the images are not blurred by diurnal motion even when the camera is stationary. But in the greater part of astronomical photography the camera must follow the object in its diurnal motion. This necessitates an equatorial mounting with an efficient clock drive and serviceable slow motions in both directions. Extraordinary feats have indeed been achieved from time to time in the way of guiding by hand (i.e. without a mechanical drive), exposures of half an hour or more having been made; but one is left wondering why such indubitable feats—like that of the man who climbed to the top of the Empire State Building twice a day for seventeen years—were thought worth performing.

Accurate guiding—which demands practice and efficiently operating equipment—is one of the first conditions of successful astronomical photography. If the clock, the setting of the polar axis, or the slow motions are individually or collectively faulty, guiding becomes very tiring and the resulting negatives correspondingly unsatisfactory. Comfort while guiding is much increased by the use of a diagonal; it is also advisable, before starting on a long exposure, to try out the limits of the observing position to see that they are feasible.

Even with a good clock drive, displacements of the image on the film will occur during the exposure. The commonest causes are:

(a) residual errors in the drive;
(b) residual error in the polar axis setting;
(c) atmospheric refraction variable with altitude;
(d) atmospheric refraction variable with temperature, humidity, and pressure at a given altitude.

Since the scale of the focal-plane image varies with F, the focal length of the guide telescope should be similar to that of the photographic instrument itself; ideally it would be longer, but, except in the case of cameras of moderate size, this may not be practical. A great deal of deep-sky photography is performed with Newtonians of $f/6$ to $f/8$, and in such cases it is not always possible to incorporate an instrument of even commensurate F on the same mounting. In these cases, a HP ocular must be used to magnify the image drift. The aperture of the guide telescope is important, because a suitably bright star is not always conveniently placed near the field being photographed. Generally speaking, difficulty in finding suitable stars on which to guide begins to be felt if the aperture falls below about half that of the main instrument, but in most instances a telescope of 75 mm to 10 cm D will be found suitable. Provision must be made for offsetting the guide telescope by a few degrees, so as to bring a suitable star to the centre of the field.

In all forms of guiding using a separate telescope, it has to be assumed that the optical axes of the two instruments remain in a fixed orientation to each other during the exposure; if this is not so, the field being photographed will suffer displacement despite the fact that the photographer painstakingly keeps his guide star on the crosswires. Absolute rigidity of linkage between the two tubes, and (more difficult to achieve with reflecting systems) immovability of the two sets of optical components, are vital to success. Two refracting telescopes form the most satisfactory combination; two reflectors, the least. Nevertheless, excellent deep-sky photographs have been achieved by, for example, guiding a 30-cm Newtonian with a 15-cm Newtonian coupled to the main tube.

The commonest form of guiding employs cross-webs in a positive ocular, which are made to bisect the slightly out-of-focus image of a fairly bright star (if possible, not fainter than about half the threshold magnitude of the instrument). To facilitate quick correction of runs-off, the ocular should be oriented so that one web is coincident with the hour circle through the star, and the other with the Dec circle; this is done by trailing the star with the clock drive disconnected.

A rather simpler arrangement to use, being more sensitive to run-off, is a photographically reduced circle located in the plane of the ocular and illuminated. The image of the guide star is enlarged until it just does not fill the circle, and placed at the centre of the circle. The narrow ring of dark sky separating the image and the ring, which should be of equal width in all position angles, provides a very sensitive criterion for guiding purposes. It is essential that each run-off should be corrected with all possible speed, as well as being halted while still small. A device using illuminated needle-points to bracket the image of a faint guide star was described in B.14.13.

When photographing an object with proper motion, the object itself must, of course, be the guide object. In the case of star-like objects, such as asteroids, this presents no difficulty apart from the additional manual operation of the slow motions and the possible faintness of the image. But comets, unless they have a sharply defined nucleus, present a different problem, methods of meeting which are described in *O.A.A.*, section 16.3.

A third method of guiding involves mounting an ocular to one side of the film frame, and following on a star just beyond the area being photographed. The separate guiding telescope, and the possibility of independent flexure, are eliminated, but at the cost of having to construct a special camera. The choice of guide stars is also lessened, and some difficulty may be experienced in obtaining coverage over a sufficiently wide field. Off-frame guiding may be accomplished by the normal use of RA and Dec slow-motions, but in large instruments it is normal practice to make the adjustments on the film-holder itself. See, further, B. 14.6.

The construction of an automatic guiding system, in which sensors hold the guide star in a fixed position, is described in B. 14.7.

20.13 Photographic seeing

Exposure of the plate when the atmosphere is prohibitively turbulent is one of the commoner causes of disappointment in astronomical photography. Where exposures can be short—the Sun, for example, and under certain circumstances the Moon—it is not only possible but essential to watch for and utilise the fleeting moments of better than average seeing. The view system (whether reflex camera or separate telescope) should have the same magnification as the photographic, since even a small amount of turbulence will distort the magnified image. Even where exposures of some length are necessitated it is still essential to exercise some choice of when to expose the plate; many nights are totally unsuitable for photography; on others the seeing improves or deteriorates either spasmodically or continuously throughout the hours of darkness.

It should be borne in mind that photographic and visual seeing are to some extent different. Visually, turbulence may either confuse the image while introducing little movement, or it may cause bodily movement of a reasonably clearly defined image (see section 26.7), or both may be present. Visually, movement is much less troublesome than confusion; photographically they are equally undesirable.

20.14 Photographic optics

Although a standard astronomical telescope is the most suitable instrument for large-scale photography of very small regions of the sky, the large bulk of amateur astrophotography is aimed at covering a fairly large field of view, and, preferably, showing stars and other deep-sky objects to the faintest possible limiting magnitude. To be done well, the optical components used must have been designed exclusively for photographic purposes. The choices facing the astrophotographer can be bracketed into three sections: (a) using new or second-hand standard lenses; (b) using second-hand or surplus photographic lenses designed, probably, for aerial photography of the Earth's surface; (c) obtaining custom-made equipment (e.g. Schmidt camera).

The choice should be dictated, ideally, by the work to be done; in practice, because of the continuous consumption of material, financial considerations may well restrict the choice of format to 35-mm film rather than cut film or plates, and this, in turn, will have a strong influence on the choice of equipment.

20.14.1 Standard 35-mm lenses: The actual negative area on a 35-mm film measures 24 mm by 36 mm. The area of sky embraced by this format depends, of course, on the focal length of the lens being used, and the angular dimensions covered using various popular focal lengths is as follows:

F (mm)	Angular coverage (°)		Scale (mm/°)
'Fisheye'	Approx. all-sky		0·13
35	39	× 59	0·61
45	31	46	0·79
55	25	37½	0·96
90	15	23	1·6
135	10	15	2·4
200	6·9	10	3·5
300	4·6	6·9	5·2
400	3·4	5·2	7·0
500	2·8	4·1	8·7
1000	1·4	2·1	17·5

The general-purpose lenses, of F around 50 mm, commonly have a full aperture of between $f/1·8$ and $f2·8$. However, as indicated in section 20.5, these lenses suffer from severe vignetting away from the centre of the frame, and to obtain an approximately constant limiting magnitude from centre to extreme corner may require a reduction of aperture to about $f/4·5$ (linear lens aperture about 11 mm); such reduction will, in any case, markedly improve the quality of the images, since it is common knowledge that most lenses give optimum performance some 2 to 3 stops below their maximum opening. The lenses of longer F contain what is, in effect, a Barlow-type amplifier to give a longer effective focal length than the barrel would otherwise permit; these are termed telephoto lenses. Lenses of 90-mm and 135-mm focal length have been found particularly useful for star charting purposes, the latter just covering a rectangle of sky equal to 10° in Dec and 1h in RA at the equator.

20.14.2 Surplus lenses: Just after the war, a number of aircraft cameras and lenses became available, the best-known being the $f/2·5$ Aero-Ektar (in 178-mm and 305-mm focal lengths); the 127-mm $f/4$ Xpres; the 203-mm $f/2·9$ Pentac; and the 508-mm Aldis in apertures of $f/5·6$ and $f/6·3$. Wray 914-mm lenses ($f/4$ and $f/6·3$) have also been on the market.

Being designed for aerial photography with red or orange filters to remove blue atmospheric haze, these lenses are not particularly well-suited to photography with a panchromatic emulsion. Furthermore, being designed for use with large-format film, critical definition has sometimes been sacrificed in the interests of a wide, flat field. Despite these drawbacks, however, these lenses and others like them have given the amateur a chance to obtain large physical aperture at very reasonable cost. The Aldis, in particular, has given good results using plates or cut film; but it should be noted that the quality sometimes differs quite markedly from one sample to another.

Occasionally, so-called 'portrait' lenses may appear in sales of old photographic equipment. With focal lengths of perhaps 300 mm to 500 mm, and apertures of $f/5$ to $f/8$, these were designed for studio photography, using very large negatives, before efficient enlargers came into use. Although it is true that Barnard's superb photographs of the Milky Way (*circa* 1890) were taken with such a lens, most types suffer from severe spherical aberration and are practically useless for astronomical work unless severely stopped down.

Many old-fashioned lenses, even when capable of giving acceptable definition, suffer from field curvatures; B. 14.32 gives details for the construction of a simple field-flattening lens to be placed in front of the sensitive emulsion.

Although the $f/4$ Xpres, in particular, has been used successfully with 35-mm format, most surplus lenses, because of their bulk and design, are better adapted to large-format work from which big enlargements are not expected.

20.14.3 Specialised photographic systems: The development of reflecting and catadioptric photographic instruments, which has had an important impact on professional researches, has been scarcely noticed by the amateur. Only recently has a commercial Schmidt camera been introduced into the amateur market, while the various Schmidt-Cassegrain and Maksutov instruments are designed principally for visual use, optimum photographic performance having been sacrificed in the process.

The Schmidt camera would seem to offer the most interesting possibilities for the amateur. Its advantage over conventional photographic lenses includes aperture and light-efficiency. This leads to a corresponding reduction in the necessary exposure time—an important consideration if the climate is a predominantly cloudy one. Disadvantages include bulk, delicacy of adjustment, and the necessity of curving the film surface. Anyone interested in the possibilities of Schmidt photography could consider the lensless version for a first attempt (see section 9.12.2).

20.15 Dewcaps and shutters

Dewcaps are even more essential for camera than for telescope objectives, since they reduce extraneous light (therefore retarding fogging), as well as dew. Particularly is this hooding function important when, for example, the Moon is above the horizon or when, during total solar eclipse, sunlight is diffused through the nearby sky outside the totality zone. The functions of hooding and dew-protection are both served best by making the dewcap as deep as possible. The wide field of view of most astro-cameras renders the tubular type of telescope dewcap inadmissible, and it should instead take the form of a funnel whose apex angle is not much larger than the angular field of view. Maximum shielding is achieved by making the cross-sectional form similar in proportion to that of the film format. Card or other porous material is not really suitable unless coated with an effective sealant such as cellulose paint, when its poor heat conductivity becomes an advantage; thin-gauge light alloy is more durable, but less satisfactory as a dew inhibitor. The interior of the hood must be painted matt black.

Since it is not possible to shield a wide-angle camera objective from all possibility of dewing (since the front surface is always more or less exposed to the sky, and radiates away its heat), an artificial source of heat should be applied to the lens mount and hence, by conduction, to the lens itself. The most convenient way is to mount four small torch bulbs, wrapped in cooking foil, around the front rim of the lens, or even around the outside of the base of the dewcap if the latter is of metal construction. It is only necessary to bring the glass to a fraction of a degree above the dewpoint, and under-running the bulbs by some 25 per cent will probably produce sufficient heat for the purpose.

The most satisfactory type of shutter depends upon the nature of the work undertaken. For long and approximately timed exposures (e.g. stellar) a hinged flap mounted on the dewcap and operated from the eye end of the telescope (by Bowden cable, electromagnet acting against a spring, or double spring) gives good results. The practice of removing by hand a cap covering the objective is not to be recommended; even when eased off as carefully as possible it is difficult to avoid vibration, and during an interval at the beginning and end of the exposure the telescope is not guided. For planetary and lunar photography, where the exposures though short, need to be precisely timed, a leaved shutter of the Compur type, situated near the enlarging eyepiece, has been found satisfactory. The focal plane shutters fitted to most modern cameras, if used to make exposures longer than a fraction of a second, involve considerable vibration. This is particularly severe if the shutter operation is linked to the mirror-lift mechanism of a single-lens reflex camera. A focal plane shutter is preferable for solar photography because of the concentrated heat near the exit pupil.

If planetary photography is being undertaken with a 35-mm camera body, it is of enormous benefit to have the shutter mechanism independent of the film winding action. This enables the film to be wound on, for example, only two sprocket holes between exposures instead of the eight corresponding to a full frame, thus allowing four times as many photographs to be accommodated within a given length of film.

Short and precisely timed exposures, as of the Sun, require a good mechanical shutter. Focal plane are to be preferred to compur shutters, and electrical to bulb or plunger operation.

It has already been mentioned (section 20.8) that most plateholders can be adapted to hold sheet film adequately flat. One positive method of ensuring the flatness of sheet film is to employ a partial vacuum to draw the film against a flat perforated surface forming the back of the film-holder. The use of a glass pressure plate in front of the film is not to be recommended.

20.16 Camera: construction and mounting

If a camera for use with sheet film or plates is being constructed, a metal tube is preferable to a wooden box on the grounds of rigidity and, probably, weight. It must, of course, be completely dust- and light-proof, and finished matt black internally. The squaring-on of the film holder in a home-made camera is a matter of importance, and may present some practical difficulty. The back of the camera, to which the runways for the dark slide are attached, should be fitted with a screw adjustment so that it can be tilted in two mutually perpendicular planes. Star images on a focusing screen, at opposite ends of two mutually perpendicular diameters of the field, are then examined with a high-power magnifying glass; when they have been made as uniform as possible (great precision is difficult owing to the comatic and astigmatic distortion) final adjustment is made from stellar exposures (see section 20.9.2).

An alternative method, which is easier to operate, employs the curtain-ring dodge already described in section 13.15 when dealing with the squaring-on of anastigmats. Load the open film holder with a sheet of plane glass and observe it through a bright ring from behind, the cover being in place over the objective. Then adjust the setting of the dark slide runners to make the reflections of the ring at the glass sheet and at the objective concentric.

Complete rigidity between objective and film, and between camera and guide telescope, is essential, and should be achieved with the minimum weight consistent with this.

Cameras which are small compared with the telescope are conveniently mounted on the telescope tube in a continuation of the Dec axis. They may be bolted to a wooden base which is itself bolted to the tube cradle, or may be fixed to the tube by metal bands, screwed tight, and widely spaced wooden U-shaped distance pieces.

For larger cameras, the best position is probably the opposite end, from the telescope, of the Dec axis. This arrangement has several advantages: the weight of the camera is subtracted from, instead of being added to, the counterpoise, thus reducing the load on the bearings; and it ensures good clearance between the camera and telescope axes—if these are too close together the telescope dewcap may obtrude on the field of a short-focus, very wide-angle lens. If, however, the telescope is mounted under a dome it entails a very wide shutter opening.

Normal exposures of up to, say, 20 minutes, using standard 35-mm cameras and lenses, will not need any guiding at all provided the polar axis is aligned within the required limits and the drive is reasonably pretentious. A synchronous motor, working via a worm and wheel, will probably provide the motive power, but excellent results have been achieved using the simple 'Scotch' mounting described in section 14.13.1; B. 14.9 describes a 35-mm camera driven by a clockwork time-switch movement.

For photographic even more than for visual work the mounting must provide:

(a) Firmness and solidity; an iron pedestal or masonry base is infinitely preferable to a tripod, even with stretcher bars. Sources of vibration (see also section 14.26) include wind, traffic, the driving motor, and the operation of the slow motions.

(b) Accuracy of the setting of the polar axis, so as to eliminate as much slow-motion adjustment as possible and to reduce the rotation of the plate about the guide star.

(c) Efficient (instantly and smoothly acting) slow motions.

20.17 Colour and cine-photography

Colour films are now of quite sufficient speed to permit their use in astrophotography, though their use is usually limited to their pictorial value. A colour film has three photographic layers, with other layers acting as colour filters. The details vary from film to film, each having its own properties, which makes evaluation rather more tricky than in the case of monochromatic film. Colour reversal films, producing transparencies, are the most popular since the image is contained on a small and consequently cheap piece of film. Negative films require prints to be made on colour paper, which, apart from adding to the expense, introduces further problems of colour balance.

The processing and printing of some types of colour materials can be undertaken by the amateur; in the case of the negative materials the colour balance of the final print is often judged by eye, trial and error. A grey calibration patch on the film, negative or reversal, using the same exposure time as the sky photograph is invaluable in assessing the correct balance.

The colours reproduced by photography are on the whole accurate, though certain anomalies may be noted. In particular, it is found that features with emission line spectra can vary in colour. For example, the Orion nebula, M42, appears red in photographs, due to the presence of a strong line of the hydrogen spectrum which falls in a sensitive region of the red layer of the emulsion. The eye is more responsive to the green lines of doubly ionised oxygen, which fall in a region of relatively poor emulsion sensitivity. Anomalous response of colour emulsions to light of low intensity will also arise from the tendency of the three layers to have different reciprocity failure characteristics.

20.18 Three common faults

Among the commonest causes of faulty and disappointing photographs (they have been mentioned already, but bear repetition) are:

(a) Inefficient slow motions. A faulty drive may be largely offset by smooth and efficient slow motions, free from backlash and lag; but if the slow motions themselves are inefficient, decent exposures are almost impossible to obtain.

(b) Insufficiently precise adjustment of the plane of the emulsion to the actinic focal plane.

(c) The exposing of films when the atmosphere is too turbulent for photography.

SECTION 21

THE MEASUREMENT OF RADIATION INTENSITIES

21.1 Introduction

The amateur observer wishing to undertake a programme of photometric work should consider very carefully what it is that he hopes to achieve. If the object is the pleasure and interest involved in constructing a measuring instrument, it matters little which type of photometer is built; a great deal of care and attention to detail can be lavished on any project. However, if the observer hopes to make a real contribution to science— and both visual and photoelectric photometry undertaken by amateurs has, and continues to do, just that—then two related factors must be considered: (a) the type of instrument to be built, assuming that the programme will not be based entirely on visual estimation; and (b) the amount of time available, and the normal observing conditions in which work will have to be undertaken. The observer who has access to a site with a large number of clear, dark skies each year can, obviously, undertake more ambitious projects than are open to the person living in a large town in a typically cloudy region. However, if the latter individual is prepared to build a more sophisticated photometer, and to use it with great care, there is no reason why his contribution should be of any less importance that that of the observer in a good site. It will, nevertheless, be a contribution of a different nature—being one dependent upon precision of measurement rather than a large amount of telescope time.

A complete novice, after a little instruction, can probably estimate the brightness of a naked-eye star to within about 0·5 mag. With practice, particularly if using a telescope, this can be improved to about 0·1 mag under favourable conditions; and, once this accuracy has been achieved, the monitoring of long-period, large amplitude variable stars is both possible and useful. To improve upon an accuracy of about 0·1 mag requires some type of photometer; but it will be worthwhile, before proceeding to a description of photometric methods, to review the frequently unrecognised difficulties inherent in the theoretically simple operation of determining precisely the apparent magnitude of a star. The apparent brightness of the image will be to some extent modified by, among others, the following factors:

(*a*) Transmission factor and wavelength sensitivity of the telescope: a function of time in the case of a reflector, as well as differing from one instrument to another.

(*b*) Wavelength sensitivity of the eye, photocell, or photographic plate; that of the eye, at least, varies widely from individual to individual, as well as from time to time with a given individual.

(*c*) Transmission factor and selective absorption of the atmosphere: both vary with place and time, depending on local meteorology, geographical position and altitude of the observing station, and the altitude of the star.

(*d*) Colour of the star.

(*e*) To these must be added, if we are concerned with magnitude as an index of the actual conditions at the star, transmission factor of interstellar space along the light-path.

In view of this host of disturbing factors, mostly uncontrollable, it is hardly surprising that many published magnitudes are infected with errors whose size there is no means of estimating.

Those types of photometer that are applicable to astronomical work can be broadly classified under five heads: extinction, equalisation, standard step, photographic, and energy conversion.

21.2 Extinction photometers

The principle is to reduce the apparent brightness of the source until it falls below the threshold of visibility. The amount of the adjustment of the extinguishing mechanism necessary to perform this is a measure of the star's brightness.

Disadvantages:

(*a*) The visibility of the star does not remain steady and continuous right down to the extinction point; close to the limit of vision it will be glimpsed intermittently. The precise point of extinction is therefore difficult to determine. In an effort to get round this difficulty some observers make their measurements from the point of 'minimum steady visibility', but it is doubtful if this can be much more consistently recognised than the point of total extinction. Alternatively a device such as a stepped wedge may be used, so that the star's magnitude may be certainly assigned to a magnitude range between two definite values, instead of being of an unknown degree of uncertainty.

(b) The value of the extinction reading varies with the seeing, the clarity of the atmosphere, the ZD of the star, the brightness or otherwise of the background, and the personal equation of the observer. With all these variables involved, series of observations by a single observer cannot be assumed to be truly consistent, nor those by different observers to be strictly comparable.

(c) Results obtained by different extinction methods are not comparable without the introduction of an adjusting factor, owing to different rates of decrease of background brightness and image brightness: the wedge extinguishes the ground before the stellar image, whereas with reduction of aperture the field darkens relatively more slowly.

(d) If the image does not always fall on the same region of the retina, results obtained at different times will not be comparable; and keeping the eye directed precisely at an image which is close to the limit of vision is nearly impossible. More uniform results are obtained if the eye is directed at another, non-vanishing, point in the field, always at the same distance and position angle from the image being measured. This procedure keeps the error well below 0·1 mag, whereas discrepancies as large as 0·5 mags can be encountered when attempts are made to keep the vanishing image on the macula.

(e) Within a certain range, the extinction point is affected by the magnification.

(f) Extinction achieved by reducing the aperture produces greatly enlarged spurious discs with the small apertures at which the brighter stars disappear,* which again makes the extinction point difficult to determine. With a reflector, this may be overcome by substituting an uncoated flat for the ordinary diagonal; or, with reflectors or refractors, a known preliminary reduction of brightness can be obtained by polarisation or other means.

(g) Eye fatigue, brought on by the length of time the adjustment requires, is probably not a factor of great importance (provided the observations are made with reasonable expedition), and in any case applies to some extent to all types of visual photometer.

For all these reasons extinction methods have now been discarded by all serious workers in photometry, and are accordingly dealt with only summarily here.

---

*Sirius, just before the extinction point is reached, resembles a small planet, so exaggerated is its spurious disc.

21.2.1 Optical wedge: An optically worked, tapering wedge of glass, neutrally and uniformly tinted, and mounted against a reversed wedge of plain glass so that the outer surfaces of the combination are parallel; or a plane parallel glass slip upon which a metallic film of graded thickness has been deposited; or a graded gelatin filter (colloidal carbon or granular silver) of the type supplied by such firms as Ilford, Kodak, or Dufay-Chromex.

The light from a star, transmitted by the wedge (which is mounted against the ocular), is then reduced by an amount which is a function of the distance of its image from the end of the wedge; this distance can be read from a scale. When the wedge constant (the magnitude difference of two stars whose scale readings at extinction differ by one division) has been determined, scale readings are directly convertible to magnitudes or differences of magnitude.

21.2.2 Reduction of aperture: An objective iris diaphragm operated from the eye end of the telescope by means of an endless screw is a convenient arrangement, since it allows an accurate reading of the scale to be made, and at the same time is quickly accomplished, thus reducing eye strain. In the case of bright stars, however, it will be found necessary to employ also a neutral wedge so that extinction can be obtained without too gross a reduction of D.

21.2.3 Deflection photometer: A N-S wire is placed in the focal plane of the eyepiece; about one-seventh of the way up the tube, a narrow-angle wedge of clear glass is arranged to occupy the f half of the field. With the telescope stationary, the star is observed as it transits the field. As it passes into the area covered by the wedge, a deviated and an undeviated image are seen, the latter steadily fading as more and more of the light is intercepted by the prism. The time taken for the star to drift from its transit across the wire to complete invisibility is noted, and compared with similar observations of one or more comparison stars. The relationship between magnitude drop and time interval, once determined for a star at a particular Dec, will be a function of cos δ for other stars. The design was first described by H. M. Parkhurst in B. 15.15.

21.2.4 Bright-field photometer: By artificially increasing the background illumination, a star can be rendered invisible through loss of contrast.

Secretan's design (Figure 172), in which the artificial illumination is focused on the entrance pupil of the telescope (i.e. in the plane of the OG) has the advantage of extremely even illumination across the field of view. By altering the aperture of the iris, a setting is reached at which the star disappears from sight, and the diaphragm reading is noted. A number

FIGURE 172

Bright-field extinction photometer. Field illumination
is usually varied by adjusting the diaphragm

of observations of the variable and comparison stars must be taken. See,
further, B. 15.23.

21.3 Equalisation photometers

The principle involved here is the reduction of the intensity of a star
image until it is judged to be equal to that of another star (natural or
artificial) in the field. The magnitude difference of the two stars is then
2·5 x the log of the ratio by which the brightness has been decreased. If
the magnitude difference is great, a preliminary known reduction to near
equalisation when the brighter is observed—by reduction of aperture,
polarisation, rotating sector, etc—may be required.

Advantages:

(*a*) Providing that the two stars are near one another in the sky, it
eliminates the following factors which modify the results of extinction
photometry: seeing, atmospheric clarity, ZD, background brightness.

(*b*) With training, the estimation of the equality or otherwise of the
brightness of two stellar images is a more accurate operation than the
judgment of the extinction point.

(*c*) The eye is naturally more sensitive to the equality or otherwise of
two point images than to the exact difference between two unequally
bright images. An equalising photometer thus has the advantage over
the usual visual methods of estimating the magnitudes of variables.

Disadvantages:

If there is a natural comparison available, well and good. But more
often than not there is no star of suitable magnitude near enough to be
included in the same field. Some designs of photometer (e.g. Danjon's)
get round this difficulty. Others rely on an artificial star, which, however,
suffers from the following drawbacks:

367

(*i*) It never exactly resembles the natural star in appearance, and is often—even if filters are used—of dissimilar colour; all of which makes accurate and consistent comparisons difficult.

(*ii*) The consistency of a series of observations depends upon the invariability of the standard source providing the artificial star image.

(*iii*) The comparison of different observers' results is complicated by the fact that their standard sources cannot be directly compared.

21.3.1 Optical wedge: If the wedge is adjusted so as to cross half the field, occluding one but not the other of two stars in the field, it can be used as an equalisation photometer to reduce the intensity of the brighter image until it is judged to be equal to that of the fainter. Results so obtained are much to be preferred to those obtained by the extinction method, but a serious source of error is the greater darkness of the half of the field covered by the wedge.

Unfortunately, so conveniently placed a comparison star of known magnitude is not always available—though in the case of well-observed variables it very often is. Two ways round the difficulty suggest themselves: to feed the image of a more remote star into the field by reflection, or to employ an artificial star. The former is the method used by, e.g. Pickering's meridian photometer (section 21.3.6) and Danjon's cat's-eye photometer (section 21.3.7); the latter by the Zöllner and flicker photometers (sections 21.3.2, 21.3.9).

21.3.2 The Zöllner photometer: A typical layout is illustrated schematically in Figure 173. The image of the observed star is formed by the telescopic objective at S, the flange A fitting into the telescope drawtube. B is an unsilvered plane parallel plate which reflects two images of the artificial source into the field, one from each surface; that from the front surface, s, is used in making the comparisons, the other being out of focus. C is a low-candlepower lamp which illuminates a pinhole in the metal-foil sheet D. Pinholes of a range of different sizes (from, perhaps, 0·1 to 0·3 mm, though this depends upon the design of the instrument) should be available, so that the appearance of the natural star under different seeing conditions can be reproduced as closely as possible. This is a point of some importance, since dissimilarity in the appearance of the images makes the estimation of equality more difficult and is one of the commonest sources of inaccuracy with a Zöllner-type photometer. Variation of the appearance of the artificial image can also be obtained by varying the distance of the ground-glass screen, E, from the pinhole. The function of F, a light blue filter, is to reduce the yellowness of the artificial star. G is the optical wedge, operated by the milled head H. The associated scale should be graduated to at least 1·0 mm in divisions large enough for tenths to be estimated. J is the projection lens.

Reduction of the brightness of the artificial star can be accomplished in other ways. For instance, by means of an iris diaphragm situated in

FIGURE 173
The Zöllner photometer

the vicinity of J. Again, polarisation (see section 21.3.3) is frequently made use of. If the angles through which the analyser has to be rotated in order to equalise two stars with the artificial image are θ and θ', b and b' their respective brightnesses, and Δm their magnitude difference, then

$$\frac{b}{b'} = \frac{\cos^2 \theta}{\cos^2 \theta'}$$

and

$$\Delta m = 5(\log \cos \theta' - \log \cos \theta)$$

Zöllner photometers in general give considerably more accurate results than can ever be obtained by visual estimation; at least ±0·05 mag should be obtainable in the mean of a series of readings. Their chief weakness, which they share with most visual photometers, is their inability to cope effectively with strongly coloured stars, which present a situation in which personal equation becomes particularly rampageous.

21.3.3 Reduction by polarisation: Commonly employed in both equalisation and extinction photometry.

Light is a wave system in which the vibration is transverse—in a plane perpendicular to the direction of propagation. Within this plane the vibrations normally occur in all possible orientations. Such a wave system, in which the directions of vibration are distributed at random, is said to be unpolarised. If, conversely, the vibrations occur predominantly or solely in one direction, then the light is said to be partially or completely plane polarised.

If an unpolarised beam is passed through a filter of some polarising material (see section 21.3.4), the emergent beam will consist of waves vibrating in a single direction. The direction, in the filter, parallel to the direction of vibration of the transmitted light, is termed the polarising axis of the filter. If the vector a represents the amplitude of a vibration inclined at an angle θ to the polarising axis, then the component of a lying in the polarising axis is given by

$$a' = a \cos \theta$$

Also the intensity of a vibration is a function of the square of its amplitude. Hence, writing I_θ for the intensity of the transmitted radiation and I for the intensity when θ is zero,

$$I_\theta = I \cos^2 \theta$$

Hence the action of the polarising filter upon an incident beam of unpolarised light is not simply to transmit those vibrations whose direction

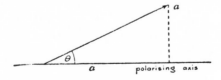

FIGURE 174
Intensity of light through crossed polaroid filters

is exactly parallel to the polarising axis, but also to transmit, of vibrations inclined to it, their components in the polarising axis.

If, now, a second filter, known as the analyser, is placed in the path of the polarised ray transmitted by the first, it will behave in an entirely analogous fashion, but now θ is the angle between the polarising axes of the two filters, and I is the intensity of the beam transmitted by the analyser when these axes are parallel. Whereas a single filter will transmit (neglecting absorption) about 50 per cent of an unpolarised incident beam irrespective of the orientation of its axis (the directions of vibration in the unpolarised beam being random), the second filter will transmit anything from 100 per cent of the polarised beam incident to it (when $\theta = 0°$) to 0 per cent (when $\theta = 90°$); see figure 175. In practice these theoretical figures are reduced by reflection at each surface of the filter, and also by absorption: the transmission factors of different Polaroid filters range from 25 to 42 per cent.

21.3.4 Polarising agents: (*a*) *Transmission polarisers.* Polaroid, a synthetic product, is a development of 'herapathite' (iodoquinine sulphate), discovered as long ago as 1852. It derives its polarising action from the linear arrangement of innumerable microscopic herapathite crystals embedded in a protective, stable, and transparent medium. It produces 90 per cent polarisation at 500 nm, is opaque to ultraviolet, and comparatively so to the shorter visible wavelengths; the infrared is transmitted unpolarised.

Various commercially produced plastic polarisers derive their action from the linear arrangement of large and complex molecule chains. An example is iodinised polyvinyl alcohol.

(*b*) *Double-refraction polarisers.* Many naturally occurring crystals (e.g. calcite) and some artificial materials (e.g. cellophane) have the property of refracting an incident ray in two directions, each of the refracted rays being completely polarised in mutually perpendicular planes. One ray arises from normal refraction, and is known as the ordinary ray. The other, the extraordinary ray, obeys neither of the laws of refraction; N_e, as measured by it, may be either greater or less than N_0, measured by the ordinary ray. The ordinary and extraordinary images may be distinguished by rotating the crystal, when the ordinary image will remain stationary while the extraordinary image will revolve round it. That the two rays are plane polarised with their axes of polarisation inclined to one another at an angle of 90° is proved by observing the two images through an analyser: when this is rotated the images will alternately appear and disappear, both never being fully visible or invisible simultaneously.

An ingenious use of the doubly refracting calcite crystal is the type of prism devised by Nicol in 1828. Two calcite crystals are cut to the shape illustrated in Figure 176, polished, and cemented together at their shaded faces with Canada balsam. The refractive index of the balsam is intermediate between N_0 and N_e, and by adjusting the angle between the balsam interface and the incident ray it is possible to eliminate one of the rays by internal reflection (its angle of incidence at the interface exceeding the critical angle) while the other is transmitted.

The advantage of the Nicol prism, which secures its use for all critical work, including photometry, resides in the fact that it is the only known agent whose polarisation and transmission are independent of wavelength,

FIGURE 175

Percentage transmission of two polarising filters
for different inclinations (θ) of their axes

i.e. its image is colourless. Its disadvantages are, firstly, cost, since sufficiently large calcite crystals are hard to come by, and secondly, for

FIGURE 176
The Nicol prism

one ray to be eliminated while the other is transmitted, the range of angles that the incident ray may make with the first face of the prism is limited, i.e. its field is small.

(c) *Polarisation by reflection.* This finds its application less in photometry than in solar work, where it is employed in solar eyepieces.* If a ray of light is directed at a plane glass surface, the angle of incidence i_p being such that

$$\tan i_p = N$$

then the reflected portion of the ray will be completely polarised in a plane perpendicular to the plane of incidence. This special value of i is known as Brewster's angle; for flint ($N=1 \cdot 615$) it is $58\frac{1}{4}°$, and for crown ($N=1 \cdot 515$) $56\frac{1}{2}°$. Polarisation is zero when $i=0°$ and $i=90°$, and partial between these limits and Brewster's angle. The refracted portion of the ray is never more than partially polarised.

Even matt surfaces partially reflect as glass, or mirror-like, surfaces, and hence the light by which they are seen is partially polarised; this fact is the basis of the use of polaroid anti-glare spectacles.

(d) *Polarisation by scattering.* Particles in suspension, whether in a liquid or in the atmosphere, both scatter and partially polarise the light passing among them, provided they are small enough. In the case of the atmosphere, maximum polarisation of the Sun's light occurs along a great circle 90° from the Sun.

*For detailed description, refer to *O.A.A.*, section 1.6.

(e) *Internal strain.* Glass is not normally doubly refracting, but under great strain, caused either by mechanical pressure or by too rapid annealing, it may become so.

(f) *Conditions in the source.* The greater part of sunlight and of light from most artificial sources is unpolarised. Electrical and magnetic fields in the source may, however, under certain conditions cause it to emit partially polarised light.

21.3.5 **The Pickering photometer:** A generic name given to a series of polarisation photometers developed by E. C. Pickering at Harvard (B. 15.20, 15.21). They share the characteristics of being restricted (like the simple wedge photometer) to the direct comparison of two stars which can be contained in the same field, and of relying on equalisation by means of a double-image prism and a rotary analyser. They differ for the most part as regards the relative positions of the components in the optical train. They all involve a considerable light loss (the brightness of the equalised images being only 30 per cent that of the primary image of the fainter star), and in an effort to circumvent this disadvantage one model incorporated a prism for reflecting the image of a bright comparison star into the field; its brightness was reductible by means of an adjustable cat's-eye diaphragm, the faint star then being viewed direct.

The general principle underlying all designs of the Pickering photometer is as follows. A doubly refracting prism, mounted near the telescopic focus, produces an ordinary and an extraordinary image of each star. By adjusting the prism's orientation and position along the axis, the ordinary image of either star can be brought close enough to the extraordinary image of the other for comparisons to be made conveniently. Either such pair of images is then equalised by rotating the analyser. If, then,

b, b' = brightness of the two stars,

$\Delta\theta = \theta - \theta'$, where θ = analyser reading when the two images are equal,

θ' = analyser reading when one image disappears,

the brightness of the two images of the first star are given by

$$b \sin^2 \Delta\theta$$
$$b \cos^2 \Delta\theta$$

and those of the second by

$$b' \sin^2 \Delta\theta$$
$$b' \cos^2 \Delta\theta$$

Of these, the second and third have been equalised,

$$\therefore \quad b \cos^2 \Delta\theta = b' \sin^2 \Delta\theta$$

$$\text{or} \qquad \frac{b}{b'} = \tan^2 \Delta\theta$$

Therefore, from equation (e), p. 10,

$$\Delta m = 5 \log \tan \Delta\theta$$

Equalisation can be produced by four positions of the analyser, all of which are used.

21.3.6 The meridian photometer: Also developed at Harvard by Pickering; it was with an instrument of this type that the Revised Harvard Photometry was compiled. Though one hesitates to say of any instrument that it is not for amateur use—now that horizontal Sun telescopes, spectrohelioscopes, Schmidt and Maksutov telescopes, and apparatus for long-wave solar study are all to be found in the hands of amateurs—it must be admitted that the meridian photometer comes very near to falling within this category.

It is a refinement on the types already described, in that for the sake of consistency the comparison star is neither an artificial star nor a casual neighbour, but in every case Polaris. Even this gain is not without its adverse side, however, for it means that the comparison star and stars in low Decs are subject to different atmospheric and meteorological modifications.

Light from Polaris and light from a star on the meridian (in which position all measurements are made) are fed into a horizontal tube, lying E and W, by two objectives; these are of equal aperture and focal length, and are inclined to one another at a small angle. The two images in the common focal plane of the objectives are then equalised as in the simple Pickering photometer. A more detailed description will be found in B. 15.22.

21.3.7 Danjon's cat's-eye photometer: Since the original description of this instrument is comparatively inaccessible (B. 15.5) it will be worthwhile treating it here in some detail, for it has outstanding advantages for the amateur. It is simple in construction and operation; it can be home-made quite cheaply by anyone used to handling tools; it introduces no false colour (cf. wedges); it employs natural stars for comparison (cf. Zöllner photometers), but these are not limited to the small area of sky covered by the telescopic field (cf. Pickering's photometer); the measures can be made quickly—as many as five a minute without difficulty; less trouble from ghosts is experienced than with polarisation photometers; and, finally, its results have been found to be of an accuracy comparable with those obtained by Stebbins with a photoelectric photometer.

The optical arrangements are extremely simple, and are illustrated in Figure 177(a). A and B are right-angled borosilicate prisms set for total internal reflection, B being mounted centrally in front of the telescope OG and A at the side of the tube. A is movable through a small arc about an axis in the plane of the diagram and perpendicular to the optical axis; B is similarly adjustable about an axis perpendicular to the diagram plane. By this means the image of a star several degrees from the axis of the OG can be reflected into the field; the controls are carried back to the eye end of the telescope. C is a cat's-eye diaphragm (Figure 177(b)) oriented so that the sides of its aperture are parallel to those of the prism faces. It is operated by remote control from the observing position through pinions engaging a ratchet on each section of the diaphragm, so that an equal and opposite movement is imparted to each. The length of the diagonal of the aperture is recorded by means of a scale mounted on one section of the diaphragm and an index on the other. The precision of the scale and the mounting and operation of the cat's-eye are determining factors in the success of the instrument. The scale is read by means of a

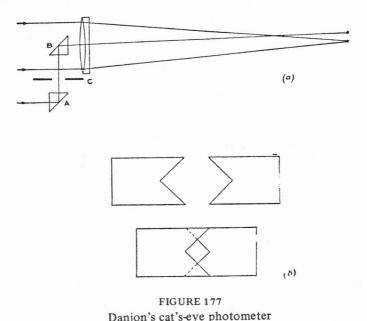

FIGURE 177
Danjon's cat's-eye photometer

small view telescope mounted on the tube near the eyepiece, being illuminated at will by a low-amp bulb and a pocket battery, and viewed (since it lies parallel to the telescopic axis) via a totally reflecting prism mounted in front of it.

The zero reading of the scale is determined by pointing the telescope towards a bright light and reducing the aperture of the diaphragm until the image of the prism is completely extinguished. Let the mean of a large number of such determinations be x_0. If, now, C = the magnitude difference of the two images of the same star for $(x-x_0) = 1$ scale unit, the mag difference of two stars whose images are equalised at a scale reading of x units is given by

$$\Delta m = C - 5 \log (x - x_0)$$

Three ranges of mag variation by means of the cat's-eye (i.e. three independent values of the constant C) may be obtained by reversing one or both of the prisms from the position of total internal reflection to that of external reflection at the hypotenuse face: C_{ii} when A and B are both reflecting internally (as in Figure 177); C_{ie} when the reflection is internal at A and external at B; C_{ee} when both A and B are reflecting externally.

These constants must be determined empirically with the greatest precision posible:

(*a*) *Two internal reflections.* A rotating sector of known angular aperture is set up in front of the instrument so that it occludes the OG but not prism A when the telescope is pointed at an artificial star some 50 yards away. From the known proportion of the open to the closed sectors the relative brightness of the two incident beams, and hence the mag difference of their respective images $\Delta m_{,}$, can be calculated. If x is the scale reading when the images are equalised,

$$C_{ii} = \Delta m + 5 \log (x - x_0)$$

(*b*) *One internal, one external, reflection.* Equalisation would now involve too great a reduction of the open/closed ratio of the sector for the accurate use of the latter alone. With both prisms set for internal reflection, reduce the used aperture of the OG by a diaphragm (whose area need not be known). Let x_1 be the scale reading at equalisation. Now reverse B and place the rotating sector (which can now have a relatively wide opening) in front of the OG diaphragm; let x_2 be the new equalisation reading. Then

$$C_{ie} = C_{ii} + \Delta m + 5 \log (x_2 - x_0) - 5 \log (x_1 - x_0)$$

(*c*) *Two external reflections.* The same differential method as in (*b*) is used, giving the value of $C_{ee} - C_{ie}$, whence

$$C_{ee} = C_{ie} + \Delta m + 5 \log (x_2 - x_0) - 5 \log (x_1 - x_0)$$

Danjon emphasises the danger of measures being invalidated by the deposit of dew on the exposed prism faces. To combat it he recommends that observations be made from inside an observatory rather than in the open air, that measures be taken to keep the temperature of the instrument slightly above that of its surroundings, and that the glass surfaces be gently wiped periodically; with borosilicate prisms he was unable to detect any systematic variation of C over long periods as a result of this practice.

21.3.8 Disc photometer: Also invented by Parkhurst (cf. 21.2.3). The image of the star being measured is defocused, and part of the expanded disc is isolated by a small aperture in the focal plane of the eyepiece. A similar, adjacent aperture is artificially illuminated. The brightness of the defocused disc is adjusted by racking the ocular in and out, until it matches that of the comparison disc. A similar reading is taken for one or more comparison stars, and the magnitude difference can be determined by measuring the movement of the ocular.

A self-luminous source, such as a betalight, could be used to illuminate the comparison disc. The device is confined to use on stars whose brightness is considerably above the threshold of the instrument.

21.3.9 Flicker photometer: Two further types of equalisation photometer must be mentioned briefly. The first, the flicker photometer, is a true equalisation photometer since it involves the adjustment of the brightness of one source until it is equal to that of another; this, however, is accomplished by a visual estimation not of equality but of the absence of 'flicker' in the images when viewed in rapid succession and in the same position.

The principle is illustrated in Figure 178. A is a circular mirror mounted centrally so as to revolve about the axis XY. Its appearance in plan is shown at A'. The convergent beam from the telescope objective is intercepted by this mirror near its edge. B is a fixed plane mirror at which the ray from an artificial star is projected via two Nicol prisms. Both rays—from the polariser and from the objective—are brought to a focus at the same point in the focal plane F. If, now, A is rotated, the images of the natural and the artificial star will be seen in rapid succession: the former when one of the mirror projections intercepts the cone from the objective, and the latter one-eighth of a revolution later, when mirror B is revealed. If the two images occupy the same position in the focal plane, they will be seen as a single stationary image, which, however, will flicker—i.e. fluctuate rapidly in brightness—so long as the images are of unequal

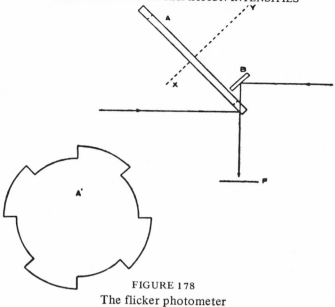

FIGURE 178

The flicker photometer

brightness. By adjusting the orientation of the analyser the flicker can be reduced and finally eliminated. At this point, to which the eye is extremely sensitive, the artificial star has been brought to equal brightness with the natural star.

The outstanding advantage offered by this type of photometer is that owing to the rapid alternation of the images, their colours, if these are different, will blend, and the task of comparing the brightness of different-coloured images, which is so productive of systematic error, is very largely obviated.

The chief practical difficulty in the construction of a flicker photometer is likely to be the elimination of vibration set up by the motor driving the rotating mirror. This, however, should not be insuperable, and the photometer would appear to offer material advantages over the Zöllner. (B. 15.14).

21.3.10 Holophane lumeter: An extremely simple device for comparing the relative intensities of extended objects without even the assistance of a telescope. A photometer of this type finds an obvious application in the observation of the Zodiacal Light. It consists (Figure 179) of a bifurcated tube, one end of which is occupied by the non-adjustable diaphragm A. The eye position is at B, just behind a circular aperture which must be larger than the pupillary aperture. The inner surface of

379

FIGURE 179
The Holophane lumeter

A is illuminated by a standard source, *C*, via a ground-glass screen, *D*, and a cat's-eye diaphragm, *E*. The interior of the lumeter is given a matt black finish except for the inner side of *A* and the space round the bulb behind *D* which are painted matt white.

Its operation is equally simple. If the intensity of an aurora, for example, is to be compared with that of the night sky, the instrument is directed so that the eye at *B* can see the aurora through the central aperture in *A*. *E* is then adjusted, so varying the illumination of *A*, until the intensity of *A* is equal to that of the aurora. At this point, providing the two are of the same colour (whence filters will have to be experimented with), the hole in *A* will disappear —i.e. the boundary between it and the diaphragm will no longer be distinguishable. The process is then repeated with the instrument directed at another part of the sky, and a second reading taken. The difference in the two screen intensities being proportional to the square of the diagonal of the cat's-eye in the two cases, the relative intensities of the aurora and of the sky can be calculated, either in terms of b/b' or of Δm.

21.4 Standard-step photometer

A device whereby each star in the field of view has a companion a known magnitude interval fainter can be of use (*a*) in establishing a new magnitude sequence, e.g. for a nova; (*b*) in estimating the magnitude of an object which is considerably brighter or fainter than nearby comparison stars.

The bar photometer, described in B. 15.8, is an example of this type of device. A weak prism, deviating the light passing through it by an amount corresponding to about $\frac{1}{4}$ of the diameter of the field of view, obstructs part of the telescope aperture. The ratio of brightness between the direct image of a star and its displaced companion therefore corresponds approximately to that of the areas of the unobstructed aperture and the prism. The latter is normally in the form of a rectangular bar mounted across a diameter of the OG; if a reflecting telescope is used, a circular prism fitted to one of the diagonal vanes may be more suitable. The precise magnitude difference, which will be constant for all stars, must be determined by observation.

The apex angle of the prism, *A*, necessary to deviate light through a small angle *x*, is given by $A = x/(N-1)$. A low-dispersion borosilicate crown glass should be used. (Figure 180.)

FIGURE 180

Two forms of standard-step prism photometer

Photocathode

Fabry lens

Focal plane diaphragm

OG

FIGURE 181

Basic layout of a photoelectric photometer

21.5 Photographic photometry

For wholesale establishment of the magnitudes of stars in a particular field (e.g. the production of a sequence for a variable star), photography offers considerable advantages over visual methods, not the least being that the sequence can be recorded in a matter of minutes while the reductions can be made at leisure. Since only one photograph (or series of photographs) is required, there is none of the continuous outlay of money and material that would be required if a star were actually being monitored photographically.

Regardless of the method of reduction employed, it is first of all necessary to consider the colour response of the film and filter combination used. When attempting to reproduce visual magnitudes, a yellow-sensitive combination, to match the acuity of the normal eye, is required; one reason for the amateur frequently wishing to take his own photographs is that professional photographic atlases are usually taken with blue- or red-sensitive combinations, which can produce disagreements of a magnitude or more with the visual appearance of strongly coloured stars. One satisfactory combination, used by the late W. E. Pennell in the production of many variable-star sequences, is Tri-X film in combination with a Wratten 8 (light yellow) filter to moderate the excess blue sensitivity of the emulsion. Such a combination, exposed for 10 minutes at the prime focus of an f/5 30·6-cm Newtonian, regularly produced measurable images down to mag 15.

The brightness of a stellar image as

382

recorded photographically is a function of D^2,* plate speed, and exposure. Various methods of determining it are available:

(*a*) Diameter of the image: for a given film and exposure, image diameter is a function of magnitude. If one or more previously-calibrated stars are in the field, they can be used as zero-points for the sequence; alternatively and less satisfactorily, a second exposure can be made, as soon as possible after the first, on a nearby field at the same altitude containing some calibrated stars, and their diameters used to set the scale. B. 25.2 contains useful photoelectric magnitudes of faint stars scattered over the sky. The relationship between magnitude and diameter, discussed in B.15.23, must be determined by a series of experimental exposures; in one instance, the prime-focal image diameter varied from 0·014 mm (mag 15·1) to 0·157 mm (mag 8·3). A microscope incorporating a focal-plane micrometer can be used to measure the negative directly; alternatively, the images on an enlargement can be measured with a simple optical micrometer incorporating a transparent scale in the focal plane. With such unpretentious equipment, visual magnitudes accurate to within about 0·2 mag, and often better, can be obtained for all but highly coloured stars. (See B. 15.16.)

(*b*) The density of the extrafocal images being a function of their brightness, this can be calibrated in terms of magnitudes and measured with a microphotometer, which makes an automatic record, via a photocell and galvanometer, of the image density. The need to defocus raises the telescope's limiting magnitude. B. 15.2 describes a combined microdensitometer and plate-measuring machine.

(*c*) A coarse grating placed in front of the objective spreads the light from each star into a series of secondary images arranged on either side of a central primary image. The number and intensity of these secondary images can be calibrated in terms of stellar magnitude, so that, as in section 21.4, the secondary image from one star can provide a known magnitude step to another, fainter star. For a description of photographic photometry using comparison prisms to provide the secondary images, see B. 15.3.

The most troublesome sources of error in photographic photometry are emulsion grain, local variations in the sensitivity of the emulsion, and non-uniform development.

---

*Stellar image brightness in theory increases proportionately to D^2. In practice, the ratio is somewhat lower, since the image is a disc and not a mathematical point. Similarly, reducing F for a given value of D should theoretically have no effect upon stellar image brightness; in practice it has.

21.6 Energy-conversion photometers

These rely on the conversion of the radiant energy of the star into a more accurately measurable form, namely electricity. Photoelectric cells fall into four main categories:

- (a) photo-emissive, or Elster Geitel, cells—either vacuum or gas-filled;
- (b) photo-voltaic, barrier layer, rectifier, or Sperrschicht cells;
- (c) photo-conductive cells;
- (d) solid-state devices.

Photoelectric photometers generally are distinguished by their great sensitivity and by the fact that their systematic and random errors are smaller than those of any other type, an accuracy of the order of ±0·001 mag being attainable. Their disadvantages are restricted applicability, cost, the complexity of the technique of their operation, and the desirability of large apertures. The threshold of Stebbins' calcium hydride cell, for example, was mag 9 with the Yerkes 1-m, and mag 7 with a 37·5-cm. However, this restriction has been to a large extent relaxed as a result of the development of the multiplier type of tube, which with a 30-cm telescope should reach to about the 10th magnitude; to say nothing of solid-state devices.

Developments in this field are rapid, with the result that more and more sophisticated equipment is becoming available to the amateur. It seems likely that, during the next few years, the enthusiast with sufficient initiative to search through the scrap boxes of local electronics laboratories will be able to build a photometer that will compare stars in real time with a precision directly comparable with the best conventional photometry of recent times.

21.6.1 Photon statistics: Comparison of the magnitudes published by professional astronomers shows that, although their results are often quoted to 0·001 mag, a typical accuracy of 0·01 mag is the best that is normally achieved. If the amateur is to approach this accuracy, it is useful to consider which of several possible factors are limiting the precision attained by professional observers.

Photon statistics are not normally a problem with relatively bright stars. The particle (photon) nature of light is a familiar contemporary concept, and many modern photometers literally count the number of photons that have been recorded by the instrument in some fixed time. The statistical nature of light being what it is, the potential accuracy of the measurement increases with the number of photons that are recorded, the relationship being expressed by

$$A \propto \frac{1}{\sqrt{n}}$$

where A is the accuracy of the result after n photons have been recorded. For example, after 100 photons have been counted, the potential accuracy is 10 per cent or about 0·1 mag; after one million, the accuracy has improved to 0·1 per cent or about 0·001 mag.

A 30-cm telescope observing a mag 6 star is, typically, capable of recording one million photons in between 10 and 100 seconds, while an objective of only 30-mm aperture will receive 1/100th (1/10²) of this number in the same time, giving potential accuracies of 0·001 mag and 0·01 mag respectively. If the latter objective were observing a star one hundred times fainter (mag 11), the accuracy would be reduced to 0·1 mag, and there would be no way of improving upon this without extending the photon-collecting interval.

It would seem, then, that the professional observer using a large telescope is not greatly hindered by photon statistics in his search for accuracy; yet, precision greater than 1 per cent (0·01 mag) is seldom achieved. Of several possible potential difficulties, the two which are most frequently to blame are (a) changes in atmospheric transparency, both from place to place in the sky and with time; and (b) mechanical instabilities in the photometer. If these causes are accepted by the amateur, it can be seen which are the aspects on which it will be most profitable to concentrate in the attempt to achieve accuracies of 0·01 mag. These are:

(a) The measurement must take place over a sufficient length of time for an adequate amount of light (number of photons) to have been measured.

(b) The star of interest, and one or more comparison stars, must be measured simultaneously; they must all be within a few degrees of each other on the sky.*

(c) The greatest care must be taken to ensure that no movement or misalignment of the mechanical parts occurs during the measurement.

21.6.2 Photo-emissive cells: These rely upon the characteristic of the alkali metals that when exposed to light they emit electrons from their surface, the rate of this emission being exactly proportional to the number of quanta in the incident radiation. The cell in its most primitive form consists of a photo-sensitive cathode mounted at the back of a glass tube whose inner surface, except for a small window opposite the cathode, is silvered. A ring anode is located centrally in the tube. The electrons

---

*Photometers which compare the star with an artificial source, or with another star at a very different zenith distance, may be interesting to construct and use but will not produce the results that a conscientious observer is capable of achieving.

emitted by the cathode when it is illuminated are collected at the anode, which is maintained at a constant potential by a H.T. battery. The current carried by the electron stream is directly proportional to the number of electrons, hence to the intensity of the radiation, and is read from a galvanometer included in the circuit; the latter may have to be at a considerable distance from the telescope, under stable physical conditions, thus involving two observers. For a given illumination the current is proportional to the applied potential up to about 25 volts, if the cell is evacuated; further increase is unaccompanied by an increase in the photoelectric current. But if the tube is filled with an inert gas (usually argon), gaseous amplification through collision ionisation will occur, and the current can be increased by raising the potential right to the point at which a glow discharge occurs. Gas-filled tubes are therefore preferred for most purposes. The external circuit should contain a resistance of about 50,000 ohms as a protective measure in the event of a glow discharge occurring. The potential can, however, be safely maintained at about 5 volts below the discharge level.

If the photoelectric current is below the threshold of the galvanometer a more sensitive meter may be used (such as a Lindemann electrometer), or the current may be amplified to a level at which it can be registered by the galvanometer. Nowadays the most usual practice is to employ the type of cell known variously as an electron-multiplier, multiplier phototube, or photomultiplier tube.* This is a vacuum tube which contains, in addition to the cathode and anode, a number (usually about ten) of dynodes. The electron stream emitted by the cathode is directed by electrostatic force to the first dynode, where it produces more electrons; thence to the remaining dynodes in turn, being amplified by secondary emission at each stage. Finally, it is collected by the anode. In its most recent development, this type of tube has a response time of a few nanoseconds, and the arrival of individual photons can be measured. The high cost (£200 to £800 each at 1979 prices) is, however, a practical deterrent to their use by amateur observers.

The tube's dark current is of little importance when the illumination is strong, but in any application of the photocell to very low intensities it becomes a factor of the first importance, since it determines the threshold to which the tube may be used. Dark current arises mainly in three ways: (*a*) ohmic leakage across incomplete insulation within the cell or

---

*The 1P21, manufactured by the Radio Corporation of America, can be used for photoelectric photometry. See B. 15.10, 11, 12, 25. Almost all the pioneering work, including the establishment of the U,B,V system, was done using tubes of this type. It still retains its usefulness where a low-cost tube is needed.

it and the base; (*b*) thermionic emission arising from the fact that the cathode emits some electrons by virtue of its own temperature; (*c*) regeneration ionisation, which occurs in a multiplier tube at about 110 volts. Of these, (*b*) is in practice usually the limiting one and can be reduced by refrigerating the cell with melting ice, or virtually eliminated with dry ice (solid carbon dioxide, $-125°C$). With dry ice refrigeration a gain of as much as 5 mags may be achieved.

Another factor affecting the performance of the cell is photoelectric fatigue. This affects all types of cell to some extent, introducing an element of uncertainty regarding their colour sensitivity as well as their absolute response. Checking with a standard source from time to time is therefore an essential precaution.

The sensitivity of the photocathode varying across its surface, it is of the utmost importance that the image of the star should illuminate as large an area of the sensitive surface as possible, and under no circumstances should movement of the star image in the field result in a movement of its image on the photocathode. The most commonly adopted system is to place a screen, carrying a hole $30''$ to $60''$ across, in the focal plane of the telescope. This restricts the field of view of the photometer, and reduces the contribution from the light of the night sky. A Fabry lens is placed behind this aperture; the purpose of this lens is to form an image of the telescope's objective on the photocathode, so that movement of the star image within the aperture, due to poor guiding or other causes, does not result in any motion across the sensitive surface itself. Behind the aperture, a sliding track bearing a totally reflecting prism allows the position of the star image to be checked. Another sliding track below the Fabry lens accepts filters for insertion into the beam, allowing the brightness of the star to be measured in different wavebands. (For a further description of a Fabry lens and its function, see B. 15.1.)

Measures should be made on the star of interest and on at least one other comparison star, which should be both nearby and of similar spectral type. A reliable routine procedure is as follows:

(*a*) Measure the sky brightness by taking alternate readings with the telescope pointing at a starless field near the star to be measured, with and without the dark slide drawn. Note the time, and repeat hourly, or more frequently if the circumstances warrant it.

(*b*) Take a set of readings of the comparison star with the dark slide alternately open and closed. Note the time of the mid-point of the set.

(*c*) Repeat with the unknown star.

(*d*) Repeat (*b*) and (*c*) twice, and end with a final set of measures of the comparison star.

387

By plotting the sky brightness readings against time, the correct value to subtract from the star readings, at the time they were made, can be determined. The mean of each set of readings of the unknown star is compared with the mean of the preceding and following sets of the comparison star. Finally b/b' is converted to Δm.

The above procedure, although used successfully by professional observers enjoying excellent sky conditions, will be less satisfactory under the rapidly-changing and far from homogeneous atmospheric veil that characterises most amateur sites. If the transparency changes even slightly between individual observations, the result will either be spurious or obviously inconsistent. Under well-lit urban conditions, increased cloud cover means not only reduced light from the star, but boosted reflected light from the lower surface of the cloud itself. The way of minimising this effect is to take readings, in rapid succession (several times per second), from each object in turn—variable, then sky, then comparison star, then sky, then variable, etc; since it would not be feasible to move the telescope itself with this rapidity, optical 'chopping' between adjacent stellar images in the focal plane is the usual solution. The adoption of a telescope similar in type to those described in sections 21.3.5, 6, 7 and 9 would allow the amateur a much better chance of competing in accuracy with the professional.

The recent emergence of fibre optics, or flexible light guides, is relevant to the problem of bringing the images of several sources to a common point for the purpose of inter-comparison. These are already so generally available that the problem is one of choice rather than availability; the quality ranges from single perspex fibres, costing a few pence per metre, intended for the automotive industry, to fused quartz fibre bundles costing perhaps £100 for the same length. Developments in this field are so rapid that no useful predictions can be made, but already (1978) there exist several 3-into-1 and 4-into-1 fibre light guide adaptors; with suitable mounting, these would seem to be the ideal way of combining the light from several stars on to the photocathode.

The sensitivity ranges and other characteristics of some photo-emissive materials which are commercially employed are summarised below:

Caesium oxide (*CsO*). Has one of the widest ranges: 340-1050 nm, maximum at about 770 nm, secondary maximum at 350 nm; high infra-red sensitivity. Unfortunately has a high dark current (average 10^{-12} amps), though this, being mostly due to thermionic emission, can be reduced by refrigeration. The caesium cell is not the most suitable for photometric purposes owing to its dark current and because the infrared must be eliminated by filters.

Potassium oxide (*KO*). Maximum sensitivity near 350 nm, falling off sharply on the short-wave side to zero at about 300 nm, and less sharply on the long-wave side to zero at about 750 nm or 800 nm. It thus covers the visible spectrum and part of the ultraviolet. One of the most satisfactory cells as regards linearity of response.

Potassium hydride (*KH*). Narrow sensitivity range: maximum at about 430 nm, falling off to zero at 600 nm, with an equally steep drop on the short-wave side. Notable for its extremely small dark current. Stebbins' Yerkes photometer (B. 15.26) was of this type.

Sodium. Individual cells usually have a range of about 200 nm between 200 nm and 550 nm. Only a slight response to visible light, therefore.

Cadmium. Increasingly sensitive down from 320 nm, limited only by absorption of the quartz cell window, i.e. sensitive to the ultraviolet only.

21.6.3 Photo-voltaic cells: These exploit the fact that the incidence of light causes the transfer of electrons across the rectifying boundary of two dissimilar materials such as copper and copper oxide. Photo-voltaic cells thus generate their own EMF, and no battery is required in the external circuit. The sensitivity ranges vary rather widely with different photovoltaic combinations.

Advantages: (*i*) simplicity; (*ii*) obviation of the necessity for an external source of potential; (*iii*) greater current sensitivity than the photo-emissive cell.

Disadvantages: (*i*) amplification of the output is not practicable, and it must therefore be read with a galvanometer; by means of an electrometer, or amplifier and galvanometer, much smaller currents can be detected with the photo-emissive cell, which will therefore be preferred when low intensities are to be measured; (*ii*) subject to fatigue; (*iii*) response is never strictly linear.

Hence the photo-voltaic cell is the more convenient when there is plenty of light and the utmost precision is not demanded; but for most astronomical requirements the photo-emissive cell is to be preferred.

21.6.4 Photo-conductive cells: These exploit the fact that the conductivity of certain materials varies with the intensity of the radiation to which they are exposed. One such substance is lead sulphide (PbS), with a sensitivity maximum in the region of 2000 nm. B. 15.7 describes in some detail the construction and performance of a sulphide cell which in the infrared showed itself to be of much superior sensitivity to the most sensitive thermopiles (section 21.7). Another photo-conductive material is selenium, whose sensitivity is superior to that of the eye (both as regards contrast and absolute values) over a range of approximately 180-1500 nm.

Its main disadvantage is that its response is not linear, being initially a function of the intensity of the radiation but finally of the square root of the intensity.

21.6.5 Solid-state devices: Many of the remarks (above) pertaining to fibre optics also apply to photosensitive solid-state devices. None of these have been developed specifically for astronomical use, yet they are likely to revolutionise instrumental astronomy over the next few years; the developments in this field are, however, so rapid (and, occasionally, innovative) that a comprehensive review of the devices would be out of place. All are direct developments from the semiconductor industry, and are, in general, intended for the mass-consumption market. Devices range from photo-resistive diodes, which change their measured resistance proportionally to the amount of light falling on them, to charge integration devices (CIDs) in which the electron-halo structure of a silicon-based compound is modified by interaction with photons. Charge is built up on the device, and can be read out in a non-destructive fashion until a meaningful signal-to-noise ratio has been accumulated. Cooling of the device to the temperature of liquid nitrogen greatly reduces noise. Quantum efficiencies as high as 80 per cent can be obtained, to be compared with about 20 per cent for the photocathode in a photomultiplier tube or generally much less than 1 per cent for a photographic emulsion.

21.7 Thermocouples and thermopiles

These—together with radiometers, bolometers, and other instruments of little interest to amateurs—are designed to detect and measure minute intensities of long-wave (heat) radiation.

They exploit a rather similar phenomenon to that made use of in the photo-voltaic cell: if two wires of dissimilar material are joined at both ends, a current (recorded by a galvanometer included in the circuit) will flow if there is a temperature gradient across the junctions; this current is proportional to the temperature difference, for a given junction. Commonly used materials for couples are iron and copper, iron and constantan, platinum and a platinum-rhodium alloy, antimony and bismuth, bismuth and a bismuth-tin alloy, copper and alumel (a nickel alloy containing a few per ccent of manganese and aluminium and traces of silicon and iron), alumel and chromel (a nickel-chromium alloy). The thermojunction is covered by a small disc of blackened metal foil on which the stellar image is projected. Evacuating the tube containing the couple increases its sensitivity: it is doubled, for example, by reducing the pressure in the tube from 10^{-3} to 10^{-6} mm.

A thermopile consists of a number of thermocouples in series: the current produced by a given intensity of radiation is then that produced by a single thermocouple multiplied by the number of couples in the pile. Alternative junctions are maintained at a constant temperature, the remainder being exposed to the radiation which it is desired to measure.

Thermocouples are less sensitive than the eye within the visual range, but in the longer wavelengths they are incomparably more sensitive than thermometers, a good thermopile and a sensitive galvanometer being able to measure temperature differences of the order of $10^{-5}\,°C$.

B. 15.17, 18, 19 describe thermocouples in use at Mt Wilson and elsewhere. The technique of making couples is described in, *inter alia*, B. 4.11, 15.29.

ACCESSORY INSTRUMENTS AND EQUIPMENT

22.1 Binoculars

A good pair of prismatic binoculars is an invaluable accessory. Not only does it provide some of the best views of the Milky Way that are to be had with any instrument, but it is frequently of use as a preliminary to the finder of the telescope, as well as in the observation of variables and comets.

The advantages of binocular over monocular vision (i.e. of both eyes using identical separate optical systems) are:

(a) faint objects are slightly more easily seen;
(b) fine detail is more quickly, if not better, seen;
(c) it is less tiring,
(d) it provides—though not in the field of astronomical work—an increased parallactic effect.

The advantages of the binocular over the naked eye are:

(a) increased light grasp;
(b) increased resolution;
(c) increased magnification.

Its advantages compared with the telescope are:

(a) wider field;
(b) increased contrast.

Prismatic binoculars—in which the rays are turned twice through 180° by Porro prisms—are, with few exceptions, the only type suitable for astronomical work; so-called opera glasses are usually Galilean systems (see section 9.2) A few small lightweight models employ a straight-through roof prism to erect the image, one well-known ultra-compact binocular of this type being the Zeiss 6 × 20.

Provision should be made for the independent focusing of each eye-piece; if, in addition, the two tubes can be adjusted together, so much the better. The connecting pieces between the tubes should also be hinged, so that the interpupillary distance can be exactly matched by the separation of the two axes. The two systems must provide identical magnifications, and their axes must be parallel. The tolerance of inaccuracy in the latter respect decreases as magnification increases, since an angle θ between the axes separates the images by an angle $\theta' = \theta(M+1)$. If the direction of separation of the images is parallel to that joining the eyes, the images are more easily superimposed than when the separation occurs at right angles to the line of eyes. The limits of θ beyond which the images can only be superimposed by conscious effort are respectively 1° and 10′, approximately. A useful pair of binoculars will secure about $1\frac{1}{2}$ to 2 mags beyond the naked-eye threshold.

Binoculars are usually described in terms of their magnification and aperture measured in mm, e.g. 7 x 50 ($M = 7$, $D = 50$ mm). Apertures in common use range from 20 mm to 80 mm, but the vast majority of models come in the 30-mm and 50-mm size, with magnifications of x7 to x10. The range of quality is equally wide; the price of instruments of the same specification can differ by four or five times, superior quality declaring itself in improved definition, better light transmission and contrast, wider and flatter field, and greater durability. Cheap binoculars are characterised by poor prisms inadequately locked into position (hence, rapid loss of collimation), narrow-field eyepieces, and generally poor colour correction. The prisms may also be too small to transmit full illumination to the useful field of the ocular, resulting in effective loss of aperture. Quite often, as well, the outer surfaces of objective and ocular are given a 'cosmetic' blooming while the numerous interior air-glass surfaces remain uncoated; the seriousness of this omission can be appreciated when it is realised that a typical binocular train contains 10 such surfaces, corresponding to a total light loss of about 34 per cent if no anti-reflection coatings are applied.

Binoculars normally being designed as wide-field instruments, magnifications near M' are usually employed; unlike astronomical telescopes, however, the value of M' is near the approximate *daylight* value, i.e. when the pupil is contracted to 3 mm or 4 mm, rather than opened to its darkness peak of 8 mm. Magnifications of around $2\frac{1}{2}D$ to $3D$ are, therefore, commonly employed in glasses designed solely for optimum daytime use (e.g. 8 x 30, 12 x 50), in order that the full value of D is usefully employed. So-called 'night glasses' are, as their name suggests, designed to give a much larger exit pupil, corresponding to the fully-dilated pupil (about 8 mm). They therefore work near the astronomical M' of about $1·25D$ (7 x 50, 11 x 80), and such instruments may lose over half of their effective aperture if employed during the day.

In between these extremes come models such as the 8 x 40 and 10 x 50 which, in practice, are excellent choices for the astronomer, since instruments in which the exit pupil and pupillary aperture are closely matched are not always comfortable to use. In addition, observers suffering from even moderate astigmatism will find that the star images formed by such a large exit pupil are intolerably deformed unless spectacles are worn. Such people will benefit from the smaller exit pupil inherent in relatively high-power types, such as the unusual 12 x 40 specification of a Russian-made binocular widely marketed in the UK (3·3-mm exit pupil). With a given aperture, as with any telescope, the limiting magnitude of a binocular increases with increasing M, due to the effective darkening of the sky background.

Large binoculars or binocular telescopes, originally made for military purposes, are now only rarely encountered; the German 25 x 105 reconnaissance model has been used by Alcock for his comet discoveries. A type of instrument to avoid is the grossly over-powered hand binocular with an aperture of 40 mm to 60 mm and magnification of x 20 or more; such instruments are rarely sufficiently well made to give satisfactory definition and to hold collimation, even if they were correctly collimated in the first place.

See also section 8.7.

22.2 Hand telescopes

The uses of a lightweight hand telescope of the coastguard variety are much the same as those of binoculars: for large and diffuse comets, 'sight-seeing', sweeping for novae, etc, and as a subsidiary finder. In addition it avoids the discontinuity between binocular and monocular vision in the observation of variables having very bright maxima. Telescopes of this type typically have apertures of 50-75 mm from about x 5 to x 8, and fields from about 5° to 8°.

An extremely useful army surplus instrument that has been available since the war is the elbow telescope. This is a remarkably compact instrument (overall length, 27·5 cm) which works at the phenomenally small focal ratio of $f/3$. The OG has an aperture of slightly less than 50 mm; the image is erected, and the convergent pencil turned through 90°, by means of a roof prism, to which is cemented the field lens of the positive achromatic ocular. It gives an 8° field with a magnification of about x 7. Other specifications of elbow telescope are also available on the surplus market, while the same source can offer separate OGs, prisms, and eyepieces, from which a useful instrument can be assembled at home.

Surplus equipment is normally above suspicion, but many modern small refractors, sold for finders or general viewing, are labelled misleadingly and should be carefully examined. It is frequently found that a telescope stated to have, for instance, a 30-mm objective, consists of a non-achromatic lens of this diameter with a stop, limiting the effective aperture to perhaps 15 mm, some way down the tube. Such instruments are, of course, useless.

22.3 Finders

There is little to say about finders except that they are usually much too small and are often mounted inconveniently far from the telescopic exit pupil. The finder of a Newtonian is apt to assume awkward positions, and it is a great convenience if more than one is fitted. As the image-reversal occasioned by a diagonal is always objectionable, the finder must be mounted well away from the telescope tube.

Low magnification, adequate light grasp and width of field are the characteristics to aim for. The method of mounting is such that the position of the finder's axis relative to that of the telescope can be adjusted, so that the two may be made parallel, The eyepiece usually contains some arrangement of cross-wires to indicate the centre of the field, though this can hardly be regarded as necessary.

22.4 Comet seekers and richest-field telescopes

The aim in designing a telescope for comet-seeking should be, primarily, width of field and light grasp. The relatively small focal ratios of comet seekers have, contrary to common belief, nothing to do with the fact that an instrument of small focal ratio is of greater photographic 'speed' than one of large. For further discussion of the qualifications of comet seekers see section 3.10.

The principles underlying the design of a richest-field telescope—i.e. one the dimensions of whose objective and ocular are adjusted so that it will show a maximum number of stars in any given field—are fully discussed by Walkden in B. 7.49.

22.5 Dielectric-filter solar telescopes

With the post-war development of the dielectric filter (see section 7.1), the modern solar observer has at his disposal a means of observing the Sun in the light of any selected wavelength, or very narrow band of wavelengths, without having to undertake the construction of a spectrohelioscope or spectroheliograph (section 22.6). A small filter, placed either before the objective or, preferably, near the solar image, replaces the arrangement of slits and grating required in the latter instrument. The main disadvantage of the narrow-band filter—its confinement to one waveband—is to a great extent mitigated by the fact that most visual and photographic monochromatic observation of the Sun is carried out at the wavelength of the $H\alpha$ line, 656·3 nm; otherwise, the only serious drawback is expense.

The waveband transmission of a dielectric filter can be made relatively wide, or extremely narrow, depending on the number of layers applied in its manufacture. There are, effectively, two classes of filter for solar work: those giving bandwidths of from about 3·0 nm down to about 0·4 nm, and very narrow-band filters with transmission widths less than 0·1 nm. The former kind are sufficiently selective to give views of prominences around the solar limb, but to see hydrogen-light details on the disc itself requires the complete removal of all adjacent wavelengths, necessitating a filter of the latter kind. Instruments using the former class of filter, known generally as 'prominence telescopes' of 'promscopes', were first popularised by Klaus in Switzerland and Dall in the UK.

Dall's design, described in B. 16.1, consists of the following features (Figure 182). A non-achromatic OG of 75 mm D, working at $f/15$, casts an image of the Sun on to a cone-shaped stop whose base diameter is just larger than that of the solar disc (since this diameter varies during the year, due to the Earth's eccentric orbit, two alternative stops with base diameters of 10·5 mm and 10·75 mm are used for summer and winter observation respectively). The cone is affixed to the centre of an achromatic field lens, whose function is to eliminate scattered light arising from diffraction around the edge of the OG; it does this by forming an image of the OG on a screen bearing a slightly under-size perforation. Both the face of the screen, and the inner side of the tube surrounding the conical stop, are dull black to absorb the scattered sunlight. A non-achromatic projection lens, positioned just behind the screen, forms an amplified image of the cone stop and surrounding prominences at the bottom of the tube, where it can be examined by an eyepiece or photographed. The final effective focal ratio is 30, and the dielectric filter is placed in the beam between the projection lens and the image. The OG must be as free from scratches and blemishes as possible, and a long lens hood, coated internally with grease to trap floating dust particles before they can land on the front surface, is helpful. Scattered light reduces the contrast of the prominences, and all optical surfaces, particularly those receiving the unfiltered sunlight, must be as clean as possible. The use of non-achromatic optical components is permissible in a system operating in monochromatic light.

On unusually clear days, when the sky is a deep blue, a 3·0 nm filter

FIGURE 182

Prominence telescope (*after Dall*)

Monochromatic solar image

Collimating 06s

Heat filter

Hα filter

Primary solar image

FIGURE 183

Arrangement of interference filter for whole-disc solar observation (*after Gabriel*)

can show prominences quite well; but a 0·4 nm filter permits observation on any day when the Sun is casting sharp shadows. At high-altitude sites, even a 12 nm filter can provide sufficient contrast for prominences to be seen. Dall emphasises the need for operating narrow-band filters at the correct temperature, a shift of about 4°C from the design specification altering the bandpass position by 0·1 nm. The filter can be 'tuned' by tilting it slightly—a facility incorporated in the promscope design. This sensitivity to the angle of incidence means that dielectric filters must be used in an approximately parallel wavefront for transmission characteristics to be similar across the field of view.

A telescope designed for whole-disc observation, using a 0·075 nm *H*α filter, has been described by Gabriël (B. 16.2). Considerably simpler, it is a standard 60-mm refractor in which the solar image is transferred to a second focus by two collimating lenses, as shown in Figure 183. The filter is placed in the parallel light between the lenses. The same effect would be achieved by placing the filter in front of the OG, but at the cost of considerable reduction of aperture. A heat-absorbing filter is placed in front of the first collimating lens to protect the unit from the Sun's direct heat. Narrow-band filters of this precision, being critically sensitive to temperature variations, are supplied in a thermostatically-controlled heat jacket or 'oven'.

The image formed in either system can be viewed visually, via a LP ocular, or else photographed. In the latter case, care must be taken to select a film with reasonable sensitivity to light at the $H\alpha$ wavelength, which occurs near the limit of red-response for most films; FP3 and HP4, for example, record much less effectively than Plus-X. For photography of the disc features, rather than the limb prominences, very high-contrast spectrographic film such as SO-392, used for professional solar patrol work, must be used (see section 40).

22.6 Spectrohelioscope

The spectroheliograph was invented independently by Hale and Deslandres in 1890. The visual form of the instrument, the spectrohelioscope, though it was adumbrated by Janssen and Lockyer as long ago as 1868, had to await effective development by Hale during the years 1924-1929. Although the development of narrow-band dielectric filters has to some extent displaced this instrument, it possesses the advantage of being tunable to any desired wavelength, and measurements of radial velocity—manifesting themselves in a slight change of wavelength of the particular feature involved—can be made with more facility. Its bandwidth can be reduced below that of a normal dielectric filter, giving better contrast. However, the image of prominences seen around the limb is considerably dimmer, aperture for aperture, than that obtained when using a filter; and the complexity of its construction, together with the amount of room its layout requires, mitigate against its use by amateurs who could, without too much difficulty, both build and accommodate a filter-type instrument. On the grounds of cost, judgment depends critically upon the maker's optical competence. A filter telescope, objective included, could be made from standard lenses, but a spectrohelioscope or spectroheliograph requires two rather large plane mirrors, two concave mirrors of very large radius of curvature, and a very long-focus OG. All of these items would have to be ordered specially, or made; if the former, the total cost of the optical components would probably be comparable in the case of the solar-disc filter telescope, but much less if a prominence telescope were being constructed.

The essential principle involved in the spectrohelioscope is the employment of the persistence of vision in the observation of a rapidly (c. 25 per sec) oscillating monochromatic image of the Sun, thus allowing a considerably wider area than a singe slit-width to be viewed.

Structurally it consists of: (a) An image-forming unit; usually an OG of about 5·5 m F (forming a solar image about 50 mm in diameter), which need not be achromatic, since in the final image only a single wavelength is used; it is mounted permanently (apart for allowance for focusing adjustment) in a position where the solar image can be fed into it by a coelostat. (*b*) A high-dispersion spectroscope, either prismatic or diffraction, of about the same focal ratio as the objective; high dispersion is necessitated by the fact that the blaze of integrated photospheric light, when the slit is overlying the disc, makes it essential that all but a single wavelength be excluded; and this, without considerable dispersion, involves impracticably narrow slit openings. (*c*) A mechanism for synthesising the monochromatic images; this may be of several types:

(*i*) oscillating slits, yielding a final image whose orientation is a mirror reversal of the direct telescopic image (i.e. left and right reversed);

(*ii*) rotating prisms, one in front of each slit; the slits now being stationary, it is the solar image and the view-point that oscillate relative to them; the orientation of the image is here the same as that of the telescopic image of the Sun;

(*iii*) rotating glass disc, coated with black laquer in which radial slits have been cut.

Hale's original model incorporated oscillating slits, and these are generally favoured, partly for cheapness and also because they transmit light more efficiently. The slits may be arranged either side by side or vertically above and below one another. The disadvantage of the latter arrangement is that the observer's head is only a few inches below the first slit, and convection results in deterioration of the definition. Hale's models have vertical slits, since this is the most convenient arrangement with Anderson prisms or the simple rocker bar mechanism he used for imparting unison motion to the slits. Slits mounted side by side require out-of-phase movement (like that of a tuning-fork, both of whose prongs move outward together and inward together, exactly in step and by precisely equal distances), which presents some problems in mechanical design. These difficulties have been overcome in Sellers' oscillating-slit mechanism, however; it is, moreover, ideally adapted to the needs of the amateur building his own instrument (B. 16.12, 13). The Sellers mechanism is employed in Hill's spectrohelioscope, which consists of the following essential features (B. 16.11):

(*i*) Coelostat mirror, diameter 13·8 cm, mounted equatorially.

(*ii*) Secondary mirror, diameter 11·2 cm, with fine adjusting controls to position the solar image correctly on the jaws of the first slit.

(*iii*) Non-achromatic OG, $D=10$ cm, $F=600$ cm (solar image diameter, about 55 mm).

(*iv*) Oscillating slits set 16·2 cm apart, effective length about 28 mm.

(*v*) Two collimating mirrors, $D=83$ mm, $F=396$ cm.

(*vi*) Bausch and Lomb reflecting grating, 76 mm×65 mm, 600 rulings/mm (*vii*) Line shifter, placed just behind the viewing slit: a piece of plane-parallel optical glass 5 mm thick, which may be rotated to displace, by refraction, the wavelength of light passing through the slit. If the light from a particular feature is Doppler-shifted through its individual velocity, measurement of the rotation necessary to restore its spectral line to the slit gives a determination of the velocity concerned.

Any amateur contemplating the construction of such an instrument should, in addition to reading Hill's paper, try to consult B. 16.5, or, more accessible, B. 16.10. A description of a miniaturised spectrohelioscope appeared in B. 16.14.

22.7 Timing devices

Some observations (e.g. lunar occultations) depend critically on being timed precisely. The following means are most commonly employed by the amateur.

22.7.1 Chronograph: An instrument which records on a continuous trace the instant at which an observation is made; either concurrently and automatically or at any required moment manually, and either on the same or a parallel trace, the instants of GMT or GST are recorded. A comparison of the two traces, or of the two sets of signals on the one trace, then allows the GMT or GST of the observation to be derived.

The arrangement of a typical chronograph is as follows. A paper strip (e.g. the roll for a print-out calculator) is drawn slowly past two pens, both of which are in contact with the paper at all times. One pen receives a regular (usually once-per-second) signal from a timing device, upon which it makes a brief kick or notch on the trace. In line with this pen, on the opposite side of the strip, is the signal pen, which makes its own separate kick on receiving a push-button signal from the observer. The machine is set in motion, the timing notches being calibrated by one or more time signals recorded by the signal pen. The observation is recorded by a further mark from the signal pen, and the interval in seconds and fractions between this and the calibration mark can be measured. A second time mark, made after the observation, provides a further check.

Two motor-driven spring-loaded rollers drawing the paper between them at a rate of about 30 cm/minute will be found satisfactory; the timing pen can be powered by an electro-magnet on the pen arm, operated by a make-and-break contact on a small synchronous motor. A similar electro-magnet operates the signal pen. If a long interval occurs between the time and signal marks, a preliminary 'stutter' of the signal pen, not long before the observation is expected to occur, helps in locating the marks. Should a series of observations be made (e.g. transits when using a cross-wire micrometer), coding signals can be employed to distinguish the various timings.

The advantages of a chronograph include: robust signals (cf. radio); visual and permanent display; limitless capacity (cf. stopwatch); and versatility. Among its disadvantages are: bulk; consumption of paper; and zero point subject to personal equation (cf. radio).

The construction of a chronograph is described in B. 16.18.

22.7.2 Stopwatch: For handy convenience, the stopwatch is by far the most useful general-purpose timing device. It can be carried in the pocket, and is ready for use at a moment's notice. If of the split-action type, in which the slave hand, which is stopped to record an observation, can be released to catch up with the master hand, repeated observations can be made in the course of a single run.

Precautions to be observed when using a stopwatch include: (*a*) when using the watch after winding it fully, allow it to run for a few minutes beforehand; (*b*) do not store the watch away fully wound; (*c*) determine its rate over the section of the winding during which it will be used; (*d*) keep the interval between observation and time-check as short as possible. See, further, section 28.19.1.

22.7.3 Tape recorder: Although a useful 'notebook' for certain types of observation, a tape recorder can be a valuable adjunct when used in conjunction with a radio receiving time signals. The time base can then be recorded together with the announcement (whether verbal or using some suitable signal) of the event. Details of stations broadcasting suitable time signals are given in section 28.18.

The advantages of the method include portability, versatility, and accuracy of the timing signals. Disadvantages are: relative complexity (cf. stopwatch); possible loss of time signal through atmospheric or other factors at the critical moment.

22.8 Blink comparator

The blink comparator is a device for viewing pairs of photographs (of the same star field, taken at different times) in such a way that the attention is quickly attracted to any difference there may be between them. Such differences are, in particular: an image whose position on the two photographs is different, relative to the general star field; an image whose brightness on the two photographs is different; an image which occurs on one photograph only.

The principle of the blink comparator is the alternating presentation of the two plates to the eye, the position of each within the visual field being identical. The vast majority of stellar images will then appear to be visible continuously and without change, since their positions and brightnesses on the two plates are identical (assuming equal exposure). But if an image occurs on plate A and not on plate B it will appear to 'blink', i.e. successively disappear and reappear as the field of vision switches back and forth between A and B. The image of an object which has moved during the interval between the exposure of the two plates will (assuming that both its images are contained in the same field of the microscope) appear to jump to and fro between its two positions, being seen alternately in each. Both these effects—blinking and jumping— are extremely eye-catching. The appearance under the blink microscope of an image which differs on the two photographs only as regards its intensity is less obviously changing, and needs more careful scrutiny: its intensity, and to some extent its size, will simply appear to vary in successive jerks between the two limits.

The labour and time spent on any work involving the comparison of stellar photographs are cut to a minimum by the use of the blink comparator. For anyone undertaking the routine photographic search for novae it may be considered indispensable. It is a great convenience in the location of minor planets, and may also be of limited usefulness to the variable-star observer. As an aid to cometary discovery it is only of value on the comparatively infrequent occasions when the cometary image is indistinguishable from that of a star. The photographic images of most comets, even at the time of discovery, are clearly non-stellar in appearance, and can be detected as easily, and more quickly, by the direct examination of the photograph with a magnifying glass as by blinking it with a paired photograph taken on an earlier occasion.

Summarily, then, the blink comparator, though a convenience in several fields of work, is only an essential part of the equipment of an observer undertaking regular search surveys—whether for novae, asteroids or variables. Such work is most economically performed by fast wide-angle cameras of the Schmidt and related families, whose pictures are so crowded with images that any other method of comparison would be impracticable.

As regards design, blink microscopes differ among themselves chiefly in the manner whereby the two plates are alternatively presented to the field of the viewing microscope or short-focus telescope. However this is arranged, there must be provision for (a) adjusting the positions of the two photographs relatively to one another, so that corresponding images can be made to occupy precisely the same position in the field of vision;

(b) moving the two negatives, or the viewing head, or both, so that the whole expanse of the photographs can be covered.

The comparator designed by Will Hay and W. H. Steavenson suffers from the disadvantage that two fairly large and specially shaped prisms are required. The Schnitzer form of comparator (B. 16.24) employs only three small mirrors and a beam-splitter, but, unlike the Hay single-lens design, a separate OG is needed to image each photograph. A simpler form, suitable for home construction, is described in B. 16.21 and illustrated in Figure 184. Field coverage is obtained by moving the optical head up and down in the y direction, while the combined negative carrier can move from side to side in the x direction.

For viewing small negatives, a very simple device has been described by Mobsby (B. 16.23). A small stereoscope-type device views the two negatives placed side by side, and a slowly-rotating 3-bladed shutter (which can be operated by hand) alternately blocks out each negative. Using binocular vision, any non-registering images can also be detected by a pseudo-stereoscopic effect; this is the basis of the professionally-used stereo-comparator.

35-mm negatives or slides may be too crowded with images for Mobsby's simple device to be able to differentiate between individual stars. An alternative method, using two identical projectors to superimpose the two photographs on a screen, is described in B. 16.22.

22.9 Ocular adaptors

In any work which involves frequent changes of magnification it becomes very tedious and time wasting to be continuously unscrewing one ocular, screwing in another, and then refocusing. Revolving ocular adaptors, working on the same principle as microscope objective mountings, obviate all this; they are part of the standard equipment of some Continental and American telescope makers, and are a great boon to the user. They are usually made to hold three or more oculars, which are set in such a manner that little or no refocusing is necessary: all one has to do to change an eye-piece is to rotate the adaptor. With telescopes mounted in the open some trouble may be experienced from dewing up, but this would rarely occur under a dome. Lindemann (B. 8.3) describes a somewhat elaborate design incorporating electrical elements to prevent the condensation of dew on the eye lenses.

The old 31·75-mm (1¼-in) screw-in ocular fittings, which were standard in the UK until the 1950s, have now been replaced by push-in fittings, either of the same diameter (American standard) or using the narrower 24·5-mm Continental barrel. The former fitting is much to be preferred,

FIGURE 184
Blink comparator (*after Mahon*)

405

allowing as it does the accommodation of a larger field stop, and hence a wider LP field. Some designs are made parfocal, so that, once the drawtube is adjusted, each ocular is automatically in focus when pushed home. If the oculars in a collection do not form a homogeneous set, they can be made parfocal by fitting small locating collars to the barrels; or, if the variation in focus is too extreme, they can be mounted in sections of brass tube which fit into the drawtube. Re-focusing can be avoided by fitting flanges on the eye-piece extensions in such a position as to give the correct focus, for a given setting of the drawtube adjustment, when they are pressed home against the end of the drawtube (Figure 185).

FIGURE 185
Parfocal ocular fitting

22.10 Verniers

A vernier is an index for the reading of a moving scale which does not consist of single pointer but of a zero pointer and a set of secondary graduations whose size is such that n graduations of the vernier are equal to $n-1$ graduations of the scale.

If $\theta =$ the required angle between zero on the scale and zero on the vernier,
   $s =$ the angular value of 1 division of the scale,
   $v =$ the angular value of 1 division of the vernier,

then
$$(n-1)s = nv$$

and
$$s - v = \frac{s}{n}$$

If, then, the vernier reads $\theta' +$, and the xth vernier graduation coincides with a scale graduation,
$$\theta = \theta' + [x(s - v)]$$

Suppose, for example, that the scale is graduated in degrees, 5 divisions of which are equal to 6 of the vernier. Then
$$s = 1°$$
$$s - v = \tfrac{1}{6}° = 10'$$

If, for example, the vernier zero lies between 30° and 31°, and the 4th vernier graduation from zero coincides with a scale graduation, then the reading of the scale is given by

$$\theta = 30° + (4 \times 10')$$
$$= 30° \, 40'$$

22.11 Observing seats

A comfortable observing position is of importance only because its lack increases the difficulty and therefore the inaccuracy of the observations; and for this reason alone it is extremely important.

The requirements of any observing seat, whatever its type, are—apart from strength, lightness, and mobility—the ability to bring the eye comfortably to any position of the ocular, and to provide for small adjustments of the height of the seat being made without frequent interruptions to observation. The type and dimensions of a suitable seat are determined to a large extent by the focal length of the telescope, its type (refractor, Newtonian, Cassegrain, etc), and its mounting, and probably no one seat will suit all positions of the telescope.

With a small altazimuth refractor on a tripod mounting the most satisfactory arrangement is to use a low observing chair of fixed height (such as a 'fireside' armchair) and to vary the height of the head from the ground by opening or closing the tripod legs; three metal or concrete strips in the hut floor or the ground, arranged radially at 120° from each other, and carrying sockets at 50-mm intervals into which the spikes of the tripod can fit, allow this to be accomplished quickly and easily.

For an altazimuth mounted on a column, or an equatorial, some variety of observing seat heights can be obtained by means of unequal-sided boxes, e.g. two boxes measuring 15 x 22 x 30 cm and 40 x 50 x 60 cm respectively. But though these may be a first approximation to a solution of the problem, they succeed in filling neither of the specifications of the ideal seat.

An improvement on them is a stepladder whose steps can be hinged upward. The disadvantage here is that either the steps must be uncomfortably narrow, or the vertical distance between steps must be undesirably great. The latter can be reduced by using a box, of height one-half the inter-step distance, as an auxiliary seat to be laid on any required step; but the arrangement is at best a makeshift. A similar arrangement, which has been strongly recommended, consists of a movable seat which can be hooked on to any desired tread of a pair of ordinary household steps.

Complete smoothness of height variation as well as ease of adjustment without moving off the ladder are provided by the Sellers observing chair (B. 16.26). A fixed step at the bottom of the ladder serves as a foot-rest during observation and a platform to stand on when adjusting the seat. The seat itself is constructed as in Figure 186, and the weight of the observer's body keeps it firmly jammed on the two uprights of the ladder. A tendency for the seat to slip if the weight is put too far back on it (also if, when the observer is standing and is taking no load, the back of his coat happens to jog it) is overcome by notching the uprights at intervals of 50 mm or so (Figure 187); this also does away with the wood-screw grips, while retaining a fine enough height adjustment for practical purposes. In one development of this type of ladder (B. 16.25) spring plungers are used to lock the seat to the uprights.

In all ladder-type observing seats, the overall dimensions are dictated by the maximum and minimum heights that the occular may assume. An angle of slope of from 60° to 70° will be found convenient; the ladder

FIGURE 186
The Sellers observing chair

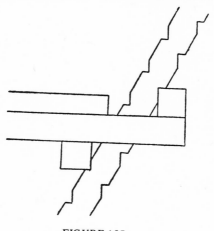

FIGURE 187
Alternative seat fitting

must not be of folding construction,* like a domestic stepladder, but solid; its back should be splayed to the base angles of a triangular base whose corners carry castors or domes of silence.

Observation near the zenith requires special arrangements. Depending on the height of the ocular from the ground when the tube is vertical, a fireside chair or a modified camp-bed may be used; or, with a star diagonal, an unequal-sided box.

---

*Or must at least be fixed in the open position by means of a strut which should be attached not less than halfway up.

CHOICE OF INSTRUMENT

23.1 Factors affecting choice

Both the reflector and the refractor possess specific advantages and weaknesses, the more important of which are listed in the two following sections. In addition, there are certain fields of work to which one is better adapted than the other, and if the observer is intending to specialise this also will affect his choice. In the case of most amateur installations, however, cost, and possibly the amount of space available, are likely to be the determining factors. The scales thus tip heavily on the side of the reflector: for even in cases where the refractor would be the superior instrument, aperture for aperture, the reflector is the superior instrument cost for cost—so important is the role of aperture in an instrument's performance.

Two words of warning should be given; the first against unfair comparisons: between telescopes of different apertures, between a good refractor and an admittedly inferior mirror, between a refractor and an improperly used reflector (e.g. unsuitable oculars). It is also wise to distinguish between those failings of the reflector that are intrinsic and those which are solely a function of focal ratio, therefore shared alike by a refractor and a reflector if their focal ratio is the same.

Secondly, it has been repeated almost *ad nauseam* in the literature of the past many decades that 'the reflector's definition is inferior to that of the refractor'. Sometimes the statement is unqualified, leaving the reader to suppose that some mysterious attribute of reflection makes it less perfect an image-former than refraction; sometimes the reflector's relative inferiority in this respect is attributed to the diffraction structure of its point image; sometimes, again, to the reflector's greater dependence on the atmospheric condition. But the supposed inferiority of the reflector—unsteadiness and poor definition of the image, which render high magnification impossible—is something that vanishes like an exorcised bogle, provided that the mirror is made of the right material, that steps are taken to eliminate tube currents, that the error of the flat is not greater than that of the parabolic figure of the mirror (and that both are small), and, finally, that suitable oculars, with or without a Barlow lens, are used. (See, further, sections 8.8, 11.2, 13.4.)

The modification of the star image by the central obstruction of the reflector has also been built up into something of a nightmare by many writers; but provided its diameter does not exceed about $D/5$ its effect is negligible (see section 2.2).

Whilst it is admittedly true that the performance of many reflectors in amateur hands is markedly inferior to that of refractors with which they should be comparable, it is nevertheless very far from inevitable that this should be so.

23.2 Advantages of the reflector

(a) Perfect achromatism. The secondary spectrum of small refractors used for visual work may not indeed be of much importance, but in the photographic field the reflector is supreme—unless a photovisual can be afforded. (The $H\delta$ focus of the Lick refractor lies 81·5 mm beyond the D_2 focus.)

(b) Relative cheapness. Even for work in a field where the refractor is superior, aperture for aperture, one would hesitate to select a 10-cm refractor in preference to the 20-cm Newtonian which the same money would buy.

(c) Where large apertures are required, the reflector has the advantage. Furthermore, the possibility of employing relatively small focal ratios makes for compactness of design, with the attendant advantages of simpler and cheaper mountings, smaller observatory, and increased stability.

(d) Ease of construction: a single reflecting surface compared with a minimum of four transmitting surfaces, which have also to be very precisely centred and squared.

23.3 Advantages of the refractor

(a) Greater independence of temperature variations, with steadier images and higher possible magnifications than with a reflector of the same aperture. Much can be done, however, to bring the reflector nearer the level of the refractor in this respect (see section 11.2).

(b) Diffraction pattern which may, but does not necessarily, favour the resolution of extended images (see sections 2.2, 2.3).

(c) Relative permanence of the optical adjustments.

(d) The inaccessible optical surfaces being also those protected from the outside air, the OG seldom needs dismantling for cleaning.

(e) A small OG is easier to mount without flexure than a mirror, since the flexures of the individual components tend to cancel one another out in their effects With very large instruments, however, the reflector has the advantage, since its mirror can be supported at as many points as required, whereas the OG can be supported at its edges only.

(f) The larger focal ratio of the refractor gives it some advantages over the Newtonian as usually constructed:

(i) highly corrected oculars are not required;

(ii) equivalent magnifications are obtained with oculars of longer focus;

(iii) larger primary image-scale, aperture for aperture;

(iv) definition at field edges tends to be better, though this is remediable in the reflector by the use of a Barlow lens;

(v) optical defects generally may be cloaked at $f/15$ while being glaringly apparent at $f/5$.

(g) Improved light grasp, at least for visual use in the apertures likely to be found in amateur hands (Figure 188). The transmission of an anti-reflection coated OG is superior to that of a freshly-aluminised Newtonian or Cassegrain combination in apertures of up to about 45 cm; if uncoated, parity occurs at about $D = 30$ cm. However, the superiority of the refractor is intimately connected with light leakage into the secondary spectrum, and hence F/D, and cannot be more than about 0·3 mag under the most favourable possible circumstances (see section 6.8).

23.4 Suitability for different types of work

As a rough guide, objects requiring light grasp rather than critical definition are best served by the reflector, those which make demands primarily on definition, by the refractor. The former choice arises principally out of aperture/cost considerations; the latter, from the reduction of scattered light, less sensitive alignment, and relative freedom from tube currents, of the typical refractor.

For work at the threshold of definition—e.g. lunar and planetary observation—a refractor is probably superior to a reflector of the same aperture, although, as already emphasised, the latter need by no means be as inferior as is often suggested. For work requiring the optical axis to remain rigidly orientated over long periods of time, as in transit instruments, reflectors are disqualified by the comparative mobility of their optical alignment. The use of refractors for double-star work, although hallowed by tradition, does not necessarily exclude the reflector entirely;

FIGURE 188

Transmission characteristics of reflecting and
refracting telescopes compared

satisfactory measures of close pairs have been secured with the 208-cm reflector of the McDonald Observatory, Texas.

For photography a reflector is to be preferred, unless a photovisual is available; this, however, is likely to be ruled out on the score of expense.

Solar work is normally carried out using a refractor, or at least with an OG fed by mirrors made of low-expansion glass. The unimportance of light-grasp should not make one forget that resolving power is also a function of aperture.

In all other fields covered by amateur observers, a moderate-sized reflector is to be preferred to a small (75- or 100-mm) refractor. The latter is perhaps chiefly of use to the beginner during his initial few months of 'star-gazing'; on the other hand, W. H. Pickering's remark that apertures smaller than 12·5 cm are useful for nothing but the observation of variables need not be taken too seriously.

The instrumental requirements in each particular field of observation are discussed more fully in *O.A.A.*, where more detailed information regarding apertures will also be found.

VISION

24.1 Structure of the eye

The eye is a closed, approximately spherical structure. Its outer protective sheath (the *sclerotic*) is transparent at the front; this window is the *cornea*, which contributes largely to the lens action of the eye. A layer of pigment cells on the inner side of the sclerotic (the *choroid layer*) forms, at the front of the eye, the *iris*; this is a diaphragm lying within the cornea, whose diameter varies from about 2·5 to 8 mm, by reflex response, according to the brightness with which it is illuminated. The central aperture of the iris is termed the *pupil*. The inner rear surface of the eye is the light-sensitive *retina*. The movements of the eye as a whole are controlled by six muscles attached to the sclerotic; by their means the whole eyeball can be moved and the eye's field of distinct vision (not more than a degree or two in diameter) may be displaced over the greater part of the whole field of vision (nearly 180°) which can be encompassed without moving the head.

Within the cornea lies the *crystalline lens*, an elastic biconvex structure whose hardness and refractive index increase towards the centre; when focused for distant vision, the radius of curvature of its outer surface is from 6 to 10 mm, that of its inner surface from 5 to 6 mm. The space between the cornea and the lens is filled with salt solution, the *aqueous humour*. The largest chamber of the eye, lying between the crystalline lens and the retina, is filled with a clear gelatinous substance known as the *vitreous humour*.

The crystalline lens is attached to the walls of the eye by the *ciliary muscles*, which are responsible for adjusting its curvature to accommodate the eye to the clearly focused vision of near and distant objects. When the ciliary muscles are completely relaxed, the eye is focused on its *far point*, which with normal eyes is infinity. To focus nearer objects, the ciliary muscles must be more or less contracted, increasing the curvature of the lens; when fully contracted, the eye is focused on its *near point*. Between the near and far points, and closer to the former, is the *distance of distinct vision,* normally about 25 cm. The distance from the near to the far point is the eye's *range of accommodation.*

The function of the cornea, the aqueous humour, the crystalline lens, and the vitreous humour is to prepare the incident light for the retina: until the light falls upon the retina there is no vision. Their refractive indices lie between 1·33 and 1·42, those of the cornea and the aqueous humour being the same. The equivalent focal length of the combination, in the case of an average adult eye, is about 15 mm, giving a retinal image scale of rather less than 4° per mm.

The retina consists of a network of nerve endings, in close contact with one another, each of which is connected with the brain by the *optic nerve.* The retinal image is therefore granular, like a photographic image, though the grain is too fine ever to be perceptible.

The point at which the optic nerve enters the retina is insensitive to light: the *blind spot.* Almost on the optical axis lies a yellow spot, about 1 mm in diameter, the *macula lutea,* centrally placed within which is a minute depression (diameter about 0·25 mm) in the retinal surface, the *fovea centralis,* which is the point of most distinct vision.

The nerve endings of the retina are of two kinds. The *rods* are the more sensitive to faint illumination and occur in greatest numbers some 20° from the centre of the retina; at the periphery they are more common than in the central region, and they are entirely absent from the fovea itself. *Cones,* which are involved in the vision of colour and of brighter illuminations, predominate within the macula lutea, comparatively few occurring outside it, and at the centre of the macula the retinal surface consists of nothing else.

Within the macula the cones are about 0·003 mm in diameter, their centres being separated by about the same distance. Within the fovea the smallest cones are only about 0·0015 mm in diameter, though larger ones, up to about 0·005 mm diameter, also occur.

The rods are bathed in rhodopsin, a substance also known as the *visual purple,* which the eye manufactures and to which the rods owe their light-sensitivity. The visual purple disappears when the retina is exposed to light, and forms again as soon as it is placed in darkness or is only faintly illuminated. Over relatively long periods the aperture of the iris is related to the amount of visual purple present, but its sudden contractions and expansions are independent of this.

24.2 Clear vision, and vision of faint illumination

Since the rods are increasingly numerous from the centre of the retina outwards, and the cones from the periphery of the retina inwards, it follows that clearest vision of a well-illuminated object is obtained when its image falls on the fovea. At $10°$ from the centre of the retina, visual acuity has already dropped to 25 per cent that provided by the fovea, while at $40°$ it is only about 3 per cent. Conversely, a faint object may be seen when its image falls on the outer region of the retina (indirect vision) although it is invisible when its image falls on the macula.

24.3 Defects of the eye and of vision

These are of various types: structural abnormalities producing defective sight which can be counteracted by suitable spectacles; defects of the normal eye, considered as an optical instrument; psychological defects of vision to which everyone is more or less prone. Some details of the commoner examples of each of these are given in the following sections.

The observer should be eye-tested as a matter of course, especially for colour blindness. Of the commoner functional defects of the eye, long and short sight can be corrected at the telescope without wearing glasses, by adjustment of the drawtube; astigmatism necessitates the wearing of glasses if clear vision is to be obtained with all magnifications, though with high powers the exit pupil is so small that they can often be dispensed with; slight or moderate colour insensitivity, providing its nature is known and can be allowed for, need not debar the observer from all work involving the estimation and comparison of stellar, lunar, or planetary colour. Both the acuity and the light-sensitivity of the two eyes are frequently different, and they should be reserved for different types of work accordingly.

24.3.1 Myopia: short sight: Due to abnormal length of the eyeball in the direction of the optical axis. The curvature of the crystalline lens cannot be reduced sufficiently to give it the long focal length necessary to throw the image of a distant object on to the retina. Hence the distance of distinct vision is less than that of a normal eye; it may be only 50 mm or so. The condition is corrected by a diverging lens of suitable curvature.

Extrafocal images of point sources a degree or more in diameter are not uncommon with myopic subjects. One such sufferer put his disability to unexpected advantage, since he found he could quite easily estimate the brightness of the totally eclipsed Moon in stellar magnitudes by making step comparisons between it and the stars!

24.3.2 Hypermetropia: long sight: Due to abnormal shortness of the eyeball in the direction of the optical axis; curvature of the crystalline lens cannot be increased sufficiently to give the short focal length necessary to throw the image of a near object on to the retina. Hence the distance of distinct vision is greater than that of the normal eye.

It is corrected by a converging lens of suitable curvature. Thus a hypermetropic eye does not see distance objects more clearly than a normal eye, but only sees near objects less distinctly. The same is true, in the reverse, of a myopic eye.

The power of the correcting lens required to focus the relaxed eye on its near point is called the amplitude of accommodation.

24.3.3 Presbyopia: The eye tends to become long-sighted with age, owing to progressive changes in the elasticity of the lens, as shown below:

Age (years)	Average amplitude of accommodation (diopters)
10	14
40	4·5
70	0·0

This type of hypermetropia is known as presbyopia.

An eye which suffers from one of these defects is said to be emmetropic.

24.3.4 Astigmatism: Typically a condition of the cornea, less frequently of the crystalline lens: due to the radii of curvature being different in different planes. Thus a point source cannot be focused to a point image.

Correction is obtained by means of a cylindrical lens so oriented that it and the cornea together form a system whose surfaces are symmetrical about its axis. (B. 17.18.)

An uncorrected astigmatic eye may produce lack of definition when low-powered oculars are used, but unless it is very marked its effect will not be noticeable with high magnifications, since the emergent pencil is then narrow, and only the central portions of the cornea and crystalline lens are involved in its transmission to the retina.

24.3.5 Spherical aberration: A condition of the normal eye. The aberration is overcorrected, and hence paraxial rays are brought to a shorter focus than peripheral (cf. biconvex lens).

24.3.6 Chromatic aberration: A condition of the normal eye. When the eye is focused for long sight, the violet rays are brought to a focus slightly nearer the crystalline lens (less than 0·5 mm) than the red.

The chromatic aberration of the eye contributes to the apparent increasing under-correction of an objective as lower and lower magnifications are used: increasingly marginal zones of the crystalline lens then being involved in the formation of the image.

24.3.7 Other defects of cornea or crystalline lens: Local abnormalities of curvature, which cannot be compensated for by glasses, have been suggested as the cause of the instinctive preference shown by some observers for particular image magnification ranges—those producing emergent pencils of diameters which render these faults least obtrusive.

The lens may also lack homogeneity, while lines of compression in the cornea or dust particles or droplets of different refractive index on its surface may be visible when very high magnifications are used.

24.3.8 Muscae volitantes: The so-called 'floating motes' which can be seen by staring at a bright sky. Their origin is uncertain, but it has been suggested that they are detached dead cells floating in the aqueous humour. Douglass estimates their diameter as of the order of 0·025 to 0·050 mm.

Telescopically, they become increasingly obtrusive as the magnification is increased, since their visibility is inversely proportional to the diameter of the emergent pencil. They also appear to be more noticeable when the head is tilted than when it is horizontal—providing an additional reason for using a star diagonal for zenith observation with a refractor or Cassegrain.

24.3.9 Irradiation: At the boundary between areas of markedly unequal brightness, the brighter appears to encroach upon the fainter to an extent that is proportional to their intensity difference. It can therefore be reduced by increasing the magnification.

Irradiation is a physiological effect, being caused by the spreading of the excitation from the retinal area which is actually stimulated by the light. It is thus somewhat analogous to photographic halation.

Examples are: the Martian caps, planetary and satellite discs, appear larger than photographicaly shown to be; dark spots, as on the Moon, appear conversely to be smaller.

24.3.10 After-images: The image of a conspicuous object which has been intently focused will survive for a few seconds after the object has been removed; it then fades gradually. After-images are of the complementary colour to that of the object. Thus, if a green ball is stared at for about half a minute, and the gaze then transferred to a sheet of white paper, a dim red disc of the same angular size as the ball will be seen.

24.3.11 Fatigue: Any operation involving concentrated visual attention fatigues the eye rapidly, so that its performance becomes increasingly unreliable. Such operations as bisecting a stellar image with a web, eliminating diffraction fringes, making visual estimates of the relative brightness of two stars, etc, should therefore be made as quickly as possible, and the eye given a chance to rest between observations.

For the same rason, movement of a faint object in the field (by tapping the telescope or even by movements of the eye itself, so that the image moves over the retina) often helps to render it visible.

A slight change of focus is also often very helpful to the eye during prolonged sessions at the telescope.

(Colour fatigue is dealt with below, see section 24.4.)

24.3.12 Mid-point error: An innate and fairly constant inability in some observers to judge accurately the mid-point of a line (vertical or horizontal), or the point midway between two other points. The tolerably constant extent of the misjudgment is shown by the constant value of the personal equation throughout a long series of such observations by a single observer. In one experimental test, the average errors of 10 observers were:

horizontal: mid-point at 0·509: 0·9 per cent error
vertical: mid-point at 0·486: 1·4 per cent error

In another test using 16 observers, where the mid-point was adjusted in increments of 1 per cent, the mid-point of a horizontal line was, on average, judged to be 0·5 per cent too far to the right, while that of a vertical line was 0·45 per cent too low.

Estimating the position of the central meridian of a disc in order to obtain transit times of planetary markings is an instance of the type of observation susceptible to this error.

24.3.13 Leading error: The tendency, when a star, perceived to be in motion across the field (by comparison with fixed webs, for instance), is being kept bisected by a moving web, to adjust the web systematically p or f the star's image. This error is to be distinguished from mid-point or bisection error, which also occurs when the image is, or appears to be, stationary. It is unlikely to be of much practical concern to amateurs.

24.3.14 Scale error: In the estimation of the proportional distance of one 'event' between two others (e.g. the trasit of a star at a micrometer web in relation to two beats of the clock, or the brightness of one star in terms of a brighter and a fainter comparison star), a systematic preference for certain proportions at the expense of others is sometimes encountered. This is termed scale error. Where present it is found that in a long series of observations, all the steps in the scale (e.g. the 10 steps of 0·1 between 0·0 and 1·0) would not occur with equal frequency.

24.3.15 Contrast error: A tendency to over-estimate the difference in brightness of two dissimilarly bright objects in the same field. Some observers suffer more than others from this source of error, but in each case it tends to be reasonably constant. If the luminosity-difference of the two points or areas is very great, decreased dark adaptation of the eye may be introduced as an additional factor tending to increase the error.

24.3.16 Other psychological pitfalls: It is vitally important to go to the telescope with a mind completely open and free from preconceptions or expectations. The insidiousness of the danger of preconceived ideas cannot be over-stressed. One should be prepared to see literally anything, and, ideally, each time an observation is made that does *not* surprise one, one should immediately be suspicious and on one's guard against a visual datum suggesting more than it it in fact states.

The study of other observers' drawings, what one has read, unconscious conditioning by one's own previous observations—all these have a strong effect, of which one is usually unaware, upon what one 'sees'. The two schools of interpretation of Martian detail, and Hargreaves' theory of 'mentors' (B. 17.9) may be mentioned in this context.

The accuracy and objectivity of one's drawings can easily be gauged by making indoor tests on meaningless drawings prepared by a second person and observed with the same relative aperture and magnification as are employed for planetary work (see B. 17.8, 17.11-16).

24.4 Colour vision, and false colour effects

Wavelengths to which the normal eye is sensitive range from 420 nm to about 700 nm; exceptionally from 320 nm to 835 nm. The latter short-wave limit approaches the region of telluric absorption which obliterates the shorter wavelengths. The greater part of colour vision depends upon the action of the cones, the rods being comparatively colour-blind. The zone of maximum sensitivity occurs in the yellow at about 560 nm when the illumination is moderate or strong, but as the illumination is decreased towards the extinction threshold it shifts towards the green (Purkinje effect, see below).

The eye's absolute discrimination of colour is comparatively poor, and, furthermore, colour fatigue occurs rapidly, causing still greater insensitivity to uncompared tints. On the other hand, the eye is extremely sensitive to differences or changes of tint; this sensitivity is greatest in the yellow and green and lower in the red, blue, and violet, the latter being largely absorbed by the crystalline lens. An average eye can distinguish tints differing by from 1 to 3 nm between 500 nm and 600 nm, the minimal perceptible difference rising to 6 nm at the ends of the visible spectrum. A normal, untrained eye can thus distinguish a hundred or more tints between the blue and the orange.

The physiology of colour vision, involving the complex interaction of rods, cones, and visual purple, is still imperfectly understood. Colour blindness exists in many forms, from a slight relative insensitivity to a single colour, to completely colourless vision. Men appear to be about ten times more prone to it than women, 4 per cent being afflicted with colour blindness of some sort or another. The colour vision of the two eyes is not infrequently different, and this also the observer should be aware of if it applies to him. If, over a long period, one eye has been continuously used for solar observation, it is likely that its colour vision will have been slightly affected.

From the macula to the periphery the retina is progressively more sensitive to short wavelengths and less sensitive to long. This is the physiological basis of the Purkinje effect, since the peripheral rods are the elements concerned in the vision of faint illuminations. This source of error may be described in various ways according to the form it takes in practice. Basically, it is that the wavelength of maximum sensitivity varies with the intensity of the illumination, shifting from the yellow (in daylight) into the green (dark-adapted night vision). Astronomically this is chiefly of interest as a source of uncertainty in the visual estimations of the brightness of faint stars when these are noticeably coloured. A source may be too dim to stimulate the colour-sensitive elements of the retina (the cones) while still being bright enough to stimulate the night-vision elements (rods), which are comparatively insensitive to colour.

Examples of the Purkinje effect are:

(a) The apparent brightness of a red star, estimated by comparing it with a white star, varies according as to whether it is relatively bright or faint. If it is faint, its estimated brightness will be too low; if it is relatively bright, or is observed against a faintly illuminated background, it will be judged too bright. More generally, if two stars of different colours are judged to be equally bright, and their brightness is then changed by an equal amount, they will no longer be judged to be equally bright.*

(b) In late twilight, red flowers lose their colour and appear black, though distinctly visible as shapes.

(c) If the brightness of a coloured light is decreased to extinction, it does not retain its colour till the last moment. Over a limited range of faint visibility immediately before extinction (the so-called achromatic interval) it appears grey; within this range the rods are alone functioning.

(d) The greyish-green colour which is the prevalent tone of objects in moonlight is an example of the Purkinje effect. Moonlight being reflected sunlight, its 'coldness' is dependent solely upon its lower intensity.

*The variable-star observer is never allowed to forget the Purkinje effect for long, and a further discussion of it in connexion with this type of work will be found in O.A.A., section 17.5.

Another effect of colour upon apparent brightness is shown by many observers who will habitually see one member of a pair of equally bright coloured stars as brighter or fainter than the other. Usually with a red and a blue pair, the blue appears to be brighter than the red. This is not, however, an example of the Purkinje effect, since it applies to stars far above the luminosity level at which the latter comes into operation.

Three well-known examples of defective colour vision which affect contiguous coloured areas are:

(*a*) The intensity of a patch of colour appears to be heightened if it is placed in contact with, or proximity to, its complementary colour.

(*b*) A white patch on a coloured ground appears to be faintly tinged with the colour complementary to that of the ground.

(*c*) When a bright and a faint object of different colours are observed together, the fainter will be tinged with the complementary colour of the brighter. For example, the white *comes* of a red primary appears bluish or greenish; dusky markings on the reddish Martian surface appear greener than they are.

24.5 Resolving and separating acuity of the eye

Depends upon the structure of the retina—i.e. upon the size and spacing of the nerve elements—and does not vary greatly from one individual to another, providing no pathological condition is present. The area of most distinct vision being the fovea, where the smallest cones are about 0·0015 mm in diameter, it follows that the centre of two cones will not be stimulated unless the diameter of the image is greater than about 20″. One function of the telescope is to spread the image over a larger area of the retina, increasing the number of cones that are stimulated, and hence the degree of resolution.

The resolution threshold of 20″ for 'point' images assumes, however, that the pupil is at its greatest aperture. Under these conditions the eye's abaxial aberrations will impair the definition of the image, and for clearly defined night vision the threshold is nearer 60″. In an extremely interesting series of experiments carried out by E. W. and A. S. D. Maunder (B. 17.14) a very constant set of results was obtained, giving the minimum diameter of a visible black spot on a white ground (eye not dark-adapted) as 34″.

A linear object may stimulate a sufficient number of cones to produce sight even though its width is 20 or 30 times less than the threshold diameter of a spot. The Maunders' results (with independently obtained results by W. H. Pickering and Lowell for comparison) for a black thread viewed against a light background were:

	Maunders	Pickering	Lowell
Invisible	—	0″72	—
Glimpsed	1″2	0″82	0″73, 0″83
Steadily held but ill-defined	1″3	—	—
Held	—	1″15	1″13
Clearly defined	1″4	—	—

Again showing remarkable uniformity considering that the conditions of the experiments were not the same in each case.

Though 1″ may be taken as, roughly, the minimum angular width of a visible line, individual observers have considerably reduced it. Barnard, for example, was able to glimpse a wire suspended against the day sky when its diameter was only 0″44. In general, a lower threshold will be reached with a white line on a dark ground if viewed in daylight, probably on account of the assistance given by irradiation: with sufficient contrast, a bright line 1″ wide looks at least 1′ wide, and often much more.

Astronomical objects offer little opportunity for observing narrow lines with a dark-adapted eye (i.e. a bright line on a dark ground), and in the case of dark linear objects against a bright ground (lunar and Martian canali, for example) the threshold is comparable with that obtained experimentally from dark lines on a bright ground, rather than from bright lines on a dark ground where irradiation tends to widen rather than narrow the image. It was discovered by Maunder that the threshold is raised if the length of wire is considerably less than the diameter of the field of view, as is normally (but not necessarily) the case in astronomical application.

The threshold for the separation of two points (black on white) is much the same as that for the practical visibility of a single point, about 1′; for the centres of two points that are to be seen separately must be more than two cone diameters apart, i.e. more than about 40″. If, however, they are stellar points, the aberrations of the dark-adapted eye with which their observation will normally be made raise the threshold by a factor of about 3, though individual eyes vary rather widely. Naked-eye separation of ε Lyrae (3′5), for instance, requires at least average acuity. Pickering, however, obtained clear resolution of pinholes in black paper held against the light when their separation was only 86″. The threshold for the resolution of parallel lines is about that for the visibility of a 'point' image— namely, the region of 1′.

The acuity of naked eyesight usually decreases with age, though there is evidence that parallel decrease in the acuity of telescopic vision need not necessarily occur. This no doubt springs from the fact that the extreme threshold of visual acuity is not in practice called into play in telescopic work, atmospheric and optical causes having a greater effect in rendering fine detail invisible.

The figures quoted above assume optimal contrast conditions. These conditions do not apply in, for example, planetary work, and this account should be read in conjunction with section 24.8 below.

24.6 Visibility of extended detail

The complexities of planetary and lunar detail test the eye's powers of resolution and separation, as well as its power of discriminating between small differences and absolute values of colour and brightness, to the limit.

Although a spot may be glimpsed if its diameter is $20''$ to $30''$, and clearly seen when its diameter is about $60''$, the shape of a visible spot does not begin to emerge until its diameter approaches $2'$, or even $3'$. Claims that, for example, the crescent of Venus ($50''$ to $60''$) or the elliptical shape of Saturn ($40'' \times 18''$) can be discerned without optical aid should therefore be treated with reserve. Differences of size between adjacent objects, on the other hand, may be detected even when as small as $10''$.

There is experimental evidence that if the eye is abnormally acute in its perception of faint linear markings, it tends to be below average in its sensitivity both to faint spots and to tone gradients. Indeed, particular acuity in any one of these directions seems often to inhibit acuity in the other two (B. 17.13).

Among the commoner types of misperception that are liable to occur near the threshold of resolution or illumination are:

(a) The appearance of two adjacent spots, each individually below the threshold of visibility, as a single spot of approximately the threshold size.

(b) The appearance of a wavy line as a straight band.

(c) The appearance of a line of spots as a continuous, though faint and ill-defined, line. Spots no more than $8''$ in diameter can produce this effect, and with a double row the diamters of the individual spots can be less than $4''$. One of the Maunders' experiments, designed to throw light upon the then mysterious nature of the Martian canali, consisted in exhibiting a 15-cm disc, on which an irregular pattern of miscellaneous spots had been drawn, to a class of schoolboys sitting at various distances from the disc; the boys then drew what they thought they saw. The results were:

4·5 m from disc (angular diameter of disc = 2°): correct delineation;

9 m from disc (angular diameter ≏ 1°): a number of canali drawn;

more than 12 m from disc (angular diameter < 43′): few details of any sort were seen or drawn.

24.7 Importance of focus

Failure to secure absolutely precise focus can produce deceptive appearances; similar conditions are provided by inferior seeing or intrinsic lack of definition. Owing to the nature of the focus there is, theoretically, no such thing as 'the' focus, the convergent and divergent cones overlapping by an amount equal to $2\frac{F}{D}.d$, where d is the diameter of the spurious disc.

The focal plane, c, is in practice impossible to locate more precisely than somewhere between a and b:

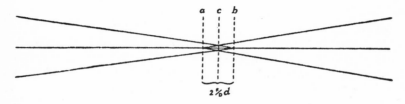

FIGURE 189

Indeterminate nature of the precise focus of an objective

Nevertheless the importance of precise focus, especially in planetary and similar work, cannot be overrated. It is therefore unfortunate that inadequate data exist concerning the effect of fatigue upon the focus of the eye, though it appears normal for it to shorten during prolonged observation, necessitating a corresponding readjustment of the telescopic focus. There is always a tendency to focus too short when striving to make out the details or existence of an object near the threshold of visibility; this imposes a strain on the eye which rapidly becomes very tiring, and it is better to start with the focus if anything too long rather than too short.

It is certain that slight variations of focus are caused by the atmosphere, but the extent and effect of these—as of the eye's focus resulting from fatigue—are very difficult to estimate. At least it appears to be fairly well established that a star at low altitude requires a slightly shorter focus than one near the zenith.

When observing objects which themselves lack sharp definition, it is advisable to establish the best telescopic focus first by means of a star, or it may be very considerably in error. A triangular objective diaphragm allows a fine adjustment of the focus to be made. The diffraction pattern of a stellar image with such a diaphragm is 6-rayed; the extrafocal images are triangular, with the angles of the inside-focus image occupying the position of the sides of the outside-focus image. A precise focus is easily obtained by adjusting the drawtube so as to make the 6 rays of the focused image equal.

Some of the subjective effects of a slight displacement of the telescopic focus have been investigated by Antoniadi (see especially B. 17.4, 5). He found, for instance, that a single band could be expanded into a pair of partically overlapping bands, of which only the darker core may be visible as an ill-defined dark line. The intersection of such bands may produce an illusion very similar to the appearance of Martian 'lakes' at the intersection of many of the line-like canali.

24.8 Importance of contrast

Adequate contrast is a vital element in the observation of extended images: even though a marking may be large enough to be resolved by the objective, the magnification sufficient for it to be resolved by the eye, and the aperture sufficient to provide a bright enough image, yet it will remain invisible unless the tonal contrast between it and its surroundings is above a certain threshold.

If the brightness of the two areas are B and b, then their contrast is measured by

$$\gamma = \frac{B-b}{B}$$

The threshold value of γ, below which B and b cannot be differentiated, may be designated γ', and is a function of B. Thus whilst a dark marking on a white ground viewed in a well day-lit room can be distinguished when $\gamma'=0.005$, γ' increases to about 0.02 in the case of the Moon and the brighter planets, to 0.3 in the case of the fainter nebulae, and to about 0.5 in the case of the night sky itself.*

These figures are 'average'. An experienced planetary observer may distinguish features two or three times less contrasted than quoted, providing the areas are not very small, and that they are contiguous. Furthermore, the contrast sensitivity of different eyes at different brightness levels varies considerably; hence one observer may be relatively

---

*This is only true if, as recommended by Danjon and Couder (B. 7.29), the magnification used on planets and the Moon is such that

$$0.1 > \left(\frac{M'}{m}\right)^2 > 0.007$$

(See section 1.3.)

426

more sensitive to faint variations of tone within a dark area, whilst to another they are most easily detected against a relatively bright ground.

The value of γ' also increases again if B is increased beyond a certain level. But this fact has little astronomical application except in the case of Venus, and possibly also of the Moon. Generally speaking, any reduction of brightness tends to increase the value of γ'. Thus lowering the magnification will make faintly contrasted detail more clearly visible, provided it covers a large enough area for the eye's resolving power not to enter the picture. Some astronomical objects, however, are so intrinsically faint that no practicable reduction of magnification will reveal their structure. A comparison of Lord Rosse's drawings of nebulae with later photographs brings this out clearly. In such circumstances there is, for visual observation, no solution but an increase of aperture.

Other agents of reduced contrast are mist, high and tenuous cloud, skylight from whatever cause, and dirty or dusty lenses.

The visibility of colour contrast, like that of tone, is also a function of image-brightness.

The artificial increase of contrast in certain cases by means of suitable filters, polarising devices, etc, is a widely practised device among planetary and lunar observers.

γ is not only a factor in the visibility or otherwise of a marking, but also in the threshold of size at which it emerges from invisibility. The subject is complex—as, indeed, is the whole field of the visibility of detail in extended images—and has never been thoroughly tackled.

The results of an investigation by Danjon and Chapeau (B. 17.6) are summarised below. Their test object was a grid of parallel equidistant lines, whose width was equal to that of the interspaces, drawn on a white ground. Though admittedly artificial, it nevertheless seems to uncover some of the factors controlling visual acuity.

Given adequate contrast between the lines and the interspaces, resolution is possible by an experienced observer when, as we have just seen, their angular separation is about $60''$. By reducing either the brightness of the illumination of the test object, or γ, this value is increased. Furthermore, without varying either of these factors, the minimum separation for resolution may be increased merely by reducing the number of lines in the grid; one set of values derived empirically was:

11 lines in grid: minimum separation for resolution, $89''$

6	”	”	”	$92''$
3	”	”	”	$101''$
2	”	”	”	$115''$

24.9 Illumination sensitivity of the eye

This varies (quite apart from real differences of sensitivity in different observers, which is partly innate and partly a function of age: night blindness tends to increase with age) with several factors:

1. degree of dark adaptation;
2. nature of the background;
3. position of image on retina;
4. nature of object;
5. colour of object.

1. Within a second or two of the eye being screened from light (or shielded by dense red goggles), the pupil dilates to a diameter of about 8 mm. The greater part of dark adaptation, however, concerns the production of the visual purple, since the area of the pupil only varies by a factor of about 16, whereas the total sensitivity range of the eye is of the order of a million. Most of the visual purple is produced during the first 20 minutes or half an hour of screening, though the final stage of adaptation may continue more slowly for as long as 2 hours. The dark adaptation of the cones, which is of little importance, is completed within the first 10 minutes.

Production of the visual purple is inhibited if the body suffers from vitamin A deficiency. Carrots and cod-liver oil, whose efficacy was proved during the war years, might profitably be included regularly in the diet of the comet-seeker. Nicotine also retards dark adaptation, even one cigarette smoked half an hour previously having a noticeable effect.

Wide variations of night-vision sensitivity are shown by different individuals, and this should be taken into account before deciding in which branch of observation to specialise. The speed with which full adaptation is reached also varies considerably between one individual and another.

2. Whether the background is dark or more or less luminous may alter the threshold of perceptible illumination by as much as two magnitudes. On a clear, moonless night the normal adapted eye can just see stars of mag 6·5. By observing the sky through a black screen which occludes all general starlight and skylight save that of the star under observation, whose light passes through a small hole, mag 8·5 can be reached. Under laboratory conditions, where all extraneous light can more easily be excluded, visibility of illuminations equivalent to mag 8·5 has also been recorded.

It will be noted that these optimum conditions approximate to those of telescopic observation with a high power ocular. For this reason, published tables relating aperture and minimum visible magnitude which are constructed by extrapolation from naked-eye observations (when the sky is considerably brighter than the field of an astronomical ocular) may be misleading by as much as 2 mags.

3. (*a*) The zone of the retina situated from about 20° to 50° from the centre is most profusely supplied with rods, and is therefore more sensitive to faint illuminations than either the macula or the extreme periphery. The focusing of an image on this outer region of the retina is termed averted, or indirect, vision. It may gain up to 2·5 mags, as compared with direct vision (the image falling on the macula), though the efficacy of indirect vision varies with individuals, and in some cases seems to offer no advantage at all; such differences are attributable to a different relative number of rods and cones in the inner and outer regions of the retinas.

(*b*) The upper part of the retina generally possesses greater sensitivity than the lower. Thus if two stars appear equally bright when the line joining them is parallel to the line joining the eyes, the lower star will appear brighter than the upper when the head is turned through 90°. It is on this fact that the variable star observer's rule is based, always to incline the head so that the line joining the variable and the comparison star is parallel to the line of eyes.

(*c*) Less noticeably, the side of the retina nearest the temple tends to be more sensitive to faint illumination that that nearer the nose.

4. Ill-defined objects respond better to indirect vision than well-defined objects (especially point sources), since the outer regions of the retina are incapable of providing sharp definition.

5. Since the outer regions of the retina are comparatively more sensitive to short- than to long-wave radiations, the observation of a blue star or a nebula (typically greenish) by indirect vision will yield a lower threshold than that of a red star. In other words, indirect vision will increase the visibility of a faint blue, green, or white object considerably, where little will be gained by observing a faint red star in this manner. It is because of this relative insensitivity of the dark-adapted eye to the longer wavelengths that a red working light is used at the telescope, and that the boredom of dark-adapting in total darkness can be obviated by the use of red goggles.

Experiments performed by Waterfield suggest that the outer side of the retina is more sensitive to all colours than the nose side; and that the lower side is more sensitive to green, but less sensitive to red and yellow, than the upper side.

As with colour and size, the eye is extremely sensitive to differences of brightness, while being comparatively insensitive to absolute values. Variable-star observers habitually make their comparisons to 0·1 mag, and under laboratory conditions sensitivity to differences of 0·01 mag has been recorded. For very bright objects, as equally for those near the limit of perception, the eye's powers of discrimination are weakened.

24.10 Tests of naked-eye acuity and illumination thresholds

The observer should determine his own threshold of each type of acuity and illumination sensitivity. A wire (measured by calipers or microscope) stretched across a window may be used. It should be approached from a distance at which it is invisible, and the distances noted at which (a) it is glimpsed, (b) it is seen steadily. From these distances and the known linear diameter of the wire its angular subtention can be calculated. A white line drawn on a black ground and observed in daylight will yield a lower threshold.

Various natural test objects have been proposed from time to time in order to provide a standard of naked-eye acuity. Pickering gives the following graduated series of tests on lunar objects:

Scale Number	Test object
1	Bright region surrounding Copernicus, the most conspicuous lunar crater
2	Mare Nectaris
3	Mare Humorum
4	Kepler
5	The Gassendi region
6	The Plinius region
7	Mare Vaporum
8	The Lubiniezky region
9	Sinus Medii: dark line running E and W
10	Faintly shaded area in the region of Sacrobosco
11	Dark spot at the foot of Mt Huyghens, across the Apennines from Mare Vaporum
12	Ripheaen Mountains

1-10 should be visible to really good eyesight, 7 representing reasonably good sight if it can be steadily held; 11 represents exceptional acuity, and 12 is possibly just beyond the reach of the best. The objects can first be identified with binoculars and a map, and then looked for with the naked eyesight.

The Pleiades furnish a convenient test of the illumination-sensitivity of the naked eye. 9 or 10 stars are normally seen by perfect, but not exceptional, sight; 6 in full moonlight. Exceptional sight will show 12, though two or three of these may only be glimpsed; even more have been claimed.

Similarly, the two outer satellites of Jupiter have occasionally been seen at elongation (B. 17.7).

Such objects, however, do not give a direct measure of the eye's acuity, since the faint objects are to some extent swamped by, in the first case, the brighter Pleiades, and, in the second, Jupiter itself. They are, much more, a test of the eye's defining power and freedom from scattered light (removed from the vicinity of Jupiter, all of the Galilean satellites would be easy naked-eye objects). Failure to pass these tests does not necessarily mean that the stars concerned would not be seen if sharply focused. In other words, naked-eye tests are not a safe guide to what can be accomplished by the same eye using a telescope. Those who fail to see 13 stars in the Pleiades need not therefore conclude that they will make inferior telescopic observers, or that their sensitivity to faint illuminations is below normal.

It should always be remembered that the sensitivities of the two eyes, both as regards resolution and the perception of faint illumination, are not necessarily the same. It is up to the individual to find out and make the best use of his natural powers.

PERSONAL EQUATION

'Personal equation' is a term applied to an observer's systematic misobservation of a particular phenomenon or situation; it also becomes, with a change of sign, the correction to be applied to his observations in order to obtain the true quantities. It is particularly applicable to estimations of the time at which an observed event occurs, but is to be encountered in most fields of observation. In some cases this systematic misobservation can be attributed to definite causes—either in the psychology or physiology of vision, or in the conditions under which the observations are made—while in others it can only be attributed to the 'personality' of the observer, which is equivalent to saying that its precise cause is unknown.

The discussion of personal equation here given will refer only to the observation of objects—stars, or the limbs of planets or of the Sun or the Moon—at a micrometer web, since it is in connexion with such work that most investigation of personal equation has been done; see also sections 24.3.12-24.3.16, 28.19, and *O.A.A.*, sections 2.7.5, 17.5, 17.7, 18.2. An observer's equation in the timing of stellar transits of webs, limb transits of webs, stellar occultations, etc, will not necessarily be the same.

The story began in 1796 when Kinnebrook, an assistant at Greenwich, was sacked by Maskelyne, the then Astronomer Royal, because his transit observations were systematically 0^s5 to 0^s8 later than his employer's. The method employed was eye-and-ear with a multiple-web eyepiece and a clock beating seconds, estimations being made to the nearest 0^s1. The problem of personal equation in transit work was dealt with, to the satisfaction of Maskelyne, by the dismissal of Kinnebrook. It was first seriously considered by Bessel, during the early years of last century. Since then it has been continuously in the front of the astronomer's consciousness, and the introduction first of the key-operated chronograph and then of the 'impersonal' micrometer has been his attempt to reduce it. The following conclusions have been reached:

1. Personal equation is normally a lag, i.e. observations are recorded too late, so that the correction to be applied to the recorded time in order to obtain the correct time carries a minus sign. This is not invariably the case, however, and to some extent it depends on the observational method used; thus the Greenwich observers (B. 18.4, 5) habitually anticipated the true time of transit when using the eye-and-ear method, but not when using the chronograph. Equations of the order of 0^s5 are common with eye-and-ear

2. For a given observer, working under uniform conditions, it may be regarded as sensibly constant for at least short periods. There is evidence of long-period variation, possibly due to the physiological effects of age. Bessel's relative eye-and-ear equation with Struve, for example, changed (probably to rather an abnormal extent in such a short period) from 0^s044 (1814) to 0^s680 (1820) to 0^s799 (1821) to 1^s021 (1823).

3. For a given observer, the personal equation depends upon the observational method employed Three varieties of equation can therefore be distinguished: (*i*) absolute personal equation—between the observed times of a single observer and the actual times; (*ii*) relative equation between two observers using the same method; (*iii*) relative equation of two methods used by a single observer. As regards methods of observation, decreasing equation is experienced along the series:

eye-and-ear: 1-sec beats;
eye-and-ear: 0·5 sec beats;
stopwatch;
hand-operated Repsold ('impersonal', or moving-web) micrometer;
mechanically operated Repsold micrometer.

The Repsold micrometer consists of a movable web which it is the observer's task to keep bisecting the stellar image; it is either hand-operated or mechanically driven with a hand-operated slow motion, and the conjunctions of the moving web with the fixed webs are automatically recorded, either by electrical contracts or by photography.

Examples of dependence of personal equation upon method are:

(*a*) The relative equation of Bessel and Argelander was reduced from 1^s22 to 0^s5 by changing from 1-sec to 0·5-sec beats.

(*b*) Average equation of the Greenwich observers over a period of 9 years was 0^s13 to 0^s20 greater by eye-and-ear than by chronograph, the former anticipating the latter.

(*c*) In another case, the average probable error of 3 observers was determined as $\pm0^s081$ for eye-and-ear, and $\pm0^s053$ with chronograph.

(*d*) Average equation of one set of 5 observers with a Repsold hand-operated micrometer was $+0^s008$, and with a mechanically-operated Repsold between $+0^s001$ and -0^s004, with a probable error of $±0^s002$.

(e) At the Cape Observatory the relative equations of experienced observers were as large as 0^s25 when the eye-and-ear and chronograph methods were in use; with a hand-operated Repsold micrometer, the relative equations of 6 observers were reduced to $<0^s06$; with a mechanically driven web they were further reduced to $<0^s02$.

4. The brightness of the star, and also therefore the aperture of the instrument, will affect the personal equation. 'Magnitude equation', i.e. variation of personal equation with the brightness of the stellar image, affects observations by the chronograph method: the usual tendency is to press the key too early with a bright star, and too late with a faint one. General order of the magnitude equation is 0^s05 to 0^s1 over a range of several magnitudes; two specific investigations arriving at equations of 0^s007 and 0^s009 per mag. Both methods of using the key (see paragraphs 5(a) and (b) below) appear to be subject to magnitude equation. The moving-web micrometer is, on the other hand, free from it.

5. 'Declination equation' is a variation of the personal equation with the Dec of the observed star, where fixed webs are used with a chronograph. There are two methods of using the latter: (a) to start the muscular action of pressing the key slightly in advance of the bisection of the star image by the web, so that the electrical contact is completed at the precise instant of bisection; (b) to start pressing the key at the observed moment of bisection of the star by the web. The former method, involving the estimation of a minute angular separation, and the relating of this to the star's angular velocity towards the web, is subject to Declination equation, since a star's velocity across the field is a function of its Dec ($\propto \sec \delta$). (b), on the other hand, is immune from Declination equation. Observations made with the travelling-web micrometer are also free from it.

6. Personal equation is affected slightly by the direction of the star's motion across the field: (a) star south of zenith, observer's head north of the telescope, star's motion (in inverted field) right to left; (b) star north of the zenith, head south of the telescope, star's motion left to right. Affects both 5(a) and 5(b). It is eliminated by making half the observations with a reversing eyepiece.

7. Personal equation with both eye-and-ear and chronometer is partly due to scale error (see section 24.3.14).

8. Observations of limb transits between parallel webs are subject to the equation, differing from that of stars, which is partially caused by bisection, or mid-point error (see section 24.3.12). Eliminated by the reversing eyepiece.

9. A given observer has different personalities in respect of the p and the f limbs in case 8 above. This equation can be reduced by the reversing eyepiece.

10. A cause of accidental error in transit observations, which is not a contributing factor to personal equation, is the phenomenon known as 'jumping the wire': an illusion that the star pauses on the f side of the web and then, an instant later, has crossed to the other side—rather in the manner that electrons are supposed to move from one orbit to another. It is particularly noticeable when the seeing is poor. The moving-web micrometer is free from this source of error.

11. The main contributions to personal equation with the Repsold micrometer is from leading error (see section 24.3.13). Not eliminated by a reversing eyepiece, since the illusion of motion is still conveyed by the fingers operating the web-driving screw. Much reduced, however, by replacing manual by mechanical web drive.

12. Personal equation is probably subject to short-term variations of small amplitude, dependent upon such factors as the observer's health, freshness or fatigue, state of digestion, even his emotional life.

13. True 'personality' is a term that may be applied to the residual tendency for an observer's reaction to lag behind (occasionally to anticipate) his reception of the stimulus, when the other factors mentioned above have been eliminated or allowed for. Factors contributing mainly to this residual lag appear to be:

(a) Reaction times vary with the sense by means of which the stimulus is received. Reaction times, when the observer is immediately pre-warned of the stimulus, are of the order of 0^s12 to 0^s18 for sound and touch, and 0^s15 to 0^s225 for sight.

(b) Reactions to multiple stimuli through more than one sense are subject to larger and less uniform errors than those to a single stimulus. Hence the superiority of the stopwatch over the eye-and-ear method. In an extreme case, when simultaneous stimuli are received by two senses, the order in which they appear to occur may depend solely upon the relative concentration that the observer devotes to each.

(c) Degree of warning of the receipt of the stimulus that is given. Immediately prior warning results in shorter and more consistent reaction times.

(d) Predisposition of the observer's attention: either mainly on the anticipated stimulus, or on the muscular response with which he will react when he receives it. Reaction times from the former are systematically longer than those from the latter by approximately 0^s1.

The determination of the relative equation of two observers is a simple enough matter: it is the mean difference between the two sets of clock corrections indicated by a series of observations made under similar conditions by each of them. The determination of an observer's absolute equation involves the employment of one of the numerous types of 'personal equation machine' that has been evolved—instruments which record both the observer's reactions and, without any equation of their own, the occurrence of the stimuli to which he is reacting.

For a general survey of the subject from the historical standpoint, see B. 18.1, which also contains an extensive bibliography; B. 18.4, 5 discuss the personal equations of the Greenwich observers from 1885 to 1894; B. 18.3 contains an interesting discussion of the errors—personal and otherwise—which are encountered in the application of the micrometer web to the measurement of planetary discs.

THE ATMOSPHERE AND 'SEEING'

In spite of the fact that astronomical observations may now be made from rockets and artificial satellites above the Earth's atmosphere, the great bulk of astronomical observations are still made through the atmosphere, which amounts to an additional element in the instrument's optical train; and its nature, behaviour, and optical effects concern the astronomer very closely.

26.1 Composition, density and extent of the atmosphere

The atmosphere consists, by weight, of

	per cent
Nitrogen	75·48
Oxygen	23·18
Carbon dioxide	0·045
Argon	1·29
Hydrogen ⎫ Helium ⎪ Neon ⎬ Krypton ⎪ Xenon ⎪ Ozone ⎭	0·005

The average pressure of the atmosphere at Mean Sea Level is 1013·2 millibars, or 760·0 mm of mercury (where 1 mb $= 0·7501$ mm of mercury $= 0·02953$ ins of mercury $= 10^3$ dynes/cm^2).

If $\rho_0 =$ atmospheric density at MSL,
$\rho_n =$ atmospheric density at $5·36n$ km above MSL,
$n =$ any integer,

then
$$\rho_n = \frac{\rho_0}{2^n}$$

Hence:

Height above MSL (km)	Density ($\times \rho_0$)
5·4	0·5
10·7	0·25
16·1	0·125
21·4	0·0625
26·8	0·0312
32·1	0·0156
37·5	0·0078
42·9	0·0039
48·2	0·0019
53·6	0·0009

Thus about nine-tenths of the atmosphere is to be found within 16 km of the Earth's surface; the highest types of cloud (cirrus, composed of ice crystals) occur between 8 and 10 km of MSL. The troposphere, or main mass of the atmosphere directly overlying the surface, is 17 km deep at the equator, 11 km in the latitudes of England, and rather less over the poles. The upper surface of the troposphere is known as the tropopause, and is characterised by a sharp change in the lapse-rate, or rate of fall of temperature with increasing altitude: throughout the troposphere the lapse-rate averages about 0·6°C per 100 m. The dry adiabatic lapse-rate, due solely to the expansion of a rising mass of air brought about by decreasing pressure, is about 1·0°C per 100 mm. The relation between the actual and the dry adiabatic lapse-rates is of some interest to the astronomer.

26.2 The atmosphere's transmission characteristics

The effects of the atmosphere upon the appearance of an object lying outside it are direct functions of the length of the light path through it from the object to the observer. If h is the effective vertical thickness of the atmosphere above the observer, the length of the light path through the atmosphere of a star whose ZD is z is given approximately by

$$l = h \sec z \dots \dots \dots \dots \dots \quad (a)$$

the approximation being acceptable provided $z < 75°$.

If it is assumed that the atmosphere above a height of 16 km exercises a negligible effect, we have:

438

Zenith distance	Length of light path
0°	16 km
45	23
55	27
65	39
75	63
⎰80	93⎱
⎱85	185⎰

The importance of observing at minimum ZD (culmination) is made abundantly clear from this table: an external object is seen through about 8 times the thickness of atmosphere when it is close to the horizon as when its altitude is 45°.

If I_o and I are respectively the original and received intensities of the radiation from an external source (i.e. before and after passing through the atmosphere to the observer), and a is the transmission coefficient representing the proportion of I_o transmitted by unit atmospheric thickness l,* then

$$I = I_o a$$

From equation (a) above,

$$I = I_o a^{\,a \, \sec \, z}$$

or

$$\log I = \log I_o + \sec z \log a \ldots \ldots \ldots \ldots (b)$$

the precise value of a being a function of wavelength (see Figure 190). Figure 191 translates the relation expressed in equation (b) into stellar magnitudes: assuming that 78 per cent of the light from a star at the zenith is transmitted by the atmosphere (the transmission at 560 nm approximately—see Figure 190), then the curve shows the light loss in magnitudes at ZDs from 0° to 85°.

The absorption of certain groups of wavelengths from mixed radiations passing through the atmosphere is responsible for the telluric lines in astronomical spectra. The most important of these lies in the ultraviolet below about 300 nm, and is due to atomic oxygen above about 100 km, and to ozone above about 40 km. Absorptions by carbon dioxide in the ultraviolet are relatively inconspicuous. The only significant tropospheric absorptions are those of water vapour, which are not conspicuous. The atmosphere's transmission curve through the visual and infrared

*Zenithal atmospheric extinction amounts to something like 0·3 or 0·4 mags on a good night.

regions to 20μ (with comparison curves for glass, rock salt, and fluorite) is shown in Figure 192.

Light loss in the terrestrial atmosphere is thus a function both of the

FIGURE 190

Percentage zenithal transmission of the atmosphere
at Mean Sea Level

altitude above MSL at which the measurements are made and also the wavelength of the radiation with which they are made.

A proportion of the incident radiation that escapes absorption is scattered by molecules, solid particles, and the smallest water droplets, the diameters of all of which are smaller than the wavelength of light.

440

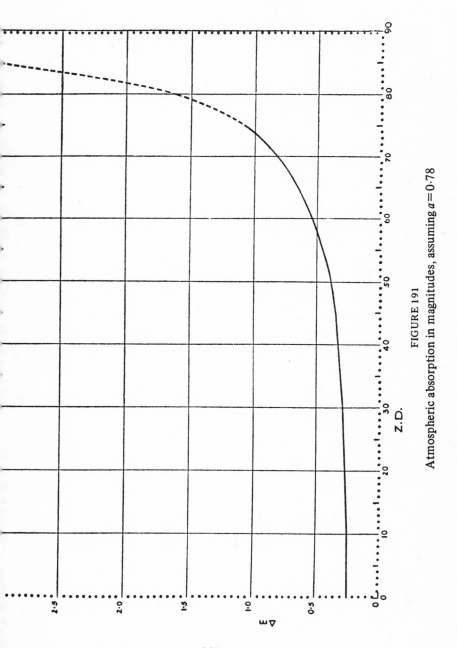

FIGURE 191

Atmospheric absorption in magnitudes, assuming $a = 0.78$

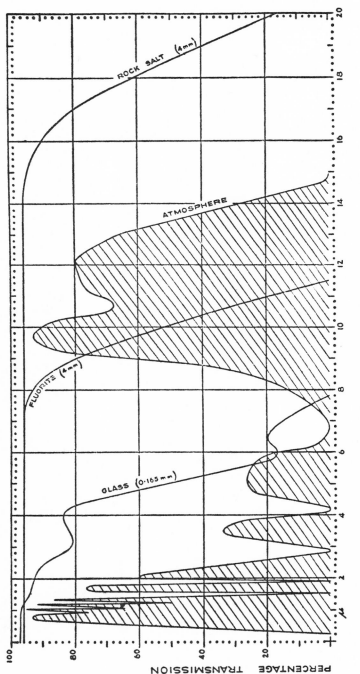

FIGURE 192

Long-wavelength transmission characteristics of the
atmosphere and some optical materials

The shorter wavelengths are scattered more than the long, whence 1 terrestrial observer has a blue sky instead of a white one. Particles t large to scatter light cause diffuse reflection which, unlike scattering independent of wavelength. Thus clouds and wet fog are white, 1 diameters of their constituent particles ranging from 0·04 to 0·1 n and 0·004 to 0·03 mm respectively. Town fog, on the other hal contains a high proportion of small solid particles which cause select scattering; hence the Sun seen through a country fog or mist is whi and through a pea-souper is red.

26.3 Atmospheric refraction

The refractive index of air at standard temperature and pressure is 1·000: with increasing altitude, and consequently decreasing pressure, N al decreases. A ray of light passing to the observer's eye through his zeni strikes the contours of equal atmospheric density normally, and is the: fore not refracted; all other rays are refracted towards the Earth's cent thus displacing the apparent source towards the zenith. Refraction the fore increases the altitude, decreases the ZD, and leaves unaffected t azimuth of a star.

Since the density of the atmosphere varies continuously with altitud atmospheric refraction—unlike that of a lens—is progressive througho its extent. Hence the angle through which the ray is refracted depen upon the thickness of the atmosphere traversed, as also upon the wav length of the light, and hence upon the ZD of the source.

Below are given two sets of values of the Mean Refraction, i.e. t refraction under standard conditions of temperature and pressure (T 10°C, b=750 mm). The rule for conversion to different conditions i for an increase of 10 mm in b, add 1·5 per cent; for a rise of 3°C, subtra 1 per cent.

The Pulkova Refraction Tables tabulate in full the factors necessary f the precise calculation of the correction to be applied to the appare ZD. These are derived from the expression

$$R = \mu \tan z \, (B.T)^A \gamma^\lambda \sigma^i$$

where z = apparent ZD,

$\left.\begin{array}{c} \mu' \\ A \\ \lambda \\ \sigma \end{array}\right\}$ are functions of z,

B is a function of the barometric pressure,

$\left.\begin{array}{c} T \\ \gamma \end{array}\right\}$ are functions of the temperature,

i is a function of the time of year.

Apparent ZD	Pulkova		Bessel		Apparent ZD	Pulkova		Bessel	
0°	0'	0.0	0'	0"	65°	2'	1.9	2'	3"
5	0	5.0			66	2	7.6		
10	0	10.1	0	10	67	2	13.8		
15	0	15.3			68	2	20.5		
20	0	20.8			69	2	27.8		
25	0	26.7	0	27	70	2	35.7	2	37
30	0	33.0			71	2	44.4		
35	0	35.7			72	2	54.0		
40	0	47.9	0	48	73	3	4.7		
45	0	57.1	0	58	74	3	16.6	3	19
46	0	59.1			75	3	30.0	3	32
48	1	3.4			76	3	45.2	3	47
50	1	8.0	1	9	77	4	2.5	4	5
52	1	13.0			78	4	22.5	4	25
54	1	18.5			79	4	45.7	4	49
55	1	21.4	1	22	80	5	13.1	5	16
56	1	24.5			81	5	46	5	49
57	1	27.8			82	6	26	6	30
58	1	31.2			83	7	15	7	20
59	1	34.8			84	8	19	8	23
60	1	38.7	1	40	85	9	40	9	47
61	1	42.8			86	11	31	11	39
62	1	47.1			87	14	7	14	15
63	1	51.7			88	17	55	18	9
64	1	56.6			89	23	53	24	25
					89 30	28	11	29	3
					90 00	33	51	34	54

In the absence of tables, Comstock's formula gives a fair approximation:

$$R = \frac{38.7}{460 + t} \tan z$$

where R is expressed in " arc,

z = apparent ZD,

t = temperature (°F),

b = barometric pressure (mm).

A still simpler approximation is given by

$$R = k \tan z$$

where $k = 57''$ at MSL (refraction at $z = 45°$ under standard temperature and pressure). The inaccuracy of R, so derived, increases with z, and becomes prohibitive when $z > 70°$ approximately.

The extent of the differential refraction, dependent on wavelength, is indicated on the following page. The two right-hand columns refer respectively to light of the wavelength of the D lines of sodium, and to the difference between the refractions of wavelengths of the F and C lines:

ZD	Mean Refraction	
	R_D	$R_F - R_C$
25°	0′ 27″	0″3
45	0 57	0·6
65	2 02	1·2
72	2 54	1·8
76	3 45	2·4
79	4 46	3·0

To obtain the true difference of ZD of two stars (difference in Dec if they are on the meridian), the difference of their refractions $(R - R')$ must be added to their observed difference in ZD $(z - z')$. $R - R'$ is tabulated below for different values of $z - z'$ at selected ZDs:

$z - z'$	ZD					
	0°	10°	20°	25°	30°	35°
0′	0″00	0″00	0″00	0″00	0″00	0″00
2	0·04	0·04	0·04	0·04	0·04	0·04
4	0·06	0·06	0·08	0·08	0·08	0·10
6	0·10	0·10	0·12	0·12	0·14	0·16
8	0·14	0·14	0·16	0·16	0·18	0·20
10	0·16	0·16	0·20	0·20	0·22	0·26
12	0·20	0·20	0·22	0·24	0·26	0·30
14	0·24	0·24	0·26	0·28	0·30	0·36
16	0·26	0·28	0·30	0·32	0·36	0·42
18	0·30	0·32	0·34	0·36	0·40	0·46
20	0·34	0·36	0·38	0·42	0·46	0·52
22	0·36	0·38	0·42	0·46	0·50	0·56
24	0·40	0·42	0·46	0·50	0·54	0·62

26.4 Atmospheric humidity

The wetness of the atmosphere is given by the proportion

$$\frac{\text{Quantity of water vapour in unit volume of air}}{\text{Quantity of water vapour required to saturate it}}$$

and is usually expressed as a percentage. Air is said to be saturated if it will stand in equilibrium over a smooth surface of water at its own temperature.

The humidity may be determined directly by means of a hygrograph, or calculated from the readings of a wet-and-dry-bulb thermometer. This consists of a pair of ordinary mercury-in-glass thermometers; the bulb of one is surrounded by air, that of the other by a piece of flannel which is kept constantly damp by means of a small water reservoir into which its other end is dipped. The rate at which water evaporates from the wet bulb is, for a given temperature, inversely proportional to the relative humidity; also, the faster it is evaporating, the more thermal energy it will utilise, and the greater will be the discrepancy between the two thermometer readings. The humidity can be obtained from tables giving the relative humidity for different values of $T_d - T_w$ at specified values of T_d.

The relative humidity is normally maximal during the night and minimal during the afternoon.

26.5 Dew

As the temperature of a volume of air is lowered, the quantity of water vapour required to saturate it falls likewise. Hence it can be saturated in either of two ways: by injecting more water vapour into it, or by cooling it. The temperature at which a given volume of air becomes saturated is known as its dewpoint: if it is cooled below its dewpoint, liquid water will condense out of it in the form of dew. A familiar example of this process—apart from the natural formation of dew, dependent upon the cooling of the ground by radiation after sunset—is the condensation of water droplets on the outside of a glass of cold water when taken into a warm room (the temperature of the glass being below the dewpoint of the air in the room. If the air is damp but unsaturated, its dewpoint temperature is lower than T_w; if it is saturated, its dewpoint temperature $= T_w = T_d$.

The condensation of dew is more marked on grass than on bare earth, since the latter's diurnal temperature change is smaller than that of grassed earth. For the same reason, certain types of soil favour the formation of dew more than others: increasing dew will be encountered along the series clay—loam—sand. Clear nights, especially if there is little or no wind, also provide the conditions most favourable for the deposit of dew.

26.6 Suspended matter

Astronomical 'seeing' does not mean, or include, clarity of atmosphere, which, indeed, is often exemplary when the seeing it at its worst. Atmospheric clarity becomes of paramount importance, however, in photometric work, the long-exposure photography of faint objects, and similar work.

Apart from such obvious drawbacks as a naturally cloudy climate, the worst enemy of atmospheric transparency is the proximity of large towns, which not only pollute the atmosphere but also illuminate the pollution. This was the cause of the removal of the Royal Observatory from Greenwich to Herstmonceux. Again, diffusion of the lights of Los Angeles by the layer of 'smog' that habitually hangs over the city reduces the maximum permissible exposure time (limited by fogging) of the Mt Wilson $f/5$ cameras from 80 minutes to about half an hour, as was proved during wartime dim-outs.

Matter in suspension in the atmosphere may be grouped under the following heads:

(a) Cloud:

Condensation occurs on small solid particles; the commonest are crystals of hygroscopic salts, such as marine brine, and oxides of nitrogen and sulphur, by-products of industrial and domestic combustion. Such particles have diameters as small as 100 nm or less.

The droplets will continue to grow until a dilution of the salt solution is reached at which no further attraction for water is exerted. Thereafter, increase in the size of the drops can only occur if the temperature falls, or more water is evaporated into the air containing the droplets, i.e. by an increase in the relative humidity of the air.

Inland, afternoons tend to be cloudier than other times of the day. There is a marked tendency for the sky to clear after sunset, and the times of minimum cloud are late evening and night. In SE England the sky is completely clear at midnight on about 122 nights per annum. During the period 1901-30 there was a daily average of 4^h9 sunlight at Herstmonceux in Sussex, equivalent to 39·9 per cent of the time the Sun was above the horizon. During the period 1927-45 the corresponding figure at Greenwich was 30·55 per cent. Measures of clear night-hours at Greenwich, by means of a photographic trace of a circumpolar star, gave 36·82 per cent of the total exposure time, over the same period. It therefore appears that there is, in SE England, little significant difference between daytime and night-time atmospheric clarity. At Greenwich in 1940-8, solar records could be obtained on an average of 261 days per annum, increasing to an average of 296 days per annum at Herstmonceux, in 1950-8.

(b) Fog and mist:

Wet, or white, fog and mist consist of water drops only, and occur typically in the country. They not only absorb light from a distant source, but also form a screen of reflected white light between the source and the observer. It is this veil that at times reduces visibility to a few metres. Sea fogs, relying for their formation on the presence of particles of extremely hygroscopic sodium chloride, can appear when the relative humidity is as low as 75 per cent.

(*c*) *Smoke fog, or pea-soup fog:*

Consists of water drops whose condensation nuclei are solid carbonaceous by-products of combustion. Unlike a white fog, it derives its opacity solely from its action in cutting down the light from a distant source.

(*d*) *Solid particles in suspension, and corrosive gases:*

Most commonly dust and the by-products of combustion. Obscuration from this cause depends solely on the existence of fires or the nature of the soil, and is independent of the relative humidity.

The larger particles resulting from combustion settle near the source, depositing a layer of fine grime over the roofs of houses in urban districts. Gaseous by-products of combustion (particularly sulphur dioxide) are largely removed by solution in cloud droplets. The most troublesome element is the smaller particles comprising smoke, of which there may be 10^6 per cubic cm of town air, and are still about 10^2 in the cleanest country air. These may remain in suspension for long periods of time and be carried for astonishingly long distances. Smoke from the Ruhr valley, for example, is brought to England by a steady SE wind; again, smoke from the industrial Midlands is detectable in Scandinavia when steady SW winds are blowing. The ultraviolet is even more drastically eliminated by smoke particles than the visible wavelengths.

Artificial pollution, both smoke and acidity, is always worst in winter, and during the late afternoons at all times of the year. Results of two years' research on atmospheric pollution carried out in Leicester (B. 19.10) may be summaried as follows:

(*i*) Street-level pollution is minimal in the early morning, rising sharply to a maximum during the mid-morning, followed by a shallow minimum during the afternoon and a secondary maximum (which may be smaller or larger than the morning maximum) in the later afternoon.

(*ii*) It is comparatively little affected by wind: an increase of wind velocity by a factor of 3 reduces mid-city pollution by only about one-third.

(*iii*) Town air is to a large extent cleaned each night; thus a pea-soup fog of the evening is a white fog by next morning.

(*e*) *Condensation nuclei:*

The much smaller particles upon which (given the requisite relative

448

humidity) cloud droplets form. In their dry state they scatter certain wavelengths of the light passing through them, causing, *inter alia*, the blueness of distances. If the relative humidity increases beyond the saturation point, droplets of increasing size form on the nuclei, as already described, and the blueness of the horizon changes to grey (haze—as often occurs on a hot day) when they become too large to scatter visible wavelengths, and finally to white (wet fog).

26.7 Astronomical seeing

The 'seeing' is the quality of the telescopic image, so far as this is dependent upon the condition of the atmosphere. Two separate factors in the state of seeing can be distinguished, either or both of which may be present at a given time and place: deformation of the image, with or without bodily motion; and motion of the image without any appreciable degree of deformation.

Motion of the image can be eliminated—or, rather, converted into deformation—by a sufficient increase of aperture. Thus with average seeing a 75–100-mm will show star images which (if the objects are not too near the horizon) are clearly defined but wandering. The movements of each image will be contained in a circle which is typically of the order of 2″ diameter. With a 150-cm telescope, on the other hand, all the star images are enlarged, blurred, and comparatively or completely stationary, the diameter of the smallest again being about 2″. When seeing is bad, even with small apertures the image may lose all resemblance to the normal diffraction pattern, becoming an enlarged, agitated, structureless blob, in which the rings are completely submerged.

This effect of atmospheric turbulence has its naked-eye counterpart in stellar scintillation or twinkling, and its extended-image counterpart in 'boiling', both of which can be reduced or eliminated by increasing the aperture (see also section 26.9). Generally, the effects of turbulence are loss of resolving power and impaired definition of the extended image.

Twinkling was at one time ascribed, erroneously, to physiological causes. It is in fact due to refractional inequalities in the atmosphere. If the surfaces of equal refractive index are resultantly convex to the eye, the intensity of a ray passing through them will be reduced; if concave, increased. Rapid deformations of the surfaces, or the horizontal passage of a deformed surface between the objective and the star, will thus produce a flickering in the perceived intensity of the star. Chromatic scintillation, visible in the neighbourhood of the horizon when the seeing is particularly bad (Sirius frequently exhibits it on frosty nights), is caused by the light being widely enough dispersed for different wavelengths to be independently affected by atmospheric turbulence. Telescopic scintillation not only jumbles the diffraction pattern but makes it 'shudder', the image's angular displacement in any direction from its mean position being, as we have seen, of the order of 1″ for average seeing, and exceptionally 10″ or, when the seeing is very bad indeed, 20″. There is scope for much more work on the correlation of seeing with local environmental conditions.

The deformations are rapid, normally between about 5 and 100 per second, only the low frequency scintillations being perceptible to the naked eye, and may affect the components of even a close pair quite independently. Atmospheric waves of different refractive index have distinguishably different effects upon the image according to their angular size. This is particularly noticeable in the case of a planetary image: comparatively small irregularities will cause a blurring of the limb, without introducing any movement; with larger waves, on the other hand, the limb may be sharply defined through wavering or 'boiling'.

Some satisfactory quantitative results concerning scintillation were obtained at Edinburgh by Ellison and Seddon (B. 19.11), using a photocell and the cine-camera recording of a cathode-ray tube.* As regards the frequency of scintillation they found that at low altitudes both low-frequency (about 5 cycles per second) and high-frequency (about 100Hz) scintillation occur, but that as the ZD decreases the former dies out. The amplitude of the brightness scintillation invariably decreases as ZD decreases, ranging (with 0·9 m aperture) from about ± 20 per cent to ±50 per cent at ZD=80° down to a normal minimum of about ±5 per cent at the zenith. A close correlation has been established between the amplitude of the scintillation and the quality of the visual telescopic image. The amplitude also varies approximately as $1/D$ for all ZDs.

Anderson (B. 19.1) has found that at moderate altitudes and with fairly good seeing, the frequency of scintillation is characteristically from a few to about 25 Hz (amplitude about 1 mag), increasing to as many as 150 Hz (amplitude 2 to 3 mags) when the seeing is bad—results in general agreement with those obtained at Edinburgh. . Anderson describes a simple technique for estimating the frequency of the scintillation visually. A small telescope (such as those used for reading laboratory scales, $D \simeq$ 25 mm, $F \simeq 20$ cm) is held near its objective by the right hand; the eye end is supported at the observer's eye with the forked fingers of the left hand, the heel of which is pressed against the observer's chin for steadiness. If now the OG end is made to describe small and rapid circles, persistence of vision will draw out the image of a star into a continuous circle. The

*See also the investigations by Butler at Dunsink (B. 19.3, 4).

speed of movement of the OG should be just rapid enough (about 5 revs/sec) to close the circle, but not so fast that successive circles overlap. If the star is scintillating it will be found that the circle consists of alternate bright and less bright segments; the number of the former can with practice be estimated quickly and reasonably accurately. This number multiplied by the number of circles described by the OG per sec gives the frequency of the scintillations.

Some light has been thrown upon the probable size and location of the atmospheric disturbances which are responsible for scintillation. That movement of the image can be converted into confusion of the image by changing from an aperture of 10 cm to one of 150 cm suggests that the distance between adjacent convexities and concavities in the theoretically plane surfaces of equal refraction lies between these limits; observations of eclipse shadow bands suggest that it lies nearer the former, and Anderson accepts about 15 cm as the order of size. Ellison and Seddon found indications that the amplitude of the scintillation reaches a maximum at about $D=75$ mm, from which they deduced this to be the order of size of the refractional irregularities causing the scintillation. As regards the height of the distorting layer above the Earth's surface, it was found at Edinburgh that Jupiter and Saturn (angular diameters 49″ and 20″) showed no trace of stellar scintillation; the scintillation of Mars (4″4) was of about half the amplitude of that of a comparison star; and that of Jupiter III (1″5) was purely stellar. They concluded that the angular size of the refractional irregularities was of the order of 3″, the angular subtention of 75 mm at a distance of about 5 km. Since the observations were made at an altitude of 30° this indicates a vertical height of about 2½ km.

The region of these refractional irregularities thus lies at a considerable height above the Earth's surface—probably between 1½ and 5 km. There is also evidence that they are confined to a comparatively thin stratum. Owing to their distance they affect fairly wide topographical areas similarly, i.e. the seeing at any moment will be equally good or bad over a wide area. Atmospheric currents themselves, if there were no temperature or pressure gradients through them, would have little or no effect upon the seeing. But owing to thermal differences, their densities and therefore refractive indices are various; constantly varying barometric pressure is another factor contributing to the optical instability of the atmosphere. Although the turbulence is directly proportional to the difference in the refractive indices of the currents and surfaces of fluctuating inclination to the ray, the number of such currents also has an important bearing on the visible effect in the telescope, there being evidence that two relatively small cross-currents have a more deleterious effect on the seeing than one conspicuous one.

Among the factors contributing to turbulence are: the settling of cold air masses during the night; the condensation of clouds; the mixing of masses of unequally heated air by convection and the wind; a lapse-rate in excess of the dry adiabatic lapse-rate.

Though turbulence increases rapidly with Zenith Distance (it is, approximately, a function of sec z), it may at any moment vary widely in different parts of the sky at the same ZD. The vicinity of clouds (where steep and erratic temperature gradients are liable to occur) in particular often provides a great variety of seeing. Also, it commonly varies by 1 or 2 divisions of the Standard Scale (see section 26.10) during the course of the night; usually, though by no means invariably, the seeing deteriorates during the latter half of the night. Very clear nights, again, are almost invariably bad; windless nights, in general, good.

A serious study of the causes and effects of atmospheric turbulence, although highly relevant to the siting of a large telescope, must be of little more than academic interest to the amateur observing from his back garden. However, a modern picture of a turbulent atmosphere envisages a random system of cells of different size and distribution scattered in an unpredictable way (just as the bubbles in a pan of boiling water are individually unpredictable as to size and position), but capable of being analysed in the mass in the form of a Kolmogrov spectrum. The very numerous small cells, of diameters up to 10 cm or so, cause small-scale destruction of definition, while the larger, comparatively rare cells, a metre or more across, cause slow bodily excursions of the whole image. The larger the value of D, the greater the number of sub-diameter cells and the worse the confusion of the image detail; see section 26.9. Work on speckle interferometry by Labeyrie and others (B. 19.14) has shown that exposures of the order of 0·01 to 0·001 sec. are required in order to 'freeze' the motion of these cells.

The origin of turbulence is, probably, very much a function of the observatory site (section 26.11), and local ground seeing may play a large part in determining conditions. It is generally agreed that the seeing will be improved considerably if the telescope can be mounted 15 m or more above its immediate surroundings.

The type of image motion known as 'jumping the wire', familiar to transit observers, is a psychological effect due to fatigue.

26.8 Direct visibility of atmospheric irregularities

Atmospheric thermal irregularities are visible to the naked eye as 'shadow bands' when the angular size of the source is reduced, as during a solar eclipse just short of the total phase.

Under certain circumstances they can also be observed telescopically. Given a sufficient aperture—not less than about 30 cm—they can be seen as more or less straight, parallel bands crossing an objective flooded with light from a bright star or planet, the ocular having been removed and the eye placed at the focus. With small apertures the objective is too small compared with the thermal waves for them to be seen individually, though their integrated effect may be seen as a flickering of the total light received. Very similar, though artificially produced, effects can be seen with a small instrument by directing it above any source of heat.

If an ocular is used, and a bright star is thrown out of focus, they may nearly always be seen, though once again the individual waves are not distinguishable as such, owing to the rapidity of their motion. The general effect is that of straight, parallel linear streams moving in the direction of the wind prevailing at the height at which they are formed. They may also be seen crossing the disc and limb of a slightly out-of-focus planetary image; they come to a sharp focus just outside that for objects at infinity, whence their distance can be roughly measured in a telescope of considerable focal length. Rarely, they are seen crossing the projected solar disc of a Sun telescope (where they are not to be confused with the commoner phenomenon of 'boiling').

More slowly moving, wavy bands, often distorted by whorls, are caused by currents in the instrument itself.

26.9 Aperture and seeing

The waves visible at the focus are the 'shadows' of the surfaces of equal refractive index, moving relative to one another, locally distorted and mutually inclined. One can look upon them as a wave system of wavelength λ (Douglass concluded from his careful investigations—B. 19.7, 9 —that the linear values lies usually between 16 and 100 mm, in general agreement with the results of Anderson, and of Ellison and Seddon); or one can consider the waves, taken altogether, as producing a residual surface, the normal to which oscillates within a cone of angle $2t$, calling t the turbulence.

For twinkling to be observed, the aperture of the image-forming lens must be small compared with λ; hence twinkling is a characteristically naked-eye rather than a telescopic phenomenon. As D increases, so that the entry pencil includes more and more complete waves, the movement of the image decreases, but the breakdown of the diffraction pattern grows progressively worse. Thus what in a small instrument appears as a well-defined but much agitated stellar image is in a large instrument a practically motionless, structureless, circular blob. The following figures are necessarily very approximate:

$D<\frac{1}{2}\lambda$: twinkling;
$D=\frac{1}{2}\lambda$ to $D=3\lambda$: image in motion, but rings and disc intact;
$D=3\lambda$ to $D=5\lambda$: decreased motion, disc destroyed, rings still intact;
$D=5\lambda$ to $D=8\lambda$: image static, rings still intact;
$D>8\lambda$: image static, neither disc nor rings visible.

The change-over from motion to confusion at about $D=5\lambda$ can be clearly witnessed by observing the limb of a planet with sufficient aperture to show it as fuzzily defined though free from 'boiling', and then progressively reducing the aperture. Conversely, these two types of poor seeing, as experienced with a given instrument, are determined by the size of the atmospheric waves at the time.

Generally speaking, image motion predominates over confusion of the diffraction pattern up to an aperture of about 10 cm; from 10 cm to 20 cm both factors are present, the former decreasing in importance; above about 20 cm deformation is always as great as is compatible with a given degree of turbulence, motion of the image being negligible. No doubt connected with this distinction is the observational fact that large apertures are at a greater disadvantage (compared with smaller) in the definition of extended images than in the resolution of stellar points. Thus under seeing conditions which prevent a 30-cm from showing any planetary detail invisible with a 15-cm it will nevertheless resolve closer doubles than the smaller instrument.

Since the quality of the image for a given turbulence, t, is a function of the aperture used, and the size of the diffraction disc is also (inversely) related to the aperture, it is possible to correlate t with r, the radius of the spurious disc. It is in this way that t, an unobservable, is derived and utilised in the third of the Standard Scales detailed below. Within the range of apertures which yield maximal image distortion for a given t-value—about 10 cm sometimes, and above 20 cm almost invariably—the extent of the deformation is a function of the relative angular magnitudes of t and r: for a given value of t, as r diminishes, so the deformation of the image increases; and conversely, for a given value of r, as t increases, so the deformation of the image increases. Hence the deformation of the image is a function of $\frac{t}{r}$.

When $\frac{t}{r}<\frac{1}{4}$ (i.e. if $D<12$ cm for average seeing, when $t=0\rlap{.}''25$), the image will suffer a little distortion, but will be displaced bodily through a total amplitude of about $2t$. With an aperture of 12 cm and the same t-value $\left(\frac{t}{r}=\frac{1}{4}\right)$ flickering and moving condensations begin to appear in the rings. With double this aperture, or double the t-value and the same aperture (in either case $\frac{t}{r}=\frac{1}{2}$), the rings begin to disintegrated, the fragments are in rapid motion, and the whole image suffers continuous distortion from the circular in random directions. Again doubling the aperture or the turbulence, giving $\frac{t}{r}=1$, turns the image into an agitated, structureless patch, the distinction between disc and rings being lost altogether. If $\frac{t}{r}>1\frac{1}{2}$, the image is not unlike a tiny, ill-defined, scintillating planetary disc of diameter $2t$ (i.e. equal in diameter to the amplitude of the image's displacements in a smaller telescope), 'boiling' but suffering no bodily displacement. The effect of this upon an extended image is seriously to impair the definition, with total suppression of the finer detail, since the image of every point on the object is a spot of radius t. Further increase of aperture will brighten the image, but will no longer reduce its angular dimensions, which remain at $d=2t$. These successive conditions can be observed with a given instrument at a given time by observing a sequence of stars of increasing ZD.

In an actual comparison of the performances of a 12-cm and a 38-cm, made by Pickering under varied atmospheric conditions, the following results appeared. They are in general agreement with the foregoing.

Poor seeing: moderately close, equally bright doubles: 12-cm as good as the 38-cm;
faint or unequal pairs: 38-cm superior;
Seeing better than average: 38-cm superior all round.

The conclusion to be drawn is that for work demanding fine definition of the image, the useful aperture is limited by the atmosphere, large apertures only coming into their own when the seeing is exceptionally good.

See further: section 27.1.

26.10 Scales of seeing

To be able to estimate the reliability of an observation, some indication of the condition of the seeing at the time must be given. Each period of observation should be preceded by an investigation of the degree of turbulence. There are numerous rough-and-ready methods of doing so—

some are mentioned below–but the only satisfactory method involves the employment of a scale of atmospheric states, dependent upon the detailed appearance of a star image.

Such a scale should possess the following characteristics: (*i*) objectivity: it must be impersonal, independent of the observer's experience and local conditions, including the aperture of his instrument, so as to allow the direct comparison of his observations with others made by different observers at different times and places, with the confidence that the scale numbers, as used by all of them, refer to identical states of seeing; (*ii*) universal applicability; (*iii*) precision: involving numerical rather than verbal description; (*iv*) simplicity.

Many approximate criteria have been proposed at one time or another, none of which is ideal in all respects. Those based on the resolution of close doubles or the definition of extended images–a planetary limb, the fine detail of the lunar or a planetary surface, etc–allow rough qualitative estimates to be made; but in general they are less precise (being ill-adapted to numerical expression), and are also more dependent on instrumental factors, than scales based upon the stability or otherwise of a bright (mag 1 or 2) stellar image. Even as a criterion of the definition of planetary detail, the condition of the limb is an untrustworthy guide: it may be clearly defined, yet the definition of the detail may be poor, or vice versa.

Five criteria, of the type described above, were devised by W. H. Pickering. In order of increasing sensitivity, they are:

(*a*) The resolution of double stars whose separation is near to the threshold of R' of the instrument.

(*b*) The visibility of elliptical outline of Jupiter I or IV. It is difficult to see what reliability or validity this criterion can have, however, since the apparent ellipticity vanishes when the satellite is observed with a large enough telescope.

(*c*) Resolution of the double 'canal' in Aristillus. The 'canals' are two fine dark lines diverging from a point on the crater floor, whence they cross the NW wall, a few miles outside which they become invisible. Length, about 40 km; width outside the crater varies with colongitude from about 100 m (0″05) to 500 m (0″25). Duplication visible between colongitudes 35° and 65°, and 110° and 140°, i.e. on only about five nights throughout the lunation. Resolution of the 'canals' inside the crater can just be glimpsed with $D = 20$ cm, and is therefore a test of seeing with this and larger apertures; the external 'canals' can be resolved with $D = 75$ mm, given the seeing.

(*d*) Visibility of a small group of 'canals' outside Aristillus, N and E of the double 'canal'.

(*e*) Visibility of a narrow bright line running parallel to the 'upper' reaches of the Mt Hadley 'river bed'; 8 km long and probably not more than about 50 m wide ($0''025$). Only visible between colongitudes about $43°$ and $57°$.

The justification for the employment of such criteria of seeing as those described above is the claim that the visibility of extended objects of comparatively little contrast is more susceptible to turbulence than the minute images of stars, heavily contrasted with their background—at any rate, that a given degree of turbulence will affect the two classes of object differently. And that therefore the visibility of planetary and lunar detail does not run parallel with the perfection or otherwise of stellar images. And that therefore a scale of seeing based on the latter is not necessarily applicable to planetary and lunar observation.

In practice it is difficult to draw this distinction. Turbulence simultaneously affects the appearance of the stellar images, reduces the telescope's resolving power as adjudged by its performance on close doubles, and reduces the definition and resolution of extended images.

Many observers find it convenient to employ 'private' scales of seeing, which they have worked out for themselves to suit their own particular requirements. Such scales should be calibrated against one of the Standard Scales, so that the equivalent Standard grading can be appended to all observations.

Below are given details of three suggested Standard Scales of Seeing, based on the appearance of the stellar image. The first two, designed respectively for 15-cm and for 12-cm objectives, suffer from the following disadvantages: (*a*) The scale number that would be ascribed to a given state of seeing will vary with the aperture used. The calibration can be adjusted to another aperture on a basis of adding 2 to the scale number when the aperture is doubled: thus perfect seeing would be rated as 14 on Scale A with a 30-cm, 12 on scale B with a 25-cm. But unfortunately even the descriptions of the several criteria vary to some extent with aperture: with small apertures an improvement of seeing causes the disc to separate out of the confused image before the rings; with larger apertures there is a progressive tendency for the rings to separate out first, the disc not becoming well defined till the rings, though they may still be in motion, are complete. (*b*) No means is provided for correcting the estimated scale number for the ZD of the star, thus converting it to standard conditions; for this reason the ZD of the star employed in the estimation of the scale number should always be given. As an example of the effect that differences of ZD can have, Douglass quotes the case of a difference of scale number of 5 when determined by a star at the zenith and one at a ZD of 80°; the amplitude of the motion of the image being 1″ in the former case and 5″7 in the latter.

SCALE A

Designed by W. H. Pickering for use with a 15-cm objective employing a magnification of from 25D to 40D.

0: disc and rings a confused mass, in violent motion; image larger than the true diffraction pattern, and varying continuously in size;

2: disc and rings a confused mass, in constant motion; image not enlarged;

4: disc well defined, rings blurred;

6: disc well defined, rings visible but broken;

8: disc well defined, rings complete but moving;

10: disc well defined, rings stationary; image moving in field;

12: disc well defined, rings stationary; image stationary in field.

SCALE B

A development of Scale A, compared with which simplicity is sacrificed for greater precision. Designed for use with a 12-cm objective; recommended magnification, not less than 25D; recommended test star, mag 1 or 2.

1: disc and rings undifferentiated; image usually about twice the size of the true diffraction pattern;

2: disc and rings undifferentiated; image occasionally twice the size of the true diffraction pattern; } Very poor

3: disc and rings undifferentiated; image still enlarged, but brighter at the centre;

4: disc often visible; also occasional short arcs of the rings;

5: disc always visible; short arcs of the rings visible for about half the time; } Poor

6: disc always visible, though not sharply defined; short arcs of the rings visible all the time;

7: disc sometimes sharply defined, and distinct from rings; } Good

8: disc always sharply defined; inner ring in constant motion;

9: disc always sharply defined; inner ring stationary; } Excellent

10: disc always sharply defined; all rings virtually stationary. } Perfect

SCALE C

The third Standard Scale to be described here is based upon the turbulence, t, and has the advantage over Pickering's scales that a correction is applied to remove the effect of ZD from the observations.

For several bright stars, differing in ZD, and observed with a magnification of not less than about $18D$, the quantity $\frac{t}{r}$ is estimated (see pp. 454-5. Let $\frac{t}{r} = x$.

Then, from equation (d) on p. 15:

$$t = \frac{5x}{D}$$

can be derived for each star. The value of t at the zenith (t_0) is given by

$$t_0 = \frac{t}{\sec z}$$

where z is the ZD. Hence for each value of t a corresponding value of t_0 is derived, the mean of which is the required criterion of the seeing.

Example: The observations of three stars and the deductions made from them are set out below, assuming $D = 10$ cm:

ZD	$\sec z$	$x = t/r$ (estimated)	$t = \dfrac{5x}{D}$	$t_0 = \dfrac{t}{\sec z}$
20°	1·064	0·75	0·94	0·88 ⎫
40	1·305	1·0	1·25	0·96 ⎬ 0·90
70	2·924	2·0	2·50	0·85 ⎭

Or t_0 can be derived in each case direct from (combining the two foregoing expressions for t and t_0),

$$t_0 = \frac{5x}{D \sec z}$$

Whether it is meaningful to correct for ZD is a matter for dispute, since the seeing affecting the observer is that which lies between his telescope and the object being viewed, rather than that which might obtain at some other altitude altogether.

26.11 Local factors

Scattered references to the dependence of seeing upon local weather and topography are to be found in the literature, but probably very little in the way of universally applicable conclusions can be arrived at. The regular observer will sooner or later discover what meteorological conditions favour observation on the subsequent night or during the subsequent period; especially if he keeps a record of the wind direction and force, type of cloud prevalent, maximum and minimum temperatures, barometric reading and humidity. Annual and diurnal variations in the seeing should also be sought.

Topographical:

1. Avoid the vicinity of valleys, since cold air flows into them at night; if it overflows, usually during the later part of the night, it may utterly ruin definition.

2. Avoid the foot or slope of a hill, since cold air flows down it at night, forming a lake at the foot which may be several hundred metres deep by the end of the night.

3. Avoid low-lying terrain (which usually contains a river or other water) owing to the high incidence of mists, especially during the winter (e.g. the Thames valley).

4. Avoid the crest of a hill: it is likely to be windy, and even on windless nights there will be constant currents owing to the sinking of cold air over the crest.

5. A high altitude is desirable, since the greater part of the dust and artificial pollution, as well as mists and fog, occur near sea level.

6. Surroundings covered by vegetation prevent the earth in the observatory's vicinity heating up badly during the day. The ever-present danger of forest fires, which may denude the mountain and impair definition over a period of years, is one of the few disadvantages of the site of the Mt Palomar Observatory.

7. The foregoing points would combine to suggest the ideal site as being a dry plateau of moderate elevation above sea level, covered with vegetation, located in a dry, clear climate, and far from valleys and the sea.

Latitude:

8. The advantage of low latitudes are: (*a*) small nightly variation of temperature, (*b*) calm nights, (*c*) they lie outside the anticyclone belt, with its constantly changing barometric pressure.

DIAPHRAGMS AND FILTERS

27.1 Diaphragms

(*a*) Anti-glare or anti-reflection stops, exactly defining the truncated cone in whose base, in the case of a refractor, is the OG and whose apex is the exit pupil; these diaphragms are normally incorporated in the tube as supplied by the telescope maker. (*b*) Field stop, limiting the visible field to the field of full illumination (see section 3.8). (*c*) Aperture stops, whose objective itself is, functionally, the aperture stop for the ocular. (*d*) Additional diaphragms, at the entry or exit pupil, are justified when it is required (*i*) to reduce any of the characteristics of telescopic vision which are dependent on aperture: in particular, light grasp, or the effects of atmospheric turbulence on definition; (*ii*) to modify the diffraction pattern of the image by changing the size or shape of the entry pupil.

An entry-pupil diaphragm will reduce the instrument's light grasp, therefore the illumination of the image; will increase the size of the spurious disc, therefore reduce the resolving power; will reduce t/r and D/λ, therefore reduce the confusion of the image due to bad seeing (see section 26.9). Thus it alters the nature, not only the illumination, of the image.

An exit-pupil diaphragm will similarly reduce both the brightness and the resolution of the image. It would be natural to suppose—and is often stated—that the resolution of the image is unaffected by interposing a diaphragm at the exit pupil, since the characteristics of the light pencil that determine the resolution of the image are already established during its passage through the objective, and cannot be altered by its subsequent passage through a diaphragm.

This reasoning is fallacious, however, and might equally be used to prove that an exit-pupil diaphragm does not reduce the brightness of the image either, since light grasp is as dependent on D as resolution is. The exit-pupil diaphragm admittedly affects neither the light grasp nor the resolving power of the objective as such—the brightness and resolution of the image in the focal plane are those appropriate to the full aperture of the objective—but it does reduce the light grasp and resolving power of the telescope as a whole. The exit pupil is the image of the objective—rays from the boundary of the latter pass through the boundary of the pupil—and to reduce its aperture is *effectively* to reduce D.

Experiments on the resolution of double stars with exit-pupil diaphragms of graded sizes may in some cases appear to indicate an improvement of resolution when the pupil is restricted. Where this occurs, however, it means that the relative difficulty of resolving the pair without the diaphragm is due so preponderantly to excessive brightness that the apparent resolution of the pair is improved by reducing the brightness even at the expense of increasing the angular sizes of the images.

Hence the reduction of image glare (e.g. of the Moon or Venus) can be effected equally by an objective diaphragm or an exit-pupil diaphragm, with the latter having the advantage of extreme convenience in use. No expensive iris diaphragm, with controls leading back to the observing position, is required. Each ocular is fitted with a slide to take a strip of aluminium, down the central line of which a series of graded holes have been drilled. Care must be taken that each hole comes centrally into position in front of the exit pupil, and that the slide does not force the eye away from the exit pupil; this will probably mean cutting away any flange or rim surrounding the eye end of the ocular.

The use of objective screens—such as perforated zinc, or even 6-mm mesh wire for an aperture of 30-cm—was first suggested by J. F. W. Herschel and Dawes, and has been advocated since by W. H. Pickering and others, both for planetary and for stellar observation. If the apertures in the screen are not very small compared with the total aperture of the objective, most undesirable modifications of the diffraction pattern of the image could result. In any case numbers of diffraction spectra are produced radially round the object. The latter, however, is seen in its natural colours with the full effective resolution, though not the full light grasp, of the whole aperture. The remarks below—on diaphragming to reduce the effects of poor seeing—apply equally to the use of perforated screens; though ultimately the proof of the pudding must always be its eating.

The position regarding the use of an objective diaphragm when the seeing is poor is an interesting one. Planetary observers in general advocate the reduction of aperture if the full aperture reveals too grossly the distortion of the image or the confusion of definition which springs from a high value of t, the turbulence; and, indeed, since the deformation of the image is greater the larger the aperture, this would seem to be reasonable.

Douglass, for example, recommended a fairly small aperture for planetary observation unless the seeing is excellent (though full aperture always for such work as micrometer measures). Jarry-Desloges, working with a 50-cm refractor under abnormally favourable conditions in Algeria, found that on 3 nights out of 4 a reduction of aperture to 20-35 cm gave better planetary images than the full aperture; and that only on about 2-5 nights in 100 did the performance of the full aperture surpass that of the 35-cm diaphragm. He considered that 50 cm is the effective limit for an instrument intended primarily for planetary work, a conclusion with which Pickering—also working under exceptionally favourable atmospheric conditions in Jamaica—agreed. With seeing 2-3 on the Standard Scale he found 12 cm to be about the maximum useful aperture, increasing to 20 cm with seeing 5-6, and from 28 to 50 cm only when the seeing was better than 6. Pickering has further stated that although a 50-cm will, when the seeing is poor, divide binaries which cannot be separated with a 25-cm, the 50-cm will nevertheless show no more planetary detail than the 25-cm; and that when t is high a 12-cm will show the planetary detail visible under the same conditions even with the Mt Wilson 2·5-m.

All this is no doubt perfectly true, but the word 'better' in the statement, 'Stopping down, when the seeing is bad, gives a better image', conceals a confusion. The image, it is true, will appear superficially better, but only because it is neither bright enough nor highly enough resolved to reveal fully the effects of the atmospheric turbulence*: it is not more highly resolved—it is indeed less so, of course. Stopping down suppresses the more glaring effects of turbulence, but suppresses also the finer detail that might otherwise have been glimpsed during the fleeting moments of improved seeing that so often occur. The full-aperture image looks horrible, but contains the resolved detail; the reduced-aperture image looks a great deal smarter, but only because its threshold of resolution is higher—the detail is actually not there, in the focal plane image.

In short, the use of an objective diaphragm reduces the distortion of the image, but at the same time reduces the resolving power (by increasing the angular size of the spurious disc) and the light grasp of the instrument. A purely superficial improvement has been gained, no real improvement. If the seeing really is so bad that observation with full aperture is 'impossible', it is better (unless the telescope is of such size that even reduced aperture is useful) to persevere or pack up, rather than to stop down. The employment of diaphragms is something of an obsession with many observers; used at random, and without a proper realisation of their

*For the same reason it is seldom necessary to employ an objective stop with apertures smaller than about 10 cm.

action, they will more often do harm than good.

It is generally agreed that in stellar observation reduction of aperture is seldom if ever a help, such authorities as Schiaparelli, Burnham and Aitken all speaking against the practice. A possible exception to this general embargo is the case of very unequal pairs, or possibly very bright pairs. A polygonal objective diaphragm, capable of being rotated about the optical axis by means of controls from the observing position, is useful in the case of unequal doubles, not because of the resultant reduction of light grasp (amounting to a little over 17 per cent in the case of a hexagon), but of the modification of the diffraction pattern that it produces. Figure 193 shows a hexagonal diaphragm and the corresponding disposition of the rays of the diffraction pattern. If the diaphragm is rotated so that the line joining the stars is parallel to one of its diagonals the *comes* will lie in one of the artificially darkened areas between the arms of the primary image, and its visibility will be slightly augmented. Another device, suggested by Sir J. Herschel, is a triangular diaphragm, producing an illumination reduction in the region of 40 per cent and a 6-rayed diffraction pattern; these rays, however, are brighter than those of the hexagon, which is therefore to be preferred. A modern device of this nature, used to reveal the comparison of Sirius, is described in B. 7.32.

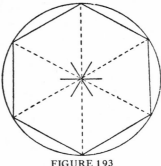

FIGURE 193

The influence of a
hexagonal diaphragm
on a star's diffraction pattern

A similar improvement can be effected by rotating the flat and ocular ring through 45°-60° if the *comes* lies on one of the arms of the diffraction cross.* Given sufficient aperture, the diffraction effects of the central obstruction can, of course, be avoided by using an eccentric diaphragm. Thus C. J. Tenukest's 50-cm Cassegrain at Sydney can still give 18 cm unobstructed aperture with an eccentric stop, and W. H. Steavenson reported that his 75-cm reflector gave excellent refractor-type images with a 29-cm eccentric stop.

27.2 Filters

There are two distinct factors involved in the improvement of the image by means of a filter:

(*a*) reduction of glare, especially with low magnifications and large

*The modifications of the image structure due to a central obstruction are described in sections 2.2, 3.

aperture, without either reducing the resolving power or increasing magnification beyond the point dictated by the state of the atmosphere.

(b) improvement of definition and augmentation of tonal and colour contrasts.

As regards (a) almost any colour filter will fill the bill, a neutral one being the obvious choice, though various tints have been used successfully.

The second factor, (b), operates partly through the filter's selective transmission (eliminating some of the secondary spectrum in refractors), and partly, it is likely, by an actual improvement of the diffraction pattern of the image (the spurious discs of bright stars can be reduced in size by suitable filters). As examples of the use of filters in planetary work may be cited: increased contrast between Jovian white spots and the belts in which they are situated, by means of a Wratten K.3; increased tonal contrast in the Martian image with the use of a very pale-red filter; wide use of filters in lunar observation.* Green and yellow filters have proved to be particularly useful in increasing contrast between extended detail.

Graded neutral filters may easily be made by exposing film for different periods—according to the degree of absorption required, and ultimately therefore on the type of observation for which they are required. Six or eight, mounted between glass slips in a circular brass disc, rotatable about a central pivot which is carried on a clip fixing it eccentrically to the ocular, provide an accessory whereby the intensity of the image can be quickly varied over as wide a range as required (Figure 194). Neutral filters (granular silver or colloidal carbon) are also supplied by the makers of colour filters, to specified densities accurate to about 5 per cent.† Complete neutrality is, however, impossible, a typical (Ilford) 'neutral' filter transmitting about 5 per cent at 400 nm, 20 per cent at 740 nm, and 10 per cent throughout the range 460-640 nm.

FIGURE 194
A rotating filter holder

*See *O.A.A.*, section 2.5.
†See also sections 21.2.1, 21.3.1.

Such firms as Eastman Kodak supply a wide range of colour filters (Kodak's Wratten series alone contains over one hundred), and will supply on request full details of the transmissions in different wavelengths, as well as the photographic exposure factors, of their products.

The transmission of a neutral filter, or of a colour filter in a particular wavelength, is the intensity of the transmitted light expressed as a percentage of the intensity of the incident light, the incident beam being parallel and normal to the surface of the filter. Hence if I_0 is the intensity of the incident radiation, and I that of the transmitted radiation, we have for the transmission of the filter in wavelength λ,

$$T_\lambda = 100 \, \frac{I}{I_0}$$

The density, D, of the filter is given by

$$D = \log \frac{I_0}{I}$$

or

$$D = \log \frac{1}{T}$$

The combined density of two or more filters is equal to the numerical sum of their individual densities, thus making D a more convenient quantity to manipulate than T.

Colour filters consist of organic dyes held in suspension in gelatin or cellulose acetate film of uniform thickness. This film, as employed for astronomical purposes, is sometimes cemented between two plane parallel (or, for the utmost accuracy, optically flat) glass slips. It should be remembered that unless the glass is replaced by quartz, filters of whatever transmission characteristics will remove the ultraviolet beyond about 330 nm. All gelatin filters, on the other hand, transmit freely in the infrared.

Gelatin swells and becomes distorted when exposed to moisture. Hence even glass-cased filters should always be kept in a dry place when not in use, and particular attention paid to the edges of the filter where, if anywhere, moisture may get at the film. Provided they are stored in a dustproof container, and their surfaces never touched with the fingers, cemented filters should require cleaning only at very long intervals. When it does become necessary, they should first be brushed with a sable brush to remove dust, and then gently polished with a piece of soft dry material such as silk or high-grade flannel; then, if this is not enough, with a screw of tissue paper dampened, but not saturated, with absolute alcohol. Undue heating will soften the balsam cement, hence filters of this type cannot be used for solar work.

Density	Transmission (%)	Density	Transmission (%)
0·00	100	0·84	14
0·02	96	0·86	14
0·04	91	0·88	13
0·06	87		
0·08	83	0·90	13
		0·92	12
0·10	79	0·94	11
0·12	76	0·96	11
0·14	72	0·98	10
0·16	69		
0·18	66	1·00	10·0
		1·02	9·6
0·20	63	1·04	9·1
0·22	60	1·06	8·7
0·24	58	1·08	8·3
0·26	55		
0·28	52	1·10	7·9
		1·12	7·6
0·30	50	1·14	7·2
0·32	48	1·16	6·9
0·34	46	1·18	6·6
0·36	44		
0·38	42	1·20	6·3
		1·22	6·0
0·40	40	1·24	5·8
0·42	38	1·26	5·5
0·44	36	1·28	5·2
0·46	35		
0·48	33	1·30	5·0
		1·32	4·8
0·50	32	1·34	4·6
0·52	30	1·36	4·4
0·54	29	1·38	4·2
0·56	28		
0·58	26	1·40	4·0
		1·42	3·8
0·60	25	1·44	3·6
0·62	24	1·46	3·5
0·64	23	1·48	3·3
0·66	22		
0·68	21	1·50	3·2
		1·52	3·0
0·70	20	1·54	2·9
0·72	19	1·56	2·8
0·74	18	1·58	2·6
0·76	17		
0·78	17	1·60	2·5
		1·62	2·4
0·80	16	1·64	2·3
0·82	15	1·66	2·2

Density	Transmission (%)	Density	Transmission (%)
1·68	2·1	2·38	0·42
1·70	2·0	2·40	0·40
1·72	1·9	2·42	0·38
1·74	1·8	2·44	0·36
1·76	1·7	2·46	0·35
1·78	1·7	2·48	0·33
1·80	1·6	2·50	0·32
1·82	1·5	2·52	0·30
1·84	1·4	2·54	0·29
1·86	1·4	2·56	0·28
1·88	1·3	2·58	0·26
1·90	1·3	2·60	0·25
1·92	1·2	2·62	0·24
1·94	1·1	2·64	0·23
1·96	1·1	2·66	0·22
1·98	1·0	2·68	0·21
2·00	1·00	2·70	0·20
2·02	0·96	2·72	0·19
2·04	0·91	2·74	0·18
2·06	0·87	2·76	0·17
2·08	0·83	2·78	0·17
2·10	0·79	2·80	0·16
2·12	0·76	2·82	0·15
2·14	0·72	2·84	0·14
2·16	0·69	2·86	0·14
2·18	0·66	2·88	0·13
2·20	0·63	2·90	0·13
2·22	0·60	2·92	0·12
2·24	0·58	2·94	0·11
2·26	0·55	2·96	0·11
2·28	0·52	2·98	0·10
2·30	0·50	3·00	0·10
2·32	0·48	4·00	0·01
2·34	0·46	5·00	0·001
2·36	0·44		

Cellulose acetate filters are more resistant to rough handling than gelatin, being little affected by fingering, damp, or grease; in particular, they are not distorted by water. Cellulose acetate filters, cemented between glass, are unaffected even by immersion in water, and may therefore be used without fear of consequences in the heaviest dew. Within the range −40°C to 70°C they are unaffected by temperature.

Most filters are affected in their transmission characteristics by exposure to light over long periods (from one to twelve months' total exposure) and should therefore be kept in a lightproof as well as dust- and damp-proof container when not in use.

TIME AND CLOCKS

Two systems of time reckoning are based upon the position of the Sun: Apparent (Solar) Time and Mean (Solar) Time.

28.1 Apparent Time (AT)

The Hour Angle of the true Sun, measured from lower meridian transit (i.e. midnight). This is the time told by a sundial..

AT is useless for civil time reckoning, since, for two reasons, it is non-uniform: (*a*) the Sun's motion in the ecliptic is not uniform, (*b*) Hour Angle is measured in the equator, which is inclined to the ecliptic.

28.2 Mean Time (MT)

To avoid these sources of non-uniformity of the time units a fictitious *mean sun* is invoked: this moves in the equator with uniform angular velocity.

Mean Time is then defined as the Hour Angle of the mean sun measured from lower meridian transit (i.e. midnight). Thus the MT at a given moment is the Hour Angle of the mean sun $\pm 12^h$.

28.3 Equation of Time (E/T)

The E/T is the difference in RA between the mean sun and the true sun, or $\Upsilon M - \Upsilon T'$ in Figure 195. It is thus the correction to be applied to MT in order to derive AT, being positive when the true Sun is W of the mean sun, negative when E of the mean sun. The value of the E/T varies continuously throughout the year, the limiting and zero values being as follows:

December	0	minutes
February	-14.25	
April	0	
May	$+ 3.75$	

470

June	0
July	− 6·5
September	0
November	+ 16·25

For Table of daily values of the E/T, see the *A.E.*

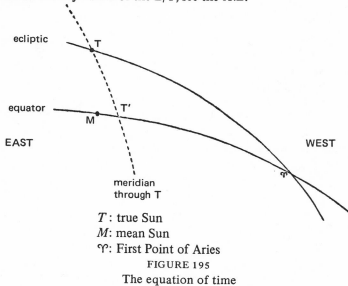

T: true Sun
M: mean Sun
♈: First Point of Aries

FIGURE 195
The equation of time

28.4 Universal Time (UT) or Greenwich Mean Time (GMT)

Mean Time for the meridian of Greenwich, England, reckoned from midnight. For civil purposes, this time system is always referred to as GMT, but astronomical usage favours UT to avoid possible confusion with GMAT (section 28.6).

UT is derived from the mean sun, the regularly moving body which returns to coincidence with the true Sun at yearly intervals. The reference point is taken as the Sun's longitude at mean noon on 1900 January 0 (i.e. 1899 December 31); UT is based on the hour angle, at Greenwich, of the calculated position of the mean sun at any instant. Since this value is related to the rotation of the Earth, UT can be derived, as it is in practice, from observations made of ST, and such time is referred to as UTO. Observations of ST are affected, however, by minute secondary effects: (*a*) varying position, relative to the Earth itself, of its axis of rotation; (*b*) small irregularities in the period of rotation. Effect (*a*) can be observed directly and allowed for; (*b*) cannot be observed directly, but may be extrapolated from previous long-continued observations. Allowance for these irregularities enable UT1 and UT2, respectively, to be derived. Co-ordinated UT (UTC), the time system actually transmitted, is a smoothed version of UT2. The difference between UTC and UT1 (DUT1) is kept within about 0·9 of a second by periodical step adjustments of 1·0 second, normally applied at the last UTC second on June 30 and December 31.

28.5 Ephemeris Time (ET) and International Atomic Time (TAI)

Secular variation in the Earth's rotation period means that true UT is not uniform; the length of each successive year is approximately one second longer than the previous one. The introduction of ET, based on observations of the position of the Moon in its orbit (hence independent of the rotation of the Earth), gives the astronomer the nearest *observational* approach to a uniform time system. These positions are compared with the orbital motion calculated from gravitational theory, and from a long series of observations ET may be calculated retrospectively from the relationship

$$ET = UT + \Delta T$$

once ΔT has been determined.

ET is of practical interest only in computing practice, the predicted difference ΔT having already been applied to all predictions (e.g. lunar occultations) where it would matter, so rendering them approximately into UT. It should be noted that ET, being based on gravitational theory, is dependent on the value of the gravitational constant. The value of ΔT for 1978 was predicted to be about +49 seconds.

The invention of the atomic clock in the 1950s has given what appears to be a time base independent of all exterior influences, at least to a precision of about one part in 10^{12} (equivalent to an uncertainty of one second in 30,000 years). All international UTC time signals are based on TAI.

28.6 Greenwich Mean Astronomical Time (GMAT)

GMT reckoned from midday. It has the advantage over UT of not involving a change of date at midnight, in the middle of the astronomer's working period; consequently employed in some fields of observation, though UT is the generally employed system, and should always be used in any rewriting of the observations. In practice it is convenient to ignore the change of date when recording observations in UT at the time they are made (thus carrying on the reckoning to 25^h, 26^h, etc), and only to

472

make the necessary and easy adjustment when reporting or publishing relevant observations.

Note that the astronomical date does not change at 24^h GMAT but at 12^h. Hence the date of the period midnight-to-midday on, say, February 10 (civil date) is still February 9.

$$UT = GMAT + 12^h$$
$$GMAT = UT - 12^h$$

28.7 Standard Time, or Zone Time

The Local Mean Time (LMT) in any of the recognised Time Zones into which the Earth's surface is divided. E.g. the 1-hour-separated Eastern Standard Time, Central Standard Time, Mountain Standard Time, and Pacific Standard Time of the North American continent. Similarly GMT is the Standard Time for all places lying between longitudes $7\frac{1}{2}°$ East and West. Standard Times normally differ from GMT by a whole number of hours.

28.8 British Summer Time (BST)

In operation between April and October; it is 1^h ahead of GMT:

$$GMT = BST - 1^h$$
$$BST = GMT + 1^h$$

28.9 Julian Date (JD)

This is a continuous calendar, useful for computing purposes involving long intervals of time. JD 0 commenced at 0^h GMAT on 4713 January 1 BC; the Julian Date on 1979 Jan 1·0 GMAT (Jan 1·5 UT), for example, was 2,443,875·0. The *Astronomical Ephemeris* tabulates JD for January 0 of every leap year from 1697 BC to 2296 AD, as well as that for day 0 of each month, 1900-2000 AD. A table covering the period 1970-1995 is given in *O.A.A.*, section 17.14.

28.10 Solar days and years

The mean solar day is the interval between successive transits of the mean sun over the same meridian:

$$24^h \text{ MT} = 24^h 3^m 56^s555 \text{ mean Sidereal Time}$$

The solar year, or Besselian fictitious year, begins when the sun's mean longitude is $280°$, which instant is written 1975·0, 1976·0, etc.

28.11 Recording the time of observations

(*a*) Employ UT rather than GMT.

(*b*) Always specify whether using UT or GMAT.

(*c*) Never use BST, or a LMT other than that of Zone 1 (i.e. GMT) for recording observations made in Great Britain.

28.12 Sidereal Time (ST)

Owing to the orbital motion of the Earth, causing the Sun to advance eastward along the ecliptic, the star sphere completes one revolution in a slightly shorter period than the Sun. ST is based upon the rotation of the star sphere, and a ST clock and a MT clock will tell the same time only once a year—at the autumnal equinox (about September 23), when the Sun and the First Point of Aries (FPA) are 12^h apart in RA.

The Local Sidereal Time (LST) at any moment and place is the RA of the meridian at that place. The difference between the RA of a star (α) and that of the meridian is known as the star's Hour Angle (*H*); it is measured westward from the meridian (in practice, for greater convenience, it is often reckoned in both eastward and westward directions to 12^h). Since RA is measured eastward from the FPA, it follows that the RA of the meridian is the Hour Angle of the FPA; also that the Hour Angle of the FPA is the ST at any moment (0^h occurring at upper meridian transit— cf. AT and MT). Hence:

LST = Hour Angle of the FPA = RA of the meridian = $\alpha+H$

These relations are illustrated in Figure 196, which represents the view of the southern sky.

It is now easy to see the relation of the Sun to ST. ST = RA + Hour Angle whether the object concerned is a star, the true Sun, the mean sun, or anything else; and since MT = Hour Angle of the mean sun,

ST = MT + RA of mean sun

In the special case of Mean Noon, when the mean sun is on the meridian,

ST = RA of mean sun = RA of true Sun + E/T

Owing to nutation, ST is not uniform, which leads to a distinction between *Mean* or *Uniform Sidereal Time* and *Apparent* or *True Sidereal Time*, analogous to MT and AT; the relation analogous to E/T is termed the *Nutation in RA*:

Nutation in RA = Apparent ST − Mean ST

Sidereal Day: the interval between successive transits of a star across the same meridian—

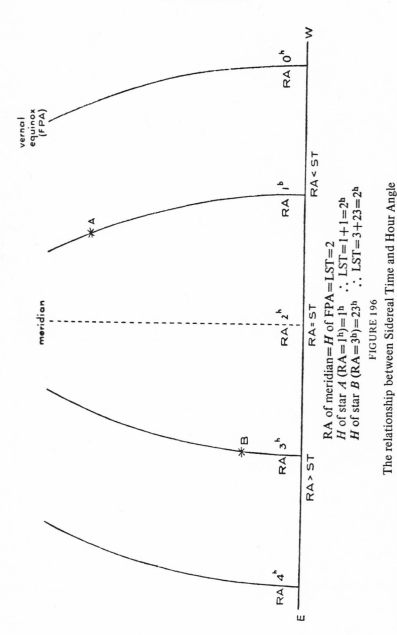

RA of meridian = H of FPA = LST = 2
H of star A (RA = 1^h) = 1^h ∴ LST = $1 + 1 = 2^h$
H of star B (RA = 3^h) = 23^h ∴ LST = $3 + 23 = 2^h$

FIGURE 196

The relationship between Sidereal Time and Hour Angle

$$24^h \text{ Mean ST} = 23^h \ 56^m \ 4\overset{s}{.}091 \text{ MT}$$

This is, on average, $0\overset{s}{.}009$ shorter than the Earth's rotation period, owing to the Precession of the Equinoxes.

28.13 Greenwich and Local Times

Greenwich Sidereal Time (GST) and Local Sidereal Time (LST) are related as follows. If H_G = Hour Angle of a star for an observer in the Greenwich meridian, and H_λ = its Hour Angle for an observer in longitude λ, then

$$\left. \begin{array}{l} H_G = H + \lambda \\ H_\lambda = H_G - \lambda \end{array} \right\} \text{ when } \lambda \text{ is West}$$

the signs of λ being reversed if it is East, and λ being converted to time by means of $1° \equiv 4^m$.

Thus
$$\text{LST} = \text{GST} - \lambda \quad \text{(West of Greenwich)}$$
$$\text{LST} = \text{GST} + \lambda \quad \text{(East of Greenwich)}$$

A similar relationship holds between GMT and LMT, the Hour Angle of the mean sun replacing that of the star:

$$\text{LMT} = \text{GMT} - \lambda \quad \text{(West of Greenwich)}$$
$$\text{LMT} = \text{GMT} + \lambda \quad \text{(East of Greenwich)}$$

28.14 Conversion of ST/MT intervals

The relation between ST and MT can be expressed in a variety of ways, of which two of the more important are:

(a) The Tropical Year contains 365·2422 mean solar days,
" " " " 366·2422 sidereal days.

(b) Since $24^h \text{ ST} = 23^h \ 56^m \ 4\overset{s}{.}09 \text{ MT}$,
$24^h \text{ MT} = 24^h \ 3^m \ 56\overset{s}{.}56 \text{ ST}$,

a sidereal clock gains on a MT clock at the rate of

$$9\overset{s}{.}8565 \text{ per MT hour,}$$
$$9\overset{s}{.}8296 \text{ per ST hour.}$$

The conversion of a ST interval to the equivalent MT interval, and vice versa, may be carried out in various ways, of which the most practical and convienient is (a) below:

(a) *A.E.* Tables give the MT equivalent of every ST hour from 1 to 23 and of every ST minute and second from 1 to 60; also the ST equivalents of the same MT intervals. Summarised, these read:

ST interval	Equivalent MT interval			MT interval	Equivalent ST interval		
	h	*m*	*s*		*h*	*m*	*s*
1ʰ	0	59	50·17	1ʰ	1	0	9·86
1ᵐ	0	0	59·84	1ᵐ	0	1	0·16
1ˢ	0	0	0·99(7)	1ˢ	0	0	1·00(3)

(*b*) A convenient approximation, involving errors not exceeding a few secs:

$$T = t + \frac{t}{360}$$

(where T is the number of sidereal h.m.s. in an interval, and t the number of mean h.m.s. in the same interval). To divide h.m.s. by 360, divide by 6 and enter the h. in the m. column, the m. in the s. column.

(*c*) A less convenient expression is

$$T = 1·003t$$

(*d*) Table showing the gain of ST on MT for every h., m., s. A summarised Table of this sort will be found in on p. 481.

28.15 Equivalent ST and MT instants

The simplest and most direct method of deriving the ST at a given MT and longitude is to make use of the two *A.E.* tables: GST of midnight, and conversion of MT to ST intervals.

Suppose we want to know the local ST at 14ʰ 30ᵐ MT on 1977 Jan 1 at a point 20° W of Greenwich. The *A.E.* gives the GST of the preceding midnight as 6ʰ 42ᵐ 07ˢ7. From the *A.E.* conversion table we find that 14ʰ 30ᵐ MT ≡ 14ʰ 32ᵐ 22ˢ9. Therefore the GST at 14.30 MT is 21ʰ 14ᵐ 30ˢ6. But we want the LST at 20° W longitude, not GST. 20° ≡ 1ʰ 20ᵐ, and since the longitude is W we subtract. Therefore the required LST = 19ʰ 54ᵐ 30ˢ6.

The derivation is the work of a moment, and slips are less likely to occur if it is laid out as below:

MT	14ʰ 30ᵐ 00ˢ0
Equivalent ST (from *A.E.*)	14ʰ 32ᵐ 22ˢ9
GST of midnight (from *A.E.)*	06 42 07·7
∴ GST at 14ʰ 30ᵐ	21 14 30·6
Less 20° W	01 20 00
. required LST	19 54 30·6

The procedure for calculating the LMT at a given ST is analogous and equally simple, the required MT being the MT of the last transit of the 0^h circle of RA, added to the MT equivalent of the given ST. For example: what is the LMT at $20^h\ 30^m\ 00^s$ ST on 1977 January 1 at a place $20°$ W of Greenwich?

ST	$20^h\ 30^m\ 00\overset{s}{.}0$
Equivalent MT (from *A.E.*)	$20^h\ 26^m\ 38\overset{s}{.}5$
MT of 0^h RA transit (from *A.E.*)	17 15 02·3
∴ GMT at given ST	37 41 40·8
Less $20°$ W	01 20 00
∴ required LMT (less 24^h)	12 21 40·8

An alternative method of deriving LST is comprised in the following formula, which obviates the need to refer to the *A.E.* and can be conveniently worked out on a pocket calculator:

$$\text{LST} = k + 1.0027379t + 0.06570982d - \lambda/15.$$

The symbol t represents UT (hours and decimals); d is the number of complete days that have elapsed since the preceding December 31, 0^h UT; λ is the observer's W longitude in degrees and decimals; and k is a constant which must be taken from the following table for the year in question:

Year	k	Year	k
1976	6·586475	1981	6·638322
1977	6·636270	1982	6·622408
1978	6·620356	1983	6·606493
1979	6·604441	1984	6·590579
1980	6·588527	1985	6·640374

The derived time, in decimals of an hour, must be converted to minutes and seconds. (See B. 20.7.)

28.16 Precision timekeepers

The history of astronomical observation has been very closely related to the development of precision timekeeping. Only in the present century has the role of the transit instrument, astrolabe and zenith tube been superseded by the superior constancy of man-made timekeepers. Whereas in earlier times the Earth's rotation was used to check clocks, modern clocks can now be used to measure the constancy of the Earth's rotation.

The limit of accuracy attainable with a mechnaical pendulum clock was achieved by the Shortt-Synchronome free pendulum timekeeper devised in 1921-4. A master pendulum, free of all escapement duties and swinging in a near-vacuum, was maintained in its vibration by $\frac{1}{2}$-minute electro-magnetically controlled impulses from a synchronised slave pendulum. The interference with the free vibration of the master pendulum was therefore kept to a minimum, and it approached very closely to the ideal of a pendulum swinging uninterruptedly through a constant arc. Accuracy of the order of one part in 10^7—equivalent to about 3 seconds per annum—was achieved, and these clocks represented the ultimate in accuracy until the introduction of electronically maintained crystals, and later atoms, which are allowed to vibrate at their own extremely constant frequency.

The quartz clock came into widespread use in the 1940s, its accuracy exceeding that of the free-pendulum clock by at least a factor of 10. A small ring of quartz, excited by an electric current, oscillates at a self-determined frequency of about 10^5 Hz (as against 0·5 Hz in the case of a standard mechanical clock with a seconds pendulum). An electronic circuit tuned to this frequency operates the clock dial. In the 1950s, the magnetic oscillation of a beam of caesium atoms—about $9·2 \times 10^9$ Hz —was found to provide a stability of about one part in 10^{12}. The current standard 'atomic time' (TAI) is based on the frequency of vibration of the ammonia atom.

28.17 The observatory clock

Clocks of the order of precision of those described above are, however, of no more than academic interest to the average amateur. A number of advanced observers used to equip themselves with a simplified single-pendulum Synchronome clock, or its equivalent (stability better than about one second per day), but the current availability of relatively cheap and extremely accurate electronic timepieces has now put pendulum clocks out of court, since UT precise to a fraction of a second is readily available. Modern domestic quartz clocks and watches are normally guaranteed to run within fairly precise limits of the true rate (e.g. gain or loss not exceeding 15 seconds per month), which is sufficiently accurate for the routine recording of observations. Even an ordinary synchronous wall clock will normally be correct to within about 20 seconds, but it has the disadvantage, for more precise purposes, that fluctuations of rate within each 24-hour period cannot be predicted accurately.

If more precise timing is required, it will be necessary to check the timepiece systematically against standard time signals (see section 28.18). It is of little importance if the clock is running noticeably fast or slow, provided that the rate is both uniform and known; the behaviour of the clock should, therefore, be kept under constant supervision, preferably by plotting the error on a chart at frequent intervals. The following terms are used:

Clock running fast: gaining rate; clock error +; correction −.
Clock running slow: losing rate; clock error −; correction +.

The timepiece should be adjusted as rarely as possible (e.g. when the error has attained 60 seconds exactly), the correction at other times being added to the reading. A good quartz timepiece, whose rate is known, should allow a confident timing to within ±0·2 second to be made within 24 hours of the last time-check.

Timings required to be more precise than this should, wherever possible, be taken directly from standard time signals.

Sidereal time, which, being local time, is unique to each longitude, must be calibrated by the observer. Extreme precision is not normally required, however, since the purpose of a sidereal clock is to allow the telescope to be set to the correct RA for location purposes, and an error of half a minute or so will still bring the object within the field of view of an LP ocular, assuming that the telescope's circles are correctly adjusted.

Synchronous or quartz-controlled sidereal clocks can be purchased, but an ordinary 12-hour dial mechanical timepiece, rated to gain about 10 seconds per MT hour (3 minutes 56 seconds per 24 MT hours), will be perfectly adequate for a single night's work. It must, of course, be remembered that ST between 12^h and 24^h is subject to a correction of $+12^h$. Alternatively, a clock running at MT rate can be used, once it has been set to the correct ST, by adding a correction, obtained from the accompanying Table, for the appropriate MT interval. The equivalent equatorial angular discrepancy is also given, showing that even during a 6-hour observing session, the equatorial angular error will amount to only $\frac{1}{4}^°$ if no correction is made at all; objects in higher Decs will be nearer to the centre of the field of view by an amount proportional to $\cos \delta$.

If ST needs to be found, and no clock is available, a quick method which is accurate enough for most practical purposes is simply to read the Hour Angle of any star whose RA (α) is known, and to set the clock accordingly from

$$LST = \alpha + \text{Hour Angle},$$

Hour Angle being negative before meridian transit and positive thereafter.

MT interval	Gain of ST on MT	Angular equatorial equivalent
10^m	1^s7	25^s5
30	5·0	1′ 15
1^h	9·8	2 25
2	19·7	4 55
6	59	14 45
12	1^m 58	29 30
24	3 56	59 0

28.18 Time signals

Since the Greenwich time service was opened in 1833 with the midday dropping of the time ball, more and more accurate time has been purveyed to the world, both from the Royal Observatory and elsewhere. Today, signals generated by atomic clocks in various countries, generally correct (at source) to about 0·0001 second, can be picked up by the appropriate receiver. In addition, domestic radio services frequently broadcast six pips to mark the last five seconds of the hour. In all cases, the commencement of the signal indicates the timing reference point. In the UK, the claimed accuracy of the regular six pips is about 0·5 second.

28.18.1 **Regular radio time signals**: Although a number of broadcasting stations operate a time-signal service, many of them, being discontinuous, may not be suitable for the direct timing of a particular event. The following stations operate for long continuous periods:

Call sign and station	Frequency (kHz)	Wavelength (metres)	Notes
MSF Rugby, England	60	5000	Continuous except for last 5^s of each hour, and 10^h to 14^h on first Tuesday of each month. Second markers 100 ms; minute markers 500 ms; both markers interruption of carrier wave.
DIZ Nauen, E. Germany	4525	66·3	Continuous except 08^h 15^m to 09^h 45^m. Second markers 100 ms; minute markers 500 ms (also last 2^s of each hour).

Call sign and station	Frequency (kHz)	Wavelength (metres)	Notes
ZUO Olifants- fontein, S. Africa	5000	60·0	Continuous. Second markers 5 ms; minute markers 500 ms; time announcement in morse every 5 minutes.
VNG Lyndhurst, Australia	7500	40·0	Continuous except 22^h 30^m to $22^h\,45^m$. Second markers 50 ms (5 ms, 55^s-58^s and 50^s-58^s every 5th minute); minute markers 500 ms.
WWV Fort Collins, Colorado, USA	2500 5000 10000 15000 20000 25000	120·0 60·0 30·0 20·0 15·0 12·0	
WWVH Kauai, Hawaii, USA	2500 5000 10000 15000 20000	120·0 60·0 30·0 20·0 15·0	Both stations continuous. Second markers 5 ms; minute markers 800 ms. Voice announcement precedes every minute.
CHU Ottawa, Canada	3330 7335 14670	90·1 40·9 20.45	Continuous. Second markers 300 ms; minute markers 500 ms; 51^s-59^s, voice announcement, station and time.

Further details of these and other stations are to be found in B. 20.9, 10.

DIZ, which can be picked up by any short-wave receiver with the appropriate waveband, is widely used by observers in the UK. MSF Rugby, although less affected by interference, requires a special receiver to convert the carrier interruptions into pulses. Since neither station identifies the individual minutes, a timepiece accurate to the nearest minute is also required. WWV and CHU signals are normally used by observers in the USA.

A design for a digital clock, synchronised by MSF signals, was published in B. 20.1.

28.18.2 Telephone speaking clock: By dialling 8081, UK subscribers can obtain time signals for each 10-second interval during the 24 hours. This so-called 'speaking clock', which is synchronised by a master clock at the Royal Greenwich Observatory, is accurate to at least 0·05 second. In the USA, the National Bureau of Standards atomic time signals are available by dialling 1-303-449-7111.

28.19 Timing observations

The instant at which an observed event occurs may be determined in a variety of ways.

28.19.1 Stopwatch and clock: The use of a clock as an intermediate step between the standard time signal and the stopwatch is necessary only if the interval between the nearest available signal and the time of the observation is too long to be accommodated with accuracy by the stopwatch. This might be the case if the observation has to be carried out at some remote site. In such a situation a quartz timepiece will be the best choice, but any reliable timekeeper with a seconds hand will serve. The time shown must be checked against a standard signal as soon as possible both before and after the observation, and any rate error noted and allowed for. Similarly, the error of the stopwatch must be known, or, preferably, reduced to zero over the short amount of time during which it must run.

The procedure is to start the watch from zero at the instant of the observed event. Carry it to the clock. Pick up the rhythm of the clock beat, and as its second hand comes up to a complete minute, stop the watch. Then if

$T_0 =$ the required UT of the event,

$T_1 =$ watch reading (secs),

$T_2 =$ clock time when watch stopped (UT),

$\Delta t =$ clock correction (+ if slow, − if fast),

$\Delta t' =$ correction for personal equation (error +, correction −, assuming normal lag),

$$T_0 = T_2 - T_1 + \Delta t + \Delta t'$$

For example: Clock rate $= +0\overset{s}{.}8$ per hour,

Clock reading $= 17^h \ 1^m \ 5\overset{s}{.}0$ at $17^h \ 0^m \ 0^s$ (TIM),

$T_2 = 17^h \ 31^m \ 0^s$,

$T_1 = 46\overset{s}{.}6$,

Clock error $= +(1^m \ 5\overset{s}{.}0 + 0\overset{s}{.}4) = +1^m \ 5\overset{s}{.}4$,

$\Delta t = -1^m \ 5\overset{s}{.}4$,

Personal equation $= +0\overset{s}{.}3$,

$\Delta t' = -0\overset{s}{.}3$,

$\quad T_0 = 17^h \ 31^m \ 0^s - 46\overset{s}{.}6 - 1^m \ 5\overset{s}{.}4 - 0\overset{s}{.}3$

$\quad\quad = 17^h \ 29^m \ 7\overset{s}{.}7$

28.19.2 Stopwatch and standard time signal: The objections to using an additional clock are (*a*) an extra synchronisation has to be made; (*b*) the rate of the clock may not be constant. For this reason, timed observations of the highest quality are always made by synchronising the stopwatch directly against a radio signal or the speaking clock. Then, if the watch is reading T_1 and the standard signal at the instant of synchronisation is T_2, the required time T_0 is obtained directly from $T_0 = T_2 - T_1 + \Delta t'$.

Since the accuracy of the observation depends as much upon the instant of stopping the watch as starting it, some thought should be given to the procedure involved. The technique usually adopted is to take up the rhythm of the pips, pressing the button to coincide with a convenient minute marker. A method advocated by Taylor is for the observer to listen to a series of pips whilst watching the second hand on the stopwatch, mentally determining as he does so on which tenth or fifth of a second the pips occur. Therefore, regardless of any error in stopping the watch, the fraction will already be known. K. Fea has advocated the use of a split-action watch. The slave hand can be stopped repeatedly against different pips to obtain the best value for the fraction of the second.

Neither MSF nor DIZ minute markers are announced by a preliminary code; the best technique is to await the marker and to stop the watch on, say, the fifth subsequent second.

28.19.3 Eye and ear: Rarely used today, this was once a common technique when observatories were always equipped with a clock beating seconds. The method is to estimate the fraction of the interval ($\frac{1}{2}/\frac{1}{2}$ or even $\frac{1}{4}/\frac{3}{4}$ is quite feasible with 1^s beats) between beats at which the observed event occurs, and start counting the ticks. 'Carrying the beat' in the head, note this down, go to the clock and identify the beat. Suppose, for example, that the event is judged to occur $\frac{3}{4}$ of the inter-beat interval after one tick and $\frac{1}{4}$ of the interval before the next. Counting starts with the latter tick (nought). On going to the clock, still mentally counting, it is found that the 31st tick occurs at $7^h 4^m 0^s$. Then the time of the observation was

$$7^h 4^m 0^s - 31^s0 - 0^s25$$
$$= 7^h 3^m 28^s75$$

An alternative method, which is safer but needs rather more practice, is to look at the clock and start counting shortly before the expected event, estimate the inter-second fraction of the instant of observation, record it, carry the beat to the clock and verify the count. The technique can be used, of course, to measure the time of an observation against radio time signals.

28.19.4 Chronograph: The advantages of the chronograph are that fractions of a second do not have to be estimated by ear, that neither a clock nor an assistant is required, that it is less distracting to the observer than the eye-and-ear method, and that it involves a more systematic personal equation than the latter.

See section 22.7.

28.19.5 Tape recorder and time signals: As already mentioned (section 22.7.3), this method possesses a number of advantages, not the least of them being that the recording can be analysed and re-measured as often as required. The likelihood of miscounting minutes or seconds is, therefore, minimal. In addition, the observer can make immediate verbal comments on the quality of the timing while the event is still fresh in his mind. The tape can either be measured directly, or else timed repeatedly using a stopwatch.

All observations are recorded in MT, sidereal clocks being used only for finding purposes. If, therefore, the sidereal clock is used in the determination of the time of an observation, the derived time must be converted to MT for the record.

SECTION 29

MATHEMATICAL TREATMENT OF OBSERVATIONS

Any set of measures of a physical quantity will exhibit a greater or smaller degree of dispersion. That is to say, the individual observations are to some extent inaccurate, and are scattered round the bull's-eye represented by the true value of the quantity.

The errors responsible for the scatter in the observational material are of two kinds, random and systematic. Random errors are the inescapable result of the humanity of the observer and the imperfection of his instrument, and occur, as the name implies, entirely fortuitously; they alone are susceptible of treatment by statistical methods based on probability theory. Systematic errors are not susceptible to such treatment, since they spring from some tendency to scatter the observed values which is not reduced by increasing the number of the observations. This tendency may be intermittent or periodic in action; when the latter, and its period happens to coincide with that of the quantity under measurement, it is often extremely difficult to disentangle.

Freedom from unknown sources of error can, as a general rule, be increased by varying the conditions under which the observations are made. This is exemplified by the old laboratory maxim 'Reverse everything reversible, and repeat.'

Three main stumbling blocks lie in the way of the application of probability considerations to observational material: the normal law is inapplicable to systematic errors, and therefore to observational data which are still tainted by such error; the number of observations must be adequate for the operation of probability among them; and the relative values of the individual observations must be correctly estimated, i.e. the system of weighting (*v. inf.*) must be carefully considered.

If a large number of readings of a certain physical quantity—micrometer readings of the separation of two stars, for example—are taken, the type of their distribution can be represented conveniently by means of a histogram (Figure 197) in which divisions of the range of the observations

486

—e.g. $0''1$, where the measures themselves are to $0''01$—are represented on the x-axis, and the y-axis level of the histogram corresponding to each such interval is determined by the number of measures that fall within it.

FIGURE 197

A histogram—normal distribution

The outline of the histogram is in fact an approximation to a curve of the form shown in Figure 198, the so-called Normal Curve, which closely represents the distribution of values obtained under conditions exemplified by, for example, those of astronomical observation.

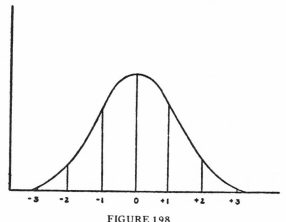

FIGURE 198

Normal Curve covering a range of observations

The aim of any theory of error and probability is at least twofold: to provide a means of estimating the most probable value of a quantity from a set of more or less discordant measures of it, and to provide a means of estimating the accuracy of the derived result and the degree of precision

487

of the observation on which it rests. The normal, or Gaussian, law of error, which is embodied in the Normal Probability Curve, has the additional advantages of simplicity, the yielding of unique solutions, and convenience and uniformity in the application of the rules derived from it. In practice these advantages are as important as the achievement of the very highest degree of probability. The Normal Curve is to be regarded as the appropriate mathematical model to use in cases of the type we have been discussing, and its value resides to a large extent in the fact that its characteristics and properties are both definite and well known. This point will be returned to later.

Faced, then, with a set of observational data, discordant measures of a given quantity, the problem resolves itself thus: what are the errors of the individual readings? what is the best value to take? what is the error of the best value? The normal law of error gives as the most probable value of a quantity q the arithmetical mean of the measures $q_1, q_2 \ldots q_n$. Thus

$$\bar{q} = \frac{\Sigma q}{n}$$

where $\bar{q}$ represents the arithmetical mean, and Σq the sum of n equally reliable observations of q. The most probable value of a quantity, understood in this sense, is applicable only to the observations themselves, and not to functions of the observed quantity. Thus, for example, the most probable value of the distance of a star is that distance corresponding to the arithmetical mean of the parallax measures, and not the mean of the derived distances, which are a function of the observable.

In more complex problems, where it is required to derive the most probable values of several different quantities, each a function of q, resort must be had to the method of least squares; this is an application of the simple proposition, deductible from the normal law, that the sum of the differences between the observed and computed values of the variables, when raised to a common power, should be minimal.* An example of such a problem is the derivation of the elements of an orbit from a set of observations of position. Providing the sum of the squared residuals is kept to a minimum, the errors of solution are confined within tolerable and known limits.

If certain members of a series of measures are considered on good grounds to be more reliable than the others, they must be given accordingly greater weight in the calculation of the most probable value. Thus if the

---

*For detailed treatments of the operation of the method, the reader should refer to the textbooks (see Bibliography).

observations of several observers are being combined, and one observation made by A is considered to be worth w observations by B—i.e. that A's observations are w times as reliable as B's—then the weight of A's observations is w when that of B's is unity.

If the weights allotted to $q_1, q_2 \ldots q_n$ are respectively $w_1, w_2 \ldots w_n$, then the weighted mean is given by

$$\bar{q} = \Sigma \frac{(wq)}{\Sigma w}^*$$

(cf. preceding expression for the corresponding unweighted mean), and the weight of $\bar{q}$ is Σw. For example, if q_1 is considered to be three times as reliable as q_2 and q_4, q_3 twice as reliable as q_2 and q_4, and q_5 and q_6 twice as reliable as q_1, then they will be weighted as follows:

q_1	3
q_2	1
q_3	2
q_4	1
q_5	6
q_6	6
$\Sigma q =$	19

If the values of $q_1 \ldots q_6$ are respectively 50·3, 49·8, 50·2, 50·1, 49·7 and 49·9, then the weighted mean is

$$\bar{q} = \frac{(3 \times 50{\cdot}3) + (49{\cdot}8) + (2 \times 50{\cdot}2) + (50{\cdot}1) + (6 \times 49{\cdot}7) + (6 \times 49{\cdot}9)}{19}$$
$$= 49{\cdot}94$$

To answer the remaining questions we must derive some expression for the scatter of the observations, i.e. discover a parameter which will convey significant and sufficient information regarding the manner in which the individual measures are dispersed about the most probable value. One such parameter of dispersion is the *range*. The range of the six quantities used in the example above is $50{\cdot}3 - 49{\cdot}7 = 0{\cdot}6$, i.e. the difference between the extreme values. This is not a very satisfactory criterion, for the very reason that it is based upon exceptional rather than on normal or typical values. Moreover it tells us nothing about the way in which the measures cluster about the best value.

A more satisfactory description of the dispersion of a set of observations is provided by the *mean deviation* (or mean error). This is defined as the average deviation of a set of measures of a quantity from the most

*Numerator expanded as $w_1q_1 + w_2q_2 + \ldots + w_nq_n$.

probable value of that quantity. If $q_1, q_2 \ldots q_n$ are the measured values of a quantity q, then the most probable value of q is given by

$$\bar{q} = \frac{\Sigma q}{n}$$

and the residuals are $q_1 - \bar{q}, q_2 - \bar{q} \ldots q_n - \bar{q}$. Call these respectively $d_1, d_2 \ldots d_n$, giving each a positive sign. Then the mean deviation is

$$\frac{\Sigma d}{n}$$

Taking the same six unweighted values as before, the most probable value of q is

$$\frac{\Sigma q}{n} = \frac{300}{6} = 50 \cdot 0$$

and the residuals are $0 \cdot 3, 0 \cdot 2, 0 \cdot 2, 0 \cdot 1, 0 \cdot 3, 0 \cdot 1$. Hence

$$\Sigma d = 1 \cdot 2$$
$$\frac{\Sigma d}{n} = 0 \cdot 2$$

The mean deviation of the six measures is thus $0 \cdot 2$, and the derived value of q would be given as

$$\bar{q} = 50 \cdot 0, \quad \text{M.D.} = 0 \cdot 2$$

Mean deviation gives a truer and more significant picture of the dispersion of $q_1 \ldots q_n$ than does the range, since it does not depend solely upon two exceptional values, but upon all the values in the set.

A word may be said at this point concerning residuals. The term has somewhat different connotations according to its context, but in general the residuals of a set of observations are the deviations of the individuals from some mean, most probable, computed, independently derived, or true value of the quantity observed. If O is the observed value of a quantity, and C is the approximately computed or arbitrarily chosen value (its true value being unknown), then $O - C$ is the *residual* (or apparent error) in the strict sense. It provides a check on the accuracy of the computed value, and thence on the computation or assumptions upon which C rests—e.g. observations of cometary positions to improve an orbit. If, on the other hand, C is the known true value of the quantity, then $O - C$ is not a residual but a *true error*, and provides a check on the accuracy of O. $C - O$ is then a *correction* to be applied to the observed value in order to obtain the true value—e.g. testing an instrument or observer for defects or personal equation.

Even better than mean deviation, as an expression of dispersion, is the so-called *standard deviation*. This is the square root of the sum of the squared residuals (deviations from the most probable value) divided by the number of measures in the set. Thus if, as before,

$$d_n = q_n - \bar{q}$$

and
$$\Sigma d^2 = d_1{}^2 + d_2{}^2 + \ldots d_n{}^2$$

then the standard deviation, σ, is given by

$$\sigma = \sqrt{\frac{\Sigma d^2}{n}}$$

Using the same set of figures as before,

$$\sigma = \sqrt{\frac{0\cdot 28}{6}}$$
$$= 0\cdot 216$$

The importance of standard deviation as an interpretation of the scatter of a set of values lies in the fact that, if the latter is assumed to be represented by a Normal Curve, then we at once know that approximately $\frac{2}{3}$ of the distribution lies within 1σ of the mean, 95 per cent within 2σ of the mean, and < 1 per cent more than 3σ from the mean. Whatever the shape of a particular Normal Curve (depending upon the parameter σ) its form is such that the area beneath it is divided as follows:

within 4σ	(+ *and* −) of the mean:	99·994%
" 3σ	" " "	99·73
" 2σ	" " "	95·4
" 1σ	" " "	68·1
" $0\cdot 6745\sigma$	" " "	50·0
" $0\cdot 25\sigma$	" " "	20·8

The name *probable error* is given to the deviation whose value of $0\cdot 6745\sigma$ is equally likely to be exceeded as not. The probable error in the arithmetical mean of a set of observations being inversely proportional to $\sqrt{n}$, there is little to gain in multiplying the number of the observations beyond a certain point. On the contrary, if a reduction of 50 per cent in the number of observations is accompanied by a similar reduction of the random errors, then the value of the result is doubled. Probable error has rather fallen into disfavour with the statisticians, who today tend to regard standard deviation as a more useful and significant guide.

REDUCTION TO EQUINOX OF DATE

The derived position of any object, determined by reference to the stars, will depend upon the epoch of the Catalogue from which the positions of these stars are taken. Similarly, the current position of a star will differ from that given in a Catalogue constructed for a past epoch.

The system of geocentric coordinates is not static, but is subject to progressive and non-uniform change. In addition, the actual position of the object, relative to the Earth, may change. The factors concerned in both these types of change are:

Precession: a secular change due to motion of the ecliptic round the equator; it affects both the RA and Dec of an object. Precession in RA is a function of both the RA and Dec of an object; precession in Dec is a function of its RA only. General Precession is the term applied to the total effect of the two components, Planetary Precession and Luni-solar Precession.

Nutation: a periodic change which can be broken down into long- and short-period terms, the latter being negligible in most cases.

Aberration: a periodic change, depending on the finite velocity of light and the changing direction of the Earth's motion.

Parallax: a periodic change due to the Earth's changing position during the course of the year.

Proper motion: a secular variation of the object's position relative to the coordinate system itself.

Over a long period, the secular variations tend to become increasingly important, as compared with the periodic ones; of them, precession is in the general run of cases much more important than proper motion. Successive refinements would therefore involve consideration and correction of the following:

(a) precession only;

(b) precession, aberration, and long-period terms of nutation;

(c) precession, aberration, long- and short-period terms of nutation;

(d) precession, aberration, nutation, proper motion and parallax.

The terms Mean Place, True Place and Apparent Place must be defined:

Mean Place: the position of an object at a given epoch, referred to the Mean Equinox. MP's of selected objects are quoted in Almanacs for the beginning of the current year, and in Catalogues for a stated epoch. MP's are affected by precession and proper motion.

True Place: the position of an object, referred to the True Equinox. The TP is equivalent to the MP corrected for nutation.

Apparent Place: the position of an object as seen from the Earth's centre. Equivalent to the TP corrected for aberration.

The quantity that it is required to derive is the AP of the object, referred to the equinox of date, i.e. at the time of observation. If its AP is quoted at intervals in the Almanac, it may be obtained for the required date by simple interpolation; if its MP on Jan 0 of the current year is given in the Almanac, its AP at date may be reduced as described below; if its MP at some past epoch is given in a Catalogue, the reduction must be performed in two stages: from MP at past epoch to MP at Jan 0 of the current year (see below), to AP at date (as before).

30.1 Derivation of MP at Jan 0 of current year (a, δ) from MP at past epoch (a_0, δ_0)

Two corrections are to be applied: for proper motion and for precession:

(i) *Proper motion:*

Given that $\Delta\alpha, \Delta\delta$ = the proper motion,

t = mean epoch of the observations on which the Catalogue is based (not the epoch of the Catalogue itself),

t' = current year,

then
$$\alpha' = \alpha_0 + \Delta\alpha(t'-t)$$
$$\delta' = \delta_0 + \Delta\delta(t'-t)$$

where α', $\delta' = \alpha_0$, δ_0 corrected for proper motion only. They are then corrected for precession, as described below.

(ii) *Precession:*

(a) Approximate corrections may be derived by graphical methods;

see Figures 199, 200; similar diagrams are to be found in *B.A.A.H.*, 1923, 1949.

FIGURE 199

Precessional corrections (RA) for intervals of 10 years

Enter the diagram with the object's RA on Scale A; take its Dec from the appropriate curve (interpolate if necessary) and read the value of $\Delta\alpha$ from Scale B or C. These are used according to the object's RA and Dec, as follows:

RA	Dec	Scale
$<12^h$	+	C
$>12^h$	+	B
$<12^h$	−	B
$>12^b$	−	C

Examples:

Position of object, 7^h, $+70°$: $\Delta\alpha = +1^m10$
20^h, $+70°$: $\Delta\alpha = -0^m02$
3^h, $-70°$: $\Delta\alpha = +0^m08$
13^h, $-70°$: $\Delta\alpha = +0^m67$

494

(*b*) Calculation: Given that the position at epoch was α_0, δ_0, and that the interval between the epoch and the current year is t years, then:

$$\alpha = \alpha_0 + t(m^s + n^s \sin \alpha_0 \tan \delta_0)$$
$$\delta = \delta_0 + t(n'' \cos \alpha_0)$$

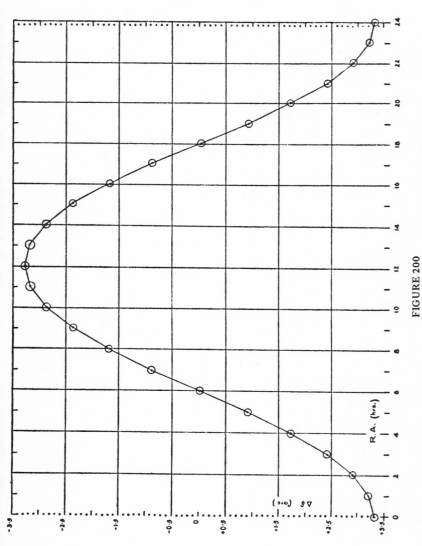

FIGURE 200

Precession corrections (Dec) for intervals of 10 years. The sign $\varDelta\delta$ is the opposite of that given if the object is in S Dec.

where m^s, n^s and n'' (m being precession in RA, n precession in Dec) are tabulated below:

	m^s	n^s	n''
1950	3.07327	1·33617	20·0426
1952	3·07331	1·33616	20·0424
1954	3·07334	1·33615	20·0422
1956	3·07338	1·33614	20·0421
1958	3·07342	1·33613	20·0419
1960	3·07346	1·33612	20·0417
1970	3·07364	1·33606	20·0409
1980	3·07383	1·33600	20·0400
1990	3·07401	1·33595	20·0392
2000	3·07420	1·33589	20·0383

30.2 Derivation of AP at date (a, δ) from MP at Jan 0 of current year (a_0, δ_0)

The complete procedure would consist of the following stages (using ρ, μ, N and A to denote the corrections for precession, proper motion, nutation, and annual aberration respectively, during the interval of τ tropical years (<1) between Jan 0 and the current date):

$$\underset{\text{Jan 0}}{MP} + p + \mu = \underset{\text{date}}{MP}$$

$$\underset{\text{date}}{MP} + N = \underset{\text{date}}{TP}$$

$$\underset{\text{date}}{TP} + A = \underset{\text{date}}{AP}$$

This procedure would be exceedingly laborious, and in practice one or other of the two methods initiated by Bessel is employed, which combine the various corrections into a single expression:

(*i*) employing Besselian day numbers and Besselian star constants: the former are taken from the *A.E.*, and the latter computed or, in practice, taken from a star catalogue; since the values of the constants vary slowly, they will not be accurately given by a catalogue of distant epoch;

(*ii*) employing independent day numbers: this method is used when the star constants are not known with sufficient accuracy.

(*i*) *Besselian day numbers:*

$$\alpha = \alpha_0 + \tau\mu_\alpha + Aa + Bb + Cc + Dd + E$$
$$\delta = \delta_0 + \tau\mu_\delta + Aa' + Bb' + Cc' + Dd'$$

where α, δ = the required AP at date,

α_0, δ_0 = the MP at Jan 0,

τ = the fraction of the tropical year between Jan 0 and date

μ_α = proper motion in RA,

$\mu_\delta=$ proper motion in Dec,

$A, B, C, D, E=$ the Besselian day numbers (derived from the constants of precession, nutation, and aberration; E, a function of nutation and precession, can usually be neglected),

$\left.\begin{array}{l} a, b, c, d \\ a', b', c', d' \end{array}\right\}=$ the Besselian star constants, quoted in the catalogue under each star. They are in fact derived as follows:

$$a=\frac{m}{15}+\frac{n}{15}\sin\alpha\tan\delta$$

$$b=\frac{\cos\alpha\tan\delta}{15}$$

$$c=\frac{\cos\alpha\sec\delta}{15}$$

$$d=\frac{\sin\alpha\sec\delta}{15}$$

$$a'=n\cos\alpha$$

$$b'=-\sin\alpha$$

$$c'=\tan\epsilon\cos\delta-\sin\alpha\sin\delta$$

$$d'=\cos\alpha\sin\delta$$

where

$\alpha, \delta=$ apparent RA, Dec (not sensibly different from α_0, δ_0)

$\epsilon=$ obliquity of ecliptic

$\left.\begin{array}{l} m= \text{precession in RA} \\ n= \text{precession in Dec} \end{array}\right\}''$ arc

Of the terms on the RHS of the above expressions for α and δ, the 2nd concerns the star's proper motion during the interval τ, the 3rd precession, the 4th the long-period terms of nutation, the 5th and 6th the annual aberration, and the 7th nutation and precession.

If required, the additional correction for (a) parallax, and (b) the short-period terms of nutation, are:

(a) $\left\{\begin{array}{l} \Delta\delta=\pi(cY-dX) \\ \Delta\delta=\pi(c'Y-d'X) \end{array}\right.$ $\left\{\begin{array}{l} \text{where} \\ X, Y \text{ are solar coordinates (A.E.),} \\ c, d, c', d' \text{ are the star constants.} \end{array}\right.$

(b) $\left\{\begin{array}{l} \Delta\alpha=A'a+B'b \\ \Delta\delta=A'a'+B'b' \end{array}\right.$ $\left\{\begin{array}{l} \text{where } A', B' \text{ are tabulated in } A.E. \text{ The short-} \\ \text{period terms of nutation in longitude and} \\ \text{obliquity are tabulated in } A.E. \end{array}\right.$

The data required for reducing α_0, δ_0 to α, δ by this method are given in the $A.E.$ for every day of the year.

(ii) *Independent day numbers:*

$$\alpha=\alpha_0+\tau\mu_\alpha+f+\frac{g\sin(G-\alpha)\tan\delta}{15}+\frac{h\sin(H-\alpha)\sec\delta}{15}$$

$$\delta=\delta_0+\tau\mu_\delta+g\cos(G-\alpha)+h\cos(H-\alpha)\sin\delta+i\cos\delta$$

where $\left.\begin{array}{l} f \\ g \\ G \\ h \\ H \\ i \end{array}\right\}$ are the independent day numbers (functions of the Besselian day numbers), the first three being concerned with precession and nutation (long-period terms), and the last three with aberration.

The additional correction for parallax, if required, is the same as before. That for the short-period terms of nutation is:

$$\Delta\alpha = f' + \frac{g' \sin (G'-\alpha) \tan \delta}{15}$$

$$\Delta\delta = g' \cos (G'-\alpha)$$

The following data, required in the operation of this method, are tabulated in the *A.E.* for every day of the year:

f
g
G
h
H
i
ST at 0^h
τ
f'', g', G'.

The *A.E.* also contains useful supplementary tables, brief descriptions of which are given below, for reference:

(*a*) Approximate reduction from standard equinox of 1950·0 to true equinox of date:

$$\alpha_{\text{date}} = \alpha_{1950} + f + 4g \sin (G + \alpha_{1950}) \tan \delta_{1950}$$

$$\delta_{\text{date}} = \delta_{1950} + g \cos (G + \alpha_{1950})$$

values of $4 \tan \delta, f, g$ and G being tabulated.

It will be seen that these approximate expressions are derived from the full equations involving the independent day numbers by the omission of the expressions for proper motion and aberration.

(*b*) Reduction of star positions from the mean equinox of the current year to the standard equinox of 1950: Two methods, with the relevant tables, are given; one accurate for all values of δ, the other inapplicable to high Decs.

(c) Differential aberration, precession and nutation: Useful in the reduction of observations of, e.g. a comet or asteroid, whose positions are obtained in the form of $\Delta\alpha$, $\Delta\delta$ relative to a fixed star. If the star's position (α', δ') is reduced to the equinox of Jan 0, then the position of the object, referred also to Jan 0 of the current year, is given by

$$\alpha = \alpha' + \Delta\alpha + [A]_\alpha + [P]_\alpha + [N]_\alpha$$
$$\delta = \delta' + \Delta\delta + [A]_\delta + [P]_\delta + [N]_\delta$$

where $\quad [A]$ = differential aberration,
$\qquad [P]$ = differential precession,
$\qquad [N]$ = differential nutation,
$\quad \Delta\alpha, \Delta\delta$ = RA/Dec of object *minus* RA/Dec of star, in units of 1^m and $1'$ respectively.

$[A]_\alpha = a\Delta\alpha + b\dfrac{\Delta\delta}{10}$ in units of $0^s\!.001$

$[A]_\delta = c\Delta\delta + d\dfrac{\Delta\delta}{10}$ in units of $0''\!.01$

$\left.\right\}$ where a, b, c, d, are derived from the tables given.

$[P]_\alpha + [N]_\alpha = e\Delta\alpha\;\dfrac{10\tan\delta}{15} - f\Delta\delta\dfrac{10\sec^2\delta}{225}$

in units of $0^s\!.001$

$[P]_\delta + [N]_\delta = f\Delta\alpha$ in units of $0''\!.01$

$\left.\right\}$ where e, f, $\dfrac{10\tan\delta}{15}$, $\dfrac{10\sec^2\delta}{225}$ are tabulated.

SECTION 31

CONVERSION OF COORDINATES

Equations for some of the more commonly required transformations are given below. The following symbols are employed:

α RA
δ Dec
h Altitude
Az ·Azimuth
H Hour Angle
ϕ Latitude (terrestrial)
ϵ Obliquity of the ecliptic
λ Longitude (celestial)
β Latitude (celestial)
z Zenith Distance
θ Sidereal Time

$$\left. \begin{array}{l} \tan H = \dfrac{\tan Az \sin X}{\cos (\phi - X)} \\[2mm] \tan \delta = \tan (\phi - X) \cos H \end{array} \right\} \tan X = \tan z \cos Az$$

$$\sin Az = \frac{\cos \delta \sin(\alpha - \theta)}{\sin z}$$

$$\cos Az = \frac{\sin \delta - \cos z \sin \phi}{\sin z \cos \phi}$$

$$\left. \begin{array}{l} \tan Az = \dfrac{\tan H \cos Y}{\sin (\phi - Y)} \\[2mm] \tan z = \dfrac{\tan (\phi - Y)}{\cos Az} \end{array} \right\} \tan Y = \frac{\tan \delta}{\cos H}$$

$$\left. \begin{array}{l} \cos z = \sin \delta \sin \phi + \cos \delta \cos \phi \cos (\alpha - \theta) \\ \sin h = \sin \delta \sin \phi - \cos \delta \cos \phi \cos H \\ \sin \delta = \sin h \sin \phi - \cos h \cos \phi \cos Az \end{array} \right\} \quad \ldots \ldots \ldots \quad (a)$$

$$\left.\begin{aligned} \tan \lambda &= \frac{\cos (Z-\epsilon) \tan \alpha}{\cos Z} \\ \tan \beta &= \tan (Z-\epsilon) \sin \lambda \end{aligned}\right\} \tan Z = \frac{\tan \delta}{\sin \alpha}$$

$$\sin \frac{H}{2} = \pm \sqrt{\frac{\sin \frac{1}{2}[Z+(\phi-\delta)] \sin \frac{1}{2}[z-(\phi-\delta)]}{\cos \phi \cos \delta}}$$

For conversion between equatorial and galactic coordinates, see B. 22.8.

OBSERVATIONAL RECORDS

It is generally true to say that an unrecorded observation is an observation wasted. Communication of results is an essential condition of the progress of any science.

When planning the form that the observational record is to take, and subsequently while making the actual notes at the telescope, the characteristics to keep constantly in mind are: clarity and unambiguity, conciseness, objectivity and comparability, orderliness, legibility. The record should be immediately intelligible to anyone familiar with the subject, without any supplementary explanation.

Whenever the nature of the observation allows it, arrange the record in tabular form: this enforces orderliness in the arrangement of the data, makes for ease and speed of reference, and is some safeguard against the omission of relevant facts. In the interests of objectivity, precision, and comparability, numerical observational data are superior to both drawings and verbal descriptions.

The particular requirements of satisfactory records in the various branches of observation are dealt with under the several sections of *O.A.A.* The miscellaneous points mentioned here are more generally applicable.

It is often convenient to record the actual observations at the telescope in one book, and subsequently to transfer the entries to a permanent record book, or possibly card or looseleaf index, in which it may be desirable to lay out the material somewhat differently. If this is so, the record made at the telescope must nevertheless be legible, ordered and sufficiently comprehensive. The entering up of the permanent record of one night's observations should not be delayed beyond the following day. Nothing must be added to the permanent record from memory: as regards observational data, the sole source for the permanent record is the observer's notebook; any amplifications or explanatory notes that have to be added to either record after the observations are ended must be distinguished from observational data recorded at the telescope—by the use, e.g. of a different-coloured ink.

The two mental processes to be most feared and shunned are expectation and memory: expectation when making the observation, and memory when recording it. The time to record the observation is whilst still at the telescope: if the nature of the object permits, it should be reobserved and compared with what has just been recorded.

It is advisable to include with the record of each observation a note of the degree of reliance placed upon it. This may take the form of a numerical weight or of a code letter (e.g. A, clearly and definitely seen; B, glimpsed at intervals, or visible only to indirect vision; C, doubtful); in either case a separate column should be set aside in the record for this note.

Never make an erasure in the record. Anything to be altered should be deleted by a single line, which allows the original as well as the amendment to be read.

In a tabular layout, draw a line through any compartment in which no entry is made.

Always quote full instrumental details and an index, as impersonal as possible, of the seeing and sky condition.

In the case of drawings, record the time at which the main outlines were laid down. The subsequent redrawing of rough sketches made at the telescope is not to be recommended: human memory is not that infallible.

Employ UT or GMAT throughout.

The indexing of photographic material falls into a rather different category. A serial number (plus a distinguishing letter for each camera if more than one is used) should be scratched on each film immediately after it is removed from the instrument for development. The record book should devote a separate page to each night's exposures, and be headed with the instrumental data and date. Against each serial number should be recorded in tabular form:

> Name or catalogue number of object
> RA and Dec of object or of film centre
> Start of exposure: UT
> > Hour angle
> > Approximate ZD
> > Sky condition
> End of exposure: UT
> Length of exposure
> UT of any interruptions to the exposure
> Focus and focal ratio
> Type of film and of any filter used
> Remarks

In the history of astronomy can be found numerous cautionary tales which illustrate the fatal consequences of messy and muddled observational records, as well as of preconceived ideas regarding what is likely or possible, and of emotional bias—expectation, disappointment, surprise, hope. No fewer than 19 pre-discovery observations of Uranus have been identified, from 1690 (by Flamsteed) onward. It is true that many of these in no way reflect upon the technique of the observers, since a single observation would quite possibly not reveal its planetary character.* But that *none* of these observations should have led to the discovery of Uranus is incredible. The case that is most relevant to the matter of observational records concerns the French astronomer Lemonnier, who in January 1796 observed Uranus six times over a period of nine days, including observations on four consecutive nights. His records of observations were kept in a particularly untidy and unsystematic fashion (one of the Uranus observations was noted down on a paper bag that had contained hair powder), and this certainly contributed to his failure: in a well-kept record the anomalies between these six observations could not have failed to strike him—and he would have anticipated Herschel in the first planetary discovery of historical times by twelve years.

Finally, having made your record, have confidence in it and be prepared to stand by it. On 1795 May 10, Lalande observed a star-like object and noted that its position had apparently changed since he observed it two days previously. But he 'assumed' that his earlier observation must have been inaccurate, or his record of it mistaken, and did not investigate further. The object was Neptune.

The lesson is clear: avoid preconceptions, avoid bias, record systematically and unambiguously, and always investigate apparent anomalies. Never accept an anomalous observation as being the result of error until investigation, not assumption, has proved it to be so. In the history of science small anomalies and unexpected discrepancies have led to more discoveries than, perhaps, any other single factor.

*On the other hand Herschel detected its non-stellar character before its motion had been established.

OBSERVATIONAL AIDS

Both the *Astronomical Ephemeris* (B. 1.1) and the *Handbook of the British Astronomical Association* (B. 1.2) contain vast quantities of information invaluable to the observer, whatever his particular field of work may be. Under sections 1, 2, 4-11, 16 of *O.A.A.* are listed the relevant data to be found in these two sources. In addition to these the *B.A.A.H.* annually gives tables of Astronomical and Physical Constants, Miscellaneous and Terrestrial Data, and notes on Time Reckoning, including a table of Julian dates. General articles dealing with some branch of observational technique, asteroid ephemerides, lists of variables especially requiring observation, etc, are included from time to time.

The B.A.A. *Circulars* (B. 1.5) are also invaluable to the practical observer. Distributed at irregular intervals to subscribers, they contain up-to-the-minute news regarding new discoveries and objects urgently requiring observation—comets, novae, planetary phenomena such as notable and ephemeral surface markings, variable stars requiring observation, outstanding solar activity, meteor showers, unusual asteroid news and discoveries, etc. The primary aim of the *Circulars* is to notify observers of new discoveries quickly enough for the objects to be adequately and widely observed before their disappearance.

As regards star charts and atlases, the reader is referred to B. 22.1-21 for details.

The British Astronomical Association is a world-wide association of amateur astronomers. Every observer should not only join such an Association but should also become an active member of the Observing Section devoted to his particular field of work. These Sections—each under the directorship of an experienced observer who collates and periodically publishes the observations of the Section, as well as giving information and advice whenever required—are as follows: Sun, Moon, Terrestrial Planets, Jupiter, Saturn, Comet, Meteor, Aurora, Variable Star, Instruments and Observing Methods, Computing, Historical, Radio Astronomy, Artificial Satellites.

The following information regarding the nature and aims of the Association is reprinted by permission:

The British Astronomical Association was originally formed in 1890 and now numbers over 5000 Members. Its leading features are as follows:

Membership—Open to all persons interested in astronomy. Nomination Forms may be had on application to the Assistant Secretary, and must be signed by one Member from personal knowledge, or by two resident householders (giving addresses) from personal knowledge.

Objects—(1) The association of observers, especially the possessors of minor instruments, for mutual help, and their organisation in the work of astronomical observation.

(2) The circulation of current astronomical information.

(3) The encouragement of a popular interest in astronomy.

Methods—(1) The arrangement of Members, for the work of observation, in Sections or departments of observation, under experienced Directors.

(2) The publication, at regular intervals, of a *Journal* containing reports of the Association's Meetings, and of its Observing Sections, papers by Members, and notes on current astronomy.

(3) The formation of Branches of the Association in the provinces and in the Commonwealth.

(4) The holding of Meetings in London, and at the seats of the Branches.

(5) The formation of a library and of collections of astronomical instruments and lantern slides, for loan to Members of the Association.

(6) The affiliation of schools and societies at a minimum annual subscription of £9·00.

Entrance Fee—£1·00 (50p under 21).

Annual Subscriptions (due on August 1 each year)—£9·00 (£7·50 under age 21 or £6·00 if over 65). The *Circulars*, giving early news of discoveries, etc, are a separate publication—annual subscription £1·00.

All cheques and postal or money orders are to be made payable to The British Astronomical Association, crossed and sent to the Assistant Secretary at the Registered Office of the Association: Burlington House, Piccadilly, London, W1V 0NL.

Meetings—The Meetings of the Association are held on the last Wednesday in each month (unless the Council shall decide otherwise), from

October to June inclusive, at 17.00 hrs at the Scientific Societies' Lecture Theatre, 23 Savile Row, London W.1.

Another national association of amateur astronomers is the Junior Astronomical Society (J.A.S.), which holds meetings four times a year in London. The J.A.S. has a variety of observing sections, publishes a journal (*Hermes*, quarterly) which is distributed free of charge to members, and also news circulars. Membership is open to all at £3·20 p.a., and enquiries should be directed to the J.A.S. Secretary, 58 Vaughan Gardens, Ilford, Essex.

Of the many overseas societies with amateur or partly amateur membership, the following may be mentioned particularly:

La Société Astronomique de France, 28 rue Saint-Dominique, Paris 7e.

Amateur Astronomers' Association, American Museum of Natural History, New York.

American Association of Variable Star Observers, 187 Concord Avenue, Cambridge, Mass. 02138.

Association of Lunar and Planetary Observers, The Observatory, New Mexico State University, Las Cruces, New Mexico 88001.

Royal Astronomical Society of Canada, 124 Merton Street, Toronto, Ontario.

SECTION 34

FIELD ORIENTATION

The normal orientation of the telescopic field—i.e. a field lying between
the zenith and the S point of the horizon, as seen in an astronomical
ocular—is

FIGURE 201
Field orientation or position angle

It is, compared with the naked-eye field, inverted (top and bottom reversed) and reversed (left and right reversed). Position angles, whatever
the orientation of the field may be, are always measured from the N point
eastwards.

Field orientation becomes complicated with the introduction of diagonals and, in the case of the Sun, of projection. An odd number of
reflections causes inversion or reversal of the image; an even number of
reflections either produces inversion and reversal or else leaves the image
as it was presented to the first reflecting surface; projection is equivalent
to one reflection.

In the following Table, the image orientation specified in the left-hand
column is obtained by a combination of those succeeding columns which
are, on any one horizontal level, marked with an x. Thus (second horizontal line) an erect image is obtained by projection with an astronomical
ocular and a diagonal turned vertically downward, the screen being viewed
from above by an observer standing behind the telescope and facing the
OG.

	Astronomical ocular	Terrestrial ocular	Diagonal		Opaque projection screen	Trans-lucent projection screen
			Turned vertically (Observer facing OG)	Turned horizontally (Observer shoulder to OG)		
Erect		×				
	×		×		×	
	×			×	×	
	×					×
Inverted	×			×		
	×		×			×
		×			×	
		×	×			
		×		×		×
Reversed	×		×			
	×			×		×
	×				×	
		×	×			×
		×		×		
Inverted and Reversed	×					
		×	×		×	
		×		×	×	
		×				×

509

The four possible orientations of a field on the meridian between the zenith and the S horizon are:

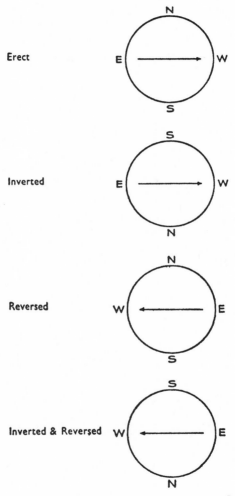

FIGURE 202

Different orientations of meridional field

ARTIFICIAL STARS

As test objects, artificial stars have the following advantages compared with natural stars:

(*a*) they are far less subject to atmospheric turbulence, owing to the reduced light path through the atmosphere;

(*b*) they are not subject to atmospheric dispersion, for the same reason;

(*c*) they are stationary relative to the telescope during the test;

(*d*) the effects of temperature gradients in the tube and in the objective are easier to eliminate.

They are also interesting demonstrators of the phenomena of diffraction.

Whatever form of artificial star is used, its minimum distance from the telescope must be such that its image is smaller than the spurious disc of a point source. It should, if possible, be greater than this minimum permissible value, since as its rays approximate to parallelism the plane in which its image is formed will approximate to the true focal plane of the objective (i.e. the plane in which the images of 'infinitely' distant astronomical objects are formed). The separation of these two planes is inversely proportional to the distance of the artificial star; thus in the case of a $f/15$ 75-mm the image is formed 13 mm from the focal plane when the star is 100 m distant, 6·5 mm from it when the star is 200 m distant, etc.

Two convenient types of artificial star may be mentioned:

(*i*) Circular aperture made with a needle in a sheet of tinfoil. If this is to be illuminated simply by an electric bulb, its diameter will have to be as large as 0·5 mm approximately. A more efficient arrangement is a very much smaller hole (diameters as small as 0·015 mm can be managed) illuminated by focusing the image of the filament of a gas-filled, unfrosted electric lamp on to it by means of a condensing lens; without this arrangement, insufficient light will pass through the hole. Holes of this calibre can only be made with a needle that has been very carefully honed on glass (its progress checked with a microscope); the punctures are made with the tinfoil glued along its edges to a sheet of glass, the needle being twisted through at least one turn while resting, with very little pressure. on the foil.

(*ii*) Reflection of the Sun in any polished convex surface of short radius, such as the field lens of a Huyghenian eyepiece, a spherical thermometer bulb, a globule of mercury on a dark non-reflecting background, or a black glass bead. The diameter of the solar image will be about 0·0025 that of the bulb or globule, which makes the calculation of the minimum permissible distance a simple matter.

ANGULAR ESTIMATES

In some branches of work (e.g. Zodiacal Light, meteors, etc) it is essential to be able to make accurate estimates of angular distances by eye; and in any case it is a useful ability to have.

An object x units long subtends an angle of $1°$ when its distance from the eye is $57 \cdot 3x$. Thus:

25 mm at $1 \cdot 43$ m subtends:	$1°$
25 mm at average arm's length (58 cm) subtends:	$2°5$
15-cm rule at arm's length subtends:	$15°$

Other approximate measurements are:

width of index nail at arm's length:	$1°$
1st-2nd knuckles of clenched fist at arm's length:	$3°$
2nd-4th knuckles of clenched fist at arm's length:	$5°$
1st-4th knuckles of clenched fist at arm's length:	$8°$
span of open hand at arm's length:	$18°$

Among the stars, the following distances are useful for training the eye:

α–η UMa	$25°$		α–β Aql	$2\frac{1}{2}°$
α–δ UMa	$10°$		α–γ Aql	$2°$
α–β UMa	$5°$		δ–ϵ Ori	$1\frac{1}{4}°$
α–β CMi	$4°$		Solar = lunar diameter $= \frac{1}{2}°$	
δ–ζ Ori	$3°$			

SECTION 37

THE NORTH POLAR SEQUENCE, AND OTHER STANDARD REGIONS

The NPS is a series of stars in the vicinity of the N Pole (chosen for visibility at all times and seasons) whose positions and other data—particularly including stellar magnitudes—have been determined with precision.

B. 25.15 contains:

(*a*) the NPS: 46 stars of mags 4·47-20·96 (photographic); positions, catalogue numbers, spectroscopic types, photometric and photographic magnitudes;

(*b*) the NPS of Red Stars: 12 stars of mags 6·69-13·34 (photographic); same data;

(*c*) Supplementary Standard Stars near the NCP: 38 stars of mags 2·71-21·00 (photographic); same data;

(*d*) photographic chart of the region of the NCP on scale $5'' = 1$ mm (Mt Wilson 1·5-m reflector, 4^h exp.).

B. 25.20 contains:

(*a*) all the B. 25.15 stars with the addition of colour indices, derived both from magnitudes and exposure ratios;

(*b*) Seares' additional list of Supplementary Stars: 56 stars of mags 11·00-16·28 (photographic); same data.

B. 25.13 is useful for users of small instruments, being a selection from B. 25.15, 25.20 in which the lower magnitudes are omitted:

(*a*) 68 stars from B. 25.15, mags 2·55-18·16 (photographic); photographic and photovisual mags, c/i from mags and exposure ratios;

(*b*) the 56 stars from B. 25.20 (*b*); same data;

(*c*) 2 photographic charts of the NCP region: $4' = 1$ mm, and $23'' = 1$ mm, approximately.

B. 25.12 gives a general account of the whole NPS work at Harvard, and contains:

(*a*) the 3 lists from B. 15.15, slightly amended;

(*b*) 36 Pleiades stars of mags 3·01-14·82 (photographic); photometric and photographic mags;

(*c*) 18 Praesepe stars of mags 6·31-10·55 (photographic); same data;

(*d*) 2 charts of the NCP region: $20'' = 1$ mm (see also B. 25.8), and $5'' = 1$ mm (see also B. 25.15).

B. 25.4 contains:

(*a*) photographic magnitudes of 262 stars in the NCP vicinity, mags 8·83-15·80 (photographic);

(*b*) chart of the area 50′ in diameter, centred on the NCP: $30'' = 1$ mm.

B. 25.6, 25.9 contain data relating to stars in the Harvard Standard Regions, which are widely distributed in both RA and Dec so that one at least may always be available near the meridian.

B. 25.8 gives no photometric information, but lists the positions of 589 stars in the vicinity of the NCP; also includes a photographic chart of the region on a scale of $20'' = 1$ mm.

The NPS is now obsolete, having been replaced by other systems of standard fields in which magnitudes have been determined photoelectrically rather than photographically. It is, however (because of its constant visibility), a most useful test field; B. 25.23 includes a chart showing telescopic stars to photovisual magnitude 14.2 within a couple of degrees of Polaris.

I.A.U. CONSTELLATION ABBREVIATIONS

The method of designating constellations throughout this book is the three-letter abbreviation system recommended by the I.A.U. This is concise, convenient, and unambiguous, and amateurs are urged to employ it in their observation records and elsewhere.

The constellation names fall into four groups, according to the method of their abbreviation:

(*a*) Sixty-five constellation names are abbreviated to the first three letters (e.g. Andromeda, And). The twenty-four exceptions to this rule are:

(*b*) Nine contractions to the first two letters of the name (or of its genitive) and a later one:

Apus	Aps	Hydrus	Hyi
Aquila	Aql	Phoenix	Phe
Aquarius	Aqr	Sculptor	Scl
Crater	Crt	Scutum	Sct
Hydra	Hya		

(*c*) Ten two-word names are contracted as follows:

Canis Major	CMa	Leo Minor	LMi
Canis Minor	CMi	Piscis Austrinus	PsA
Corona Australis	CrA	Triangulum Australe	TrA
Corona Borealis	CrB	Ursa Major	UMa
Canes Venatici	CVn	Ursa Minor	UMi

Note: Coma (Berenices) is contracted to Com

(*d*) Five contractions are formed irregularly:

Cancer	Cnc	Sagitta	Sge
Corvus	Crv	Sagittarius	Sgr
Pisces	Psc		

Note: Caelum is the modern spelling of Coelum, hence Cae. Malus (once part of Argo) is replaced by Pyxis. Puppis, Vela, Carina (once part of Argo) are now autonomous.

WAVELENGTHS AND FREQUENCIES

Wavelengths are commonly expressed in one of the following units:

micron	μ	10^{-4}	cm
nanometre (nm)	nm	10^{-7}	cm
Ångström unit	Å	10^{-8}	cm
X-ray unit	XU	10^{-11}	cm

If $\lambda =$ the wavelength of a radiation (cm),

$\nu' =$ its wave number (number of waves per cm),

$\nu =$ its frequency number (number of waves per sec or Hz),

$T =$ its period (interval between successive occurrences of the same phase at a given point),

$c =$ its velocity,

then
$$\nu = \frac{1}{T} = \frac{c}{\lambda}$$

$$\lambda = \frac{1}{\nu'} = \frac{c}{\nu} = cT$$

Electromagnetic wavebands of most interest are (approximately):

30,000	..	1000 m	long-wave radio
1000	..	100 m	medium-wave radio
100	..	10 m	short-wave radio
10	..	0·1 m	TV
0·1	..	0·001 m	radar
20,000	..	7700Å	infrared
7700	..	6500Å	red
6500	..	6000Å	orange
6000	..	5500Å	yellow
5500	..	5000Å	green
5000	..	4500Å	blue
4500	..	3600Å	violet

(red, orange, yellow, green, blue, violet) } visible light

3600	..	200Å	ultraviolet
32	..	1·0Å	X-radiation
1·0	..	0·1Å	γ-radiation
0·001	..	0·00001Å	cosmic radiation

By international convention (International Solar Union, 1907), the red line of cadmium at 6438·3696Å is taken as the fundamental reference point in the solar spectrum.

Spectral lines are denoted by their wavelengths in Å or nm. Some of the more important solar lines are also specified by letters assigned to them by Fraunhofer. The lines in the Balmer series of the hydrogen spectrum are also denoted by the letter H followed by letters from the Greek alphabet (e.g. $H\alpha$).

O.A.A., section 1.11, contains a table and map of the more important lines of Fraunhofer's spectrum.

SECTION 40

MANUFACTURERS AND SUPPLIERS

This short list makes no claim to be exhaustive, and refers wherever possible to suppliers in the UK. Inclusion in this list does not necessarily imply editor's recommendation.

A. E. Optics Ltd., 39-43 York Street, Luton, Beds: manufacturers of all types of astronomical optics.

Astro-Systems Ltd., 5b Old Bedford Road, Luton, Beds: manufacturers and suppliers of complete instruments and accessories.

Astrobits, 50 High Street, Luton, Beds: suppliers of accessories.

Astro Books & Supplies, 47 Riddlesdown Lane, Purley, Surrey: books and accessories.

Astronomical Equipment Ltd., 45a Roundwood Lane, Harpenden, Herts: manufacturers of astronomical telescopes.

Austin Roberts Ltd., Tarran Way South, Tarran Industrial Estate, Moreton, Wirral, Merseyside: observatory domes.

Bedford Astronomical Supplies, 5b Old Bedford Road, Luton, Beds: manufacturers of catadioptric systems, binocular telescopes, etc.

Bretmain Ltd., 2 Station Road, Rayne, Essex: accessories; agents for Questar telescopes.

Brunnings (Holborn) Ltd., 133 High Holborn, London WC1: new and surplus scientific equipment.

Cambridge Astronomical Telescopes, Bucklebury House, The Green, Weston Colville, Cambs: telescopes and accessories.

Carson Astronomical Instruments, Inc., PO Box 5566, Valencia, CA 91355, USA: dielectric filters.

Cave-Astrola, 62 Caversham Road, Reading, Berks: Newtonian and Cassegrain telescopes.

Chance-Pilkington Ltd., Glascoed Road, St Asaph, Clwyd: optical glass manufacturers.

DayStar Filter Corp., PO Box 1290, 1513 E 3rd Street, Pomona, CA 91769, USA (Europe: Postfach 56 A-9601, Arnoldstein, Austria): dielectric filters.

Eastman Kodak Co., Rochester, NY 14650, USA: films and plates for astronomical use.

H. W. English, 469 Rayleigh Road, Hutton, Brentwood, Essex: surplus and new scientific and optical equipment.

Charles Frank Ltd., 144 Ingram Street, Glasgow 1: surplus and new scientific equipment; agents for Celestron telescopes.

Fullerscopes (incorporating Broadhurst, Clarkson & Co. Ltd.), 63 Farringdon Road, London EC1: manufacturers and suppliers of astronomical telescopes and accessories.

Gooch & Housego Ltd., Cornhill, Ilminster, Somerset: glass milling, fine-annealing, and aluminising.

Grubb Parsons Ltd., Walkergate, Newcastle upon Tyne 6: large observatories and their equipment.

David Hinds Ltd., 2 Wolsey Road, Hemel Hempstead, Herts: manufacturer of astronomical optics; mirror kits; aluminising.

Ilford Ltd., Basildon, Essex: films, papers, etc.

H. N. Irving & Son, 258 Kingston Road, Teddington, Middlesex: manufacturer of astronomical telescopes and accessories.

Kodak Ltd., Hemel Hempstead, Herts: films, papers, filters, etc.

Edward Marcus Ltd., Moor House, 7 Moorfields, London EC2; and 46 Liverpool Street, London EC2: large binoculars.

Observatory Dome Co., 150 West Avenue, Maylandsea, Chelmsford, Essex: observatory domes.

H. V. Skan, 425/433 Stratford Road, Shirley, Solihull, West Midlands: agents for Schott optical and low-expansion glass.

UK Optical Bausch & Lomb, Bittacy Hill, Mill Hill, London NW7: optical accessories.

Vacuum Coatings Ltd., 25 Lea Bridge Road, Clapton, London E5: aluminising.

BIBLIOGRAPHY

Abbreviations used in this Bibliography are listed below; wherever possible the style of the *Astronomischer Jahresbericht* or the *World List of Scientific Periodicals* has been followed. Bibliographical data are enclosed in round brackets ( ); my own comments, if any, in square brackets [ ]. Each division of the Bibliography is arranged alphabetically by authors unless some other arrangement (e.g. chronological) is specified.

A.E. Astronomical Ephemeris.

A.J. Astronomical Journal.

A.N. Astronomische Nachrichten.

Ann. Cape Obs. Annals of the Cape Observatory.

Ann. d'Astrophys. Annales d'Astrophysique.

Ann. Lowell Obs. Lowell Observatory Annals.

Ann. Obs. Strasbourg. Annales de l'Observatoire de Strasbourg.

Ap. J. Astrophysical Journal [previously 'Sidereal Messenger', 1883-92; 'Astronomy and Astro-Physics', 1892-95].

Appl. Opt. Applied Optics.

Astr. Mitt. Göttingen. Astronomische Mittheilungen der Königlichen Sternwarte zu Göttingen.

Astr. Pap., Wash. Astronomical Papers of the American Ephemeris & Nautical Almanac.

A.T.M. Amateur Telescope Making.

B.A.A.H. Handbook of the British Astronomical Association.

Brit. J. Physiol. Optics. British Journal of Physiological Optics.

B.S.A.F. Bulletin de la Société Astronomique de France ['L'Astronomie']

Connais. Temps. Connaissance des Temps.

Contr. Dunsink Obs. Dunsink Observatory Contributions.

Contr. Lick Obs. Contributions from the Lick Observatory.

Contr. Perkins Obs. Perkins Observatory Contributions.

Contr. Princeton Obs. Princeton University Observatory Contributions.

C.R. Comptes Rendus des Séances de l'Académie des Sciences.

Engl. Mech. English Mechanics [formerly 'The English Mechanic'].

Harv. Ann. Annals of Harvard College Observatory.

Harv. Bull. Harvard Bulletins.

Harv. Circ. Harvard College Observatory Circulars.

Harv. Monogr. Harvard Observatory Monographs.

Harv. Repr. Harvard Reprints.

H.d.A. Handbuch der Astrophysik (Julius Springer, Berlin).

H.M.S.O. Her Majesty's Stationery Office.

I.A.U. International Astronomical Union.

Internat. Ass. Acad. International Association of Academies.

Irish Astr. J. Irish Astronomical Journal.

J. Appl. Phys. Journal of Applied Physics.

J.B.A.A. Journal of the British Astronomical Association.

J. Can. R.A.S. Journal of the Royal Astronomical Society of Canada.

J. Opt. Soc. Amer. Journal of the Optical Society of America.

J. Sci. Instr. Journal of Scientific Instruments.

L.O.B. Lick Observatory Bulletins.

Lund. Medd. Meddelande från Lunds Astronomiska Observatorium.

Mem. Amer. Acad. Arts Sci. Memoirs of the American Academy of Arts and Sciences.

Mem. B.A.A. Memoirs of the British Astronomical Association.

Mem. R.A.S. Memoirs of the Royal Astronomical Society and Memoirs of the Astronomical Society of London [consecutive volume numbering].

M.N. Monthly Notices of the Royal Astronomical Society.

M.W.C. Mount Wilson Contributions.

N.A. The Nautical Almanac and Astronomical Ephemeris (H.M.S.O., annually).

Nat. Nature.

Obs. The Observatory.

Perkins Obs. Repr. Perkins Observatory Reprints.

Phil. Trans. Philosophical Transactions of the Royal Society.

Photogr. J. Photographic Journal.

Phys. Rev. Physical Review [U.S.A.].

Pop. Astr. Popular Astronomy.

Proc. Amer. Acad. Arts Sci. Proceedings of the American Academy of Arts and Sciences.

Proc. Amer. Phil. Soc. Proceedings of the American Philosophical Society.

Proc. Phys. Soc. Lond. Proceedings of the Physical Society.

Proc. R. Irish Acad. Proceedings of the Royal Irish Academy.

Proc. Roy. Soc. Proceedings of the Royal Society.

Publ. A.A.S. Publications of the American Astronomical Society.

Publ. A.S.P. Publications of the Astronomical Society of the Pacific.

Publ. Tartu. Publications de l'Observatoire Astronomique de l'Université de Tartu (Dorpat).

Publ. Astrophys. Obs. Potsdam. Publikationen des Astrophysikalischen Observatoriums zu Potsdam.

Publ. Carneg. Instn. Carnegie Institution Publications, Washington.

Publ. Cincinn. Obs. Cincinnati Observatory Publications.

Publ. Dom. Astrophys. Obs. Dominion Astrophysical Observatory Publications.

Publ. Lick Obs. Lick Observatory Publications.

Publ. L. McC. Obs. Publications of the Leander McCormick Observatory of the University of Virginia.

Publ. Michigan Obs. Michigan Observatory Publications.

Publ. Univ. Pa. Pennsylvania University Publications (Astronomical Series).

Publ. Yale Obs. Yale University Observatory Publications.

Q.J. Quarterly Journal of the Royal Astronomical Society.

Result. Obs. Nac. Argent. Resultados del Observatorio Nacional Argentino.

Rev. d'Opt. Revue d'Optique Théorique et Instrumentale.

Rev. Mod. Phys. Review of Modern Physics.

Rev. Sci. Instr. Review of Scientific Instruments.

Sci. Amer. Scientific American.

Sci. Progr. Science Progress.

Sky & Tel. Sky and Telescope [amalgamation of 'Sky' and 'Telescope'].

Smithson. Contr. Smithsonian Contributions to Knowledge.

Trans. I.A.U. Transactions of the International Astronomical Union.

Veröff. Univ.Berlin-Babelsberg. Veröffentlichungen der Universitätssternwarte zu Berlin-Babelsberg.

Veröff. König. Astron. Rechen-Institut. Veröffentlichungen des Königlichen Astronomischen Rechen-Instituts zu Berlin.

Wash. Nat. Ac. Proc. Proceedings of the National Academy of Science of the United States of America.

1. OBSERVATIONAL AIDS

1.1   *Astronomical Ephemeris* (H.M.S.O., annually). [General ephemerides for Sun, Moon and major planets and their brighter satellites, as well as much other information; identical in content to the earlier *Nautical Almanac.*]

1.2   *Handbook* of the British Astronomical Association (B.A.A., annually). [Ephemerides for Sun, planets and satellites; lunar phases and occultations, as well as risings and settings; expected comet returns; meteors showers, etc.]

1.3   *Sky & Telescope* (Sky Publishing Corp., 49 Bay State Road, Cambridge, MA 02138, USA, monthly). [Articles, news notes, observations, telescope-making forum, etc.]

1.4   *The Astronomer* (12 Clare Close, Earls Barton, Northampton, monthly). [Rapid publication of observational reports from UK and overseas amateurs.]

1.5   *Circulars* of the British Astronomical Association (B.A.A., at irregular intervals). [Immediate announcement of comet and nova discoveries, etc.]

1.6   *I.A.U. announcement cards* (Central Bureau for Astronomical Telegrams, Smithsonian Astrophysical Observatory, Cambridge, MA 02138, USA). [Telegram, telex and airmail circulars detailing discoveries, etc.]

2. AMATEUR OBSERVATIONAL ASTRONOMY

2.1   J. GRIBBIN, *Astronomy for the Amateur* (Macmillan, 1976).

2.2   R. H. LAMPKIN, *Naked Eye Stars* (Gall & Inglis, 1972).

2.3   D. H. MENZEL, *Field Guide to the Stars and Planets* (Collins, 1977).

2.4   H. G. MILES, *Artificial Satellite Observing* (Faber & Faber, 1974).

2.5   P. A. MOORE (ed), *Practical Amateur Astronomy* (Lutterworth, 1978).

2.6 P. A. MOORE (ed), *Yearbook of Astronomy* (Eyre & Spottiswoode, annually).
2.7 J. MUIRDEN, *Astronomy with Binoculars* (Faber & Faber, 1976).
2.8 G. D. ROTH, *Astronomy: A Handbook* (Springer-Verlag, Berlin, 1975).
2.9 T. W. WEBB, *Celestial Objects for Common Telescopes* (Dover Books, New York, 1962).

3. OPTICS, INCLUDING TELESCOPIC FUNCTIONS AND ABERRATIONS

3.1 M. A. AINSLIE, Presidential Address to the B.A.A. (*J.B.A.A.*, **40**, No. 1, 5).
3.2 H. W. J. BLÖTE, Star Images in the Presence of Aberrations (*Sky & Tel.*, **55**, No. 4, 347).
3.3 M. BORN & E. WOLF, *Principles of Optics* (Pergamon Press, 1959).
3.4 B. A. CARTER, Limiting Magnitudes (*The Astronomer*, **6**, No. 63, 58).
3.5 F. C. CHAMPION, *University Physics* (Blackie, 1941).
3.6 A. E. CONRADY, *Applied Optics and Optical Design* (Dover Books, 1957, 2 vols).
3.7 W. R. DAWES, Catalogue of Micrometrical Measurements of Double Stars (Introduction) (*Mem. R.A.S.*, **35**, No. 4, 137). [The 'Introduction' is reprinted in *M.N.*, **27**, No. 6, 217.]
3.8 M. R. JARRY-DESLOGES, Influence des divers éléments d'un objectif (ouverture, distance focale, grossissement) sur la qualité des images télescopiques (*C.R.*, **177**, 248).
3.9 B. K. JOHNSON, *Practical Optics* (Hatton Press, Lond., 1947).
3.10 [T. LEWIS] On the Class of Double Stars which can be observed with Refractors of Various Apertures (*Obs.*, **37**, No. 479, 378).
3.11 L. C. MARTIN, *An Introduction to Applied Optics* (Pitman, 1930, 2 vols).
3.12 G. S. MONK, *Light: Principles and Experiments* (McGraw-Hill, 1937).
3.13 W. H. PICKERING, Definition and Resolution (*Pop. Astr.*, **28**, No. 9, 510).
3.14 W. H. PICKERING, Artificial Disks (*Harv. Ann.*, **32**, 149).
3.15 S. ROSIN, Optical Systems for Large Telescopes (*J. Opt. Soc. Amer.*, **51**, 331).
3.16 F. E. ROSS, Astrometry with Mirrors and Lenses (*Ap. J.*, **77**, 243).
3.17 F. E. ROSS, Lens System for Correcting Coma of Mirrors (*Ap. J.*, **81**).
3.18 R. W. SINNOTT, Limiting Magnitudes with Amateur Telescopes (*Sky & Tel.*, **45**, No. 6, 401).
3.19 J. P. C. SOUTHALL, *Mirrors, Prisms and Lenses—A Text-Book of Geometrical Optics* (Macmillan, 1933).
3.20 W. H. STEAVENSON, Note on the Light-grasp of Refractors (*J.B.A.A.*, **25**, No. 4, 186).

3.21 W. H. STEAVENSON, Further Note on the Light-grasp of Refractors (*J.B.A.A.*, **25**, No. 5, 229).
3.22 H. DENNIS TAYLOR, An Experiment with a 12½-in. Refractor, whereby the Light lost through the Secondary Spectrum is separated out and rendered approximately measurable (*Mem. R.A.S.*, **51**, 77). [Reprinted, B. 9.24, 3rd ed.]
3.23 H. DENNIS TAYLOR, The Secondary Colour Aberrations of the Refracting Telescope in Relation to Vision (*M.N.*, **54**, No. 2, 67). [Reprinted, B. 9.24, 3rd ed.]
3.24 H. DENNIS TAYLOR, Description of a Perfectly Achromatic Refractor (*M.N.*, **54**, No. 5, 328). [Reprinted, B. 9.24, 3rd ed.]
3.25 P. J. TREANOR, On the Telescopic Resolution of Unequal Binaries (*Obs.*, **66**, No. 831, 255).
3.26 C. G. WYNNE, A Field Corrector for Paraboloid Mirrors (*Proc. Phys. Soc. Lond.*, **62**, 772).
3.27 [VARIOUS] (*J.B.A.A.*, **83**, No. 4, 306 & **84**, No. 3, 220). [Light-grasp]

4. OPTICAL MATERIALS, METAL-ON-GLASS FILMS

4.1   D. A. CAMPBELL, Aluminium Films (*J.B.A.A.*, **73**, No. 7, 249).
4.2   CARTWRIGHT, HAWLEY & TURNER, Reducing the Reflection from Glass by Evaporated Films (*Phys. Rev.*, **55**, 595).
4.3   CARTWRIGHT, HAWLEY & TURNER, Reducing the Reflection from Glass by Multilayer Films (*Phys. Rev.*, **55**, 675).
4.4   CARTWRIGHT, HAWLEY & TURNER, Multilayer Films of High Reflecting Power (*Phys. Rev.*, **55**, 1128).
4.5   R. E. COX, More Reports on Sandwiched Glass Mirror Blanks (*Sky & Tel.*, **28**, No. 6, 376).
4.6   E. G HILL, A Simplified and Inexpensive Silvering Technique (*J.B.A.A.*, **75**, No. 4, 277).
4.7   D. G. HINDS, The Care of Aluminized Surfaces (*J.B.A.A.*, **71**, No. 4, 159).
4.8   E. J. HYSOM, The Long-term Stability of Telescope Discs (*J.B.A.A.*, **75**, No. 5, 332).
4.9   E. G. H. MOBSBY, Experiments with Cemented Mirrors (*Sky & Tel.*, **28**, No. 5, 305).
4.10 T. I. PUTNER & B. J. WILLIAMS, The Design of Telescope Mirror Aluminizers (*J.B.A.A.*, **83**, No. 5, 341).
4.11 J. STRONG, *et al.*, *Modern Physical Laboratory Practice* (Blackie, 1949).
4.12 A. F. TURNER, Anti-reflection Films on Glass Surfaces (*J. Appl. Phys.*, **12**, 351).
4.13 V. VAND, The Cleaning of Mirrors before Aluminising (*J.B.A.A.*, **53**, No. 7, 209).
4.14 [VARIOUS] Instructions for Silvering Glass (*A.T.M.*, **2**, 397). [Very comprehensive account of classical methods.]

5. OBJECT GLASSES

5.1   J. G. BAKER, Optical Systems for Astronomical Photography (*A.T.M.*, **3**, 1).

5.2   R. J. DONNEL, A Cooke Triplet Astrographic Lens for the Amateur (*Sky & Tel.,* **33**, No. 5, 312).

5.3   R. E. HOPKINS, Telescope Doublets (*J. Opt. Soc. Amer.,* **49**, 200).

5.4   O. R. KNAB, Making a 6-inch Air-spaced Visual Objective (*Sky & Tel.,* **40**, No. 1, 46).

5.5   H. D. KORONES & R. E. HOPKINS, Some Effects of Glass Choice in Telescope Doublets (*J. Opt. Soc. Amer.,* **49**, 896).

5.6   H. E. PAUL, Lenses for Astronomical Photography (*A.T.M.,* **3**, 307).

5.7   R. G. QUADE, Making an 8-inch Refractor Objective (*Sky & Tel.,* **41**, No. 4, 243).

6. OCULARS

6.1   L. BELL, Ghosts and Oculars (*Proc. Amer. Acad. Arts Sci.,* **56**, No. 2, 45).

6.2   B. BRIXNER, Barlow Lens Designs for a Spherical Primary Mirror (*Sky & Tel.,* **32**, No. 2, 103).

6.3   A. C. CURTIS, The 'Dynameter' and Eyepiece Focal Lengths (*J.B.A.A.,* **81**, No. 1, 24).

6.4   H. E. DALL, Telescope Eyepieces (*J.B.A.A.,* **79**, No. 5, 349).

6.5   F. J. HARGREAVES, A Modified Newtonian Telescope (*J.B.A.A.,* **52**, No. 7, 226).

6.6   C. R. HARTSHORN, The Barlow Lens (*A.T.M.,* **3**, 277).

6.7   R. T. JONES. A Wide-field Telescope with Spherical Optics (*Sky & Tel.,* **16**, No. 11, 548).

6.8   H. H. SELBY, Telescope Eyepieces (*A.T.M.,* **3**, 479).

6.9   E. W. TAYLOR, The Inverting Eyepiece and its Evolution (*J. Sci. Instr.,* **22**, No. 3, 43).

6.10   F. H. THORNTON, Mounting the Barlow Lens (*J.B.A.A.,* **60**, No. 3, 77).

6.11   [J. WALL] The Crayford Eyepiece Mounting (*Sky & Tel.,* **48**, No. 3, 182).

7. TELESCOPES, AND GENERAL WORKS ON OPTICAL INSTRUMENTS

7.1   C. ASHCRAFT, The Lensless Schmidt Camera for Astrophotography (*Sky & Tel.,* **47**, No. 5, 333).

7.2   B. BARLOW, *The Astronomical Telescope* (Wykeham Publ., 1975).

7.3   J. G. BAKER, A Method of Making Aspherical Surfaces of Revolution by means of Spherical Surfaces alone (*Pop. Astr.,* **48**, 78).

7.4   J. G. BAKER, The Solid-Glass Schmidt Camera and a New Type Nebular Spectrograph (*Proc. Amer. Phil. Soc.,* **82**, No. 3, 323).

7.5   J. G BAKER, A Family of Flat-Field Cameras, Equivalent in Performance to the Schmidt Camera (*Proc. Amer. Phil. Soc.,* **82**, No. 3, 339).

7.6   L. BELL, *The Telescope* (McGraw-Hill, 1922).

7.7 I. S. BOWEN, Future Tools of the Astronomer (*Q.J.*, **8**, No.1, 9).

7.8 E. B. BROWN, *Optical Instruments* (Chemical Publishing Co., N.Y., 1946).

7.9 R. A. BUCHROEDER, An Improved Buchroeder Catadioptric Design (*Sky & Tel.*, **36**, No. 5, 336).

7.10 R. A. BUCHROEDER, A New Three-mirror Off-axis Amateur Telescope (*Sky & Tel.*, **38**, No. 6, 418).

7.11 C. R. BURCH, On the Optical See-Saw Diagram (*M.N.*, **102**, No. 3, 159). [Advanced treatment of the Schmidt-Cassegrain.]

7.12 H. J. COOPER (ed), *Scientific Instruments* (Hutchinson's Scientific and Technical Publications, 1946; vol. 2, 1948).

7.13 H. W. COX, The Construction of Schmidt Cameras (*J. Sci. Instr.*, **16**, No. 8, 257).

7.14 H. W. & L. A. COX, The Construction of a Schmidt Camera (*J.B.A.A.*, **48**, No. 8, 308).

7.15 H. W. & L. A. COX, Further Notes on Schmidt Cameras (*J.B.A.A.*, **50**, No. 2, 61).

7.16 H. W. COX, Schmidt Camera Tests (*J.B.A.A.*, **51**, 152, 174).

7.17 H. W. COX, New Astronomical Reflectors (*J.B.A.A.*, **55**, No. 6, 137).

7.18 R. E. COX, Notes on Newtonian Reflector Alignment (*Sky & Tel.*, **31**, No. 3, 170).

7.19 R. E. COX, Methods of Alignment (*Sky & Tel.*, **35**, No. 5, 319).

7.20 R. E. COX, CHT: A Catadioptric Herschelian Telescope with Tilted Components (*Sky & Tel.*, **41**, No. 1, 46).

7.21 R. E. COX, Telescopes with Unobstructed Light Paths (*Sky & Tel.*, **43**, No. 2, 117).

7.22 R. E. COX, Temperature Effects on Mirror Cells (*Sky & Tel.*, **46**, No. 4, 256).

7.23 H. E. DALL, Diffraction Effects due to Axial Obstructions in Telescopes (*J.B.A.A.*, **48**, No. 4, 163).

7.24 H. E. DALL, Why not an Erecting System? (*Sky & Tel.*, **23**, No. 1, 48).

7.25 H. E. DALL, A Dall-Maksutov Telescope with Erecting Lens (*Sky & Tel.*, **23**, No. 2, 109). [Reprinted in B. 7.62.]

7.26 H. E. DALL, Maksutov Telescopes (*J.B.A.A.*, **73**, No. 7, 256).

7.27 H. E. DALL, A Pocketable 6-inch Multipurpose Cassegrain Telescope (*Sky & Tel.*, **35**, No. 3, 182).

7.28 H. E. DALL, Cassegrain Reflectors for Amateurs (*Yearbook of Astronomy 1966*, Eyre & Spottiswoode, 1965).

7.29 A. DANJON & A. COUDER, *Lunettes et Télescopes* (Editions de la Revue d'Optique Théorique et Instrumentale, Paris, 1935).

7.30 A. S. DeVANY, A Schmidt-Cassegrain Optical System with a Flat Field (*Sky & Tel.*, **29**, No. 5, 318 & **29**, No. 6, 380).

7.31 C. DÉVÉ (Trans. T. L. TIPPELL), *Optical Workshop Principles* (Hilger, Lond., 1945).

7.32 D. DI CICCO, Some Sightings of Sirius' Companion (*Sky & Tel.*, **49**, No. 6, 407).

7.33 D. C. DILWORTH, A New Catadioptric Telescope (*Sky & Tel.*, **54**, No. 5, 425 & **54**, No. 6, 521).

7.34 G. Z. DIMITROFF & J. G. BAKER, *Telescopes and Accessories* (Blakiston, Philadelphia, 1945).

7.35 T. DUNHAM, The Use of Schmidt Cameras in Plane Grating Spectrographs (*Phys. Rev.*, 2nd Series, **46**, 326).

7.36 T. DUNHAM, Schmidt Spectrograph Cameras (*Publ. A.A.S.*, **8**, 110).

7.37 F. J. EASTMAN, The Aberrations of a Prism Diagonal (*Sky & Tel.*, **38**, No. 4, 261).

7.38 W. F. A. ELLISON, *The Amateur's Telescope* (Belfast, 1932).

7.39 J. GREGORY, A Cassegrain-Maksutov Design for the Amateur (*Sky & Tel.*, **16**, No. 5, 236). [Reprinted in B. 7.62 and 7.91.]

7.40 F. J. HARGREAVES, A New Plea for the Reflector (*J.B.A.A.*, **46**, No. 5, 193).

7.41 F. J. HARGREAVES, *The Optical Adjustment of a Newtonian reflector* (B.A.A. Sectional Notes, No. 2).

7.42 D. O. HENDRIX, An Extremely Fast Schmidt Camera (*Publ. A.S.P.*, **51**, No. 301, 158).

7.43 D. O. HENDRIX & W. H. CHRISTIE, Some Applications of the Schmidt Principle in Optical Design (*Sci. Amer.*, **161**, 118).

7.44 E. G. HILL, A Home-made 24-inch Reflector and Warmed Observatory (*J.B.A.A.*, **73**, No. 8, 328).

7.45 J. H. HINDLE, Mechanical Flotation of Mirrors (*A.T.M.*, **1**, 229).

7.46 D. F. HORNE, *Optical Production Technology* (Crane, Russak & Co., New York, 1973).

7.47 N. E. HOWARD, *Handbook for Telescope Making* (Faber & Faber, 1970).

7.48 A. G. INGALLS (ed.), *Amateur Telescope Making* (Scientific American, Inc., New York, 1970). [Referred to as *A.T.M.*, **1**.]

7.49 A. G. INGALLS (ed.), *Amateur Telescope Making Advanced* (Scientific American, Inc., New York, 1968). [Referred to as *A.T.M.*, **2**.]

7.50 A. G. INGALLS (ed.), *Amateur Telescope Making, Book Three* (Scientific American, Inc., New York, 1969). [Referred to as *A.T.M.*, **3**.]

7.51 H. C. KING, *History of the Telescope* (Charles Griffin & Co., 1955).

7.52 O. R. KNAB, An Improved 4¼-inch Unobstructed Oblique Reflector (*Sky & Tel.*, **22**, No. 4). [Reprinted in B. 7.62.]

7.53 A. KUTTER, A New Three-mirror Unobstructed Reflector (*Sky & Tel.*, **49**, No. 1, 46 & **49**, No. 2, 115).

7.54 E. H. LINFOOT & HAWKINS, An Improved Type of Schmidt Camera (*Obs.*, **66**, No. 831, 239).

7.55 E. H. LINFOOT, On Some Optical Systems Employing Aspherical Surfaces (*M.N.*, **103**, No. 4, 210). [Advanced theory.]

7.56 E. H. LINFOOT, The Schmidt-Cassegrain Systems and Their Application to Astronomical Photography (*M.N.*, **104**, No. 1, 48). [Advanced theory.]

7.57 E. H. LINFOOT, Some Recent Applications of Optics to Astronomy (*M.N.*, **108**, No. 1, 81).

7.58   E. H. LINFOOT, Achromatised Plate-Mirror Systems (*Proc. Phys. Soc. Lond.*, **57**, No. 3, 199).

7.59   H. A. LOWER, Notes on the Construction of a $f/1$ Schmidt Camera (B. 7.49).

7.60   D. D. MAKSUTOV, New Catadioptric Meniscus Systems (*J. Opt. Soc. Amer.*, **34**, No. 5, 270).

7.61   J. MUIRDEN, *Beginner's Guide to Astronomical Telescope Making* (Pelham Books, 1975).

7.62   T. PAGE & L. W. PAGE (eds.), *Telescopes* (Collier-Macmillan, 1966).

7.63   H. E. PAUL, *Telescopes for Skygazing* (Sky Publications, 1966).

7.64   J. R. PAWLICK, A Folded Herschelian Off-axis Reflector (*Sky & Tel.*, **39**, No. 3, 191).

7.65   W. T. PETERS & R. PIKE, The Size of the Newtonian Diagonal (*Sky & Tel.*, **53**, No. 3, 220).

7.66   H. N. RUSSELL, The Schmidt Camera—Introductory (B. 7.49).

7.67   R. A. SAMPSON, On a Cassegrain Reflector with Corrected Field (*Phil. Trans., A* **213**, 27).

7.68   N. J. SCHELL & T. G. BEEDE, [A 'criss-cross' off-axis telescope] (*Sci. Amer.*, **162**, No. 5, 314).

7.69   B. SCHMIDT, Ein lichtstarkes komafreies Spiegelsystem (*Central Zeitung fur Optik und Mechanik*, **52**, 1931).

7.70   C. H. SMILEY, The Schmidt Camera (*Pop. Astr.*, **44**, No. 438, 415).

7.71   C. H. SMILEY, The Schmidt Camera—Chromatic Aberration and Other Factors Governing Design (*Pop. Astr.*, **48**, No. 474, 175).

7.72   C. H. SMILEY, The Construction of a Schmidt Camera (*Publ. A.A.S.*, **9**, Nos. 19-20, 1937).

7.73   C. H. SMILEY, A Note on the Schmidt Camera (*J.B.A.A.*, **49**, No. 1, 34).

7.74   O. STRUVE, A New Slit Spectrograph for Diffuse Galactic Nebulae (*Ap. J.*, **86**, No. 5, 613).

7.75   C. J. TENUKEST, R. SCHAEFER & H. PINNOCK, Notes on the Construction of a 6-inch Maksutov Telescope (*J.B.A.A.*, **56**, No. 7, 130).

7.76   J. TEXEREAU, *How to Make a Telescope* (Interscience Publishers, New York, 1963).

7.77   A. J. THOMPSON, *Making Your Own Telescope* (Sky Publishing Corp., Cambridge, Mass., 1947).

7.78   F. TWYMAN, *Prism and Lens Making* (Hilger & Watts, London, 1952).

7.79   T. J. WAINEO, Fabrication of a Wright Telescope (*Sky & Tel.*, **38**, No. 2, 112).

7.80   R. L. WALAND, Note on Figuring Schmidt Correcting Lenses (*J. Sci. Instr.*, **15**, No. 10, 339).

7.81   C. H. WERENSKIOLD, A Note on Curved Spiders (*Sky & Tel.*, **38**, No. 4, 262).

7.82   D. H. WHITEHEAD, An Australian 11½-inch Wright Telescope (*Sky & Tel.*, **43**, No. 5, 320).

BIBLIOGRAPHY

7.83 R. R. WILLEY, Cassegrain-type Telescopes (*Sky & Tel.*, **23**, No. 4, 191). [Reprinted in B. 7.62.]
7.84 R. R. WILLEY, Characteristics and Testing of Cassegrain-type Telescopes (*Sky & Tel.*, **23**, No. 4, 226).
7.85 D. O. WOODBURY, *The Glass Giant of Palomar* (Heinemann, 1940).
7.86 F. B. WRIGHT, An Aplanatic Reflector with a Flat Field Related to the Schmidt Telescope (*Publ. A.S.P.*, **47**, No. 280, 300).
7.87 F. B. WRIGHT, Theory and Design of Aplanatic Reflectors Employing a Correcting Lens (*A.T.M.*, **2**, 401).
7.88 F. B. WRIGHT, The Maksutov Lens Applied to Herschelian and Newtonian Telescopes (*A.T.M.*, **3**, 574).
7.89 C. G. WYNNE, New Wide-aperture Catadioptric Systems (*Obs.*, **68**, No. 843, 52).
7.90 [VARIOUS]*Bulletin A* (Sky Publishing Corp., Mass.). [Articles on the schiefspiegler, reprinted from *Sky & Tel.*]
7.91 [VARIOUS]*Bulletin C* (Sky Publishing Corp., Mass.). [Articles on the Maksutov telescope, reprinted from *Sky & Tel.*]

8. DEWING-UP AND TUBE CURRENTS

8.1 J. FRANKLIN-ADAMS, Prevention of Dew-deposit on Glass Surfaces (*M.N.*, **72**, No. 2, 165).
8.2 E. J. HYSOM, Air Currents Above Mirrors (*J.B.A.A.*, **82**, No. 4, 274).
8.3 A. F. LINDEMANN, A Revolving Eyepiece, electrically warmed (*M.N.*, **58**, 362).
8.4 W. H. STEAVENSON, The Dewing of Optical Surfaces (*A.T.M.*, **2**, 576).

9. TESTING

9.1 J. A. ANDERSON & R. W. PORTER, Ronchi's Method of Optical Testing (*Ap. J.*, **70**, No. 3, 175).
9.2 M. G. BIGOURDAN, *Les méthodes d'Examen des Lunettes et des Téléscopes* (Gauthier-Villars, Paris, 1915).
9.3 C. R. BURCH, On the Phase-Contrast Test of F. Zernike (*M.N.*, **94**, No. 5, 384).
9.4 C. R. BURCH, On Reflection Compensators for Testing Paraboloids (*M.N.*, **96**, No. 5, 438).
9.5 C. R. BURCH, Note on Compensation of Spherical Aberration in Telescopes (*J.B.A.A.*, **48**, No. 3, 115).
9.6 H. W. COX, The Testing of Wide-Angle Mirrors (*J.B.A.A.*, **56**, No. 6, 110).
9.7 H. E. Dall, A Null Test for Paraboloids (*A.T.M.*, **3**, 149).
9.8 P. FOX, An Investigation of the Forty-inch Objective of the Yerkes Observatory [by the Hartmann test] (*Ap. J.*, **27**, No. 4, 237).
9.9 F. J. HARGREAVES, A Spurious Aberration in Reflecting Telescopes (*J.B.A.A.*, **59**, No. 6, 180).
9.10 F. J. HARGREAVES, A Note on the Auto-Collimation Test for Concave Mirrors (*J.B.A.A.*, **45**, No. 2, 72).

9.11 F. J. HARGREAVES, A Further Note on the Auto-Collimation Test (*J.B.A.A.*, **45**, No. 8, 320).

9.12 F. J. HARGREAVES, A Note on the Hartmann Criterion (*J.B.A.A.*, **51**, No. 4, 121).

9.13 J. HARTMANN, An Improvement of the Foucault Knife-edge Test in the Investigation of Telescope Objectives (*Ap. J.*, **27**, No. 4, 254).

9.14 J. F. KIELKOPF, An Easily Constructed Caustic Tester (*Sky & Tel.*, **45**, No. 5, 315).

9.15 E. G. H. MOBSBY, A Ronchi Null Test for Paraboloids (*Sky & Tel.*, **48**, No. 5, 325).

9.16 W. H. NEWMAN, A Method of Measuring the Focal Lengths of Zones of a Reflecting Telescope (*J.B.A.A.*, **49**, No. 5, 186).

9.17 B. A. NORMAN, A New Test for Cassegrain Secondaries (*Sky & Tel.*, **17**, No. 1, 38).

9.18 H. E. PAUL, Schmidt Camera Notes (*A.T.M.*, **3**, 323).

9.19 J. S. PLASKETT, The Character of the Star Images in Spectrographic Work (*Ap. J.*, **25**, No. 3, 195). [Hartmann test.]

9.20 J. L. RICHTER, A Test for Figuring Cassegrain Secondary Mirrors (*Sky & Tel.*, **39**, No. 1, 49).

9.21 G. W. RITCHEY, On Methods of Testing Optical Mirrors during Construction (*Ap. J.*, **19**, 59).

9.22 R. W. SINNOTT, Calculating a Mirror's Surface Accuracy (*Sky & Tel.*, **54**, No. 2, 144).

9.23 D. E. STOLTZMANN & M. HATCH, Extensions of the Dall Null Test (*Sky & Tel.*, **52**, No. 3, 210).

9.24 H. DENNIS TAYLOR, *The Adjustment and Testing of Telescope Objectives* (Sir Howard Grubb, Parsons, 4th edn., 1946). [Last revised edn., 1921.]

9.25 J. V. THOMSON, A Compact Knife-Edge Apparatus for the Foucault Test (*J.B.A.A.*, **54**, No. 4/5, 87).

9.26 F. ZERNIKE, Diffraction Theory of the Knife-edge Test and its Improved Form, the Phase-Contrast Method (*M.N.*, **94**, No. 5, 377).

10. TELESCOPE MOUNTINGS, INCLUDING DRIVES AND POLAR AXIS ALIGNMENT

10.1 M. A. AINSLIE, A Compact 'Reflex' Mounting for a Refractor (*J.B.A.A.*, **46**, No. 9, 324).

10.2 [D. P. BAIN] A Mounting from the Past (*Sky & Tel.*, **52**, No. 2, 138). [Description of the Sellers mounting.]

10.3 J. J. BROOKS, Structural Considerations for Telescope Makers (*Sky & Tel.*, **51**, No. 6, 423).

10.4 A. COLEMAN, Equatorial Head and Clock Drive for Camera for Small Scale Photographs of the Eclipse (*J.B.A.A.*, **37**, No. 6, 216).

10.5 E. C. COLLINSON, [Report of B.A.A. Meeting] (*J.B.A.A.*, **37**, No. 5, 157).

10.6 M. A. CORNU, On the Law of Diurnal Rotation of the Optical Field of the Siderostat and Heliostat (*Ap. J.*, **11**, No. 2, 148).

10.7 H. W. COX, A Variable-Frequency Telescope Drive (*J.B.A.A.*, **48**, No. 6, 248).

10.8 R. E. COX, Balancing the Tube of a Reflecting Telescope (*Sky & Tel.*, **18**, No. 1, 47).

10.9 R. E. COX, Balancing a German Equatorial Mounting (*Sky & Tel.*, **18**, No. 2, 107).

10.10 D. W. DEWHURST, Automatic Guiding of Solar Telescopes (*J.B.A.A.*, **56**, No. 8, 144).

10.11 W. FELLOWS, A Fixed-eyepiece Refractor with Heated Observatory (*Sky & Tel.*, **41**, No. 5, 310).

10.12 R. M. FRY, The Adjustment of a Polar Axis by Star Trails near the Pole (*J.B.A.A.*, **45**, No. 6, 234).

10.13 D. R. GLASSPOOL, An Altazimuth Stand for 10½-inch Reflector (*J.B.A.A.*, **55**, No. 1, 18).

10.14 G. Y. HAIG, A Stellar Spectrograph (*J.B.A.A.*, **85**, No. 5, 408). [Description of the 'Scotch' mount.]

10.15 G. Y. HAIG, A Simple Camera Mounting for Short Exposures (*Sky & Tel.*, **49**, No. 4, 263).

10.16 A. HAMON, A Tangent Arm with a Specially Cut Cam (*Sky & Tel.*, **55**, No. 6, 530).

10.17 [W. HARVEY] A Homemade Toothed Sector (*Sky & Tel.*, **52**, No. 2, 135).

10.18 N. HERRETT, A Split-ring Mounting for a 12-inch Reflector (*Sky & Tel.*, **16**, No. 7, 341).

10.19 J. HISCOTT, Polar Axis Clutch Elimination (*J.B.A.A.*, **82**, No. 6, 436).

10.20 [G. A. HOLE], Mr G. A. Hole's Observatory at Patcham, Sussex (*J.B.A.A.*, **51**, No. 9, 329).

10.21 H. C. HUNT, A Member's Observatory (*J.B.A.A.*, **76**, No. 1, 5). [Modified English mounting.]

10.22 O. R. KNAB, [Description of a Grubb polar mounting] (*Sky & Tel.*, **25**, No. 5, 292).

10.23 R. A. KNOX, Invertor Circuits and Motors for Clock Drives (*J.B.A.A.*, **82**, No. 3, 182).

10.24 T. P. KOHMAN, Polar Alignment of Portable Equatorials (*Sky & Tel.*, **51**, No. 2, 135).

10.25 C. L. KUSSNER, An Inexpensive Declination Slow-motion Control (*Sky & Tel.*, **36**, No. 1, 46).

10.26 F. B LINDSAY, A New Mounting and Housing for a Telescope (*Pop. Astr.*, **39**, No. 8, 497).

10.27 D. G. McCARTAN & A. DE SA, A Stepping Motor Drive for Small Telescopes (*J.B.A.A.*, **85**, No. 4, 324).

10.28 P. R. McMATH, The Tower Telescope of the McMath-Hulbert Observatory (*Publ. Michigan Obs.*, **7**, No. 1, 1).

10.29 G. W. MOFFITT, Frequency Requirements and the Control of Frequency for Synchronous Motor Operation of Astronomical Telescopes (*Rev. Sci. Instr.*, **3**, No. 9, 499).

10.30 C. NASH, An Amateur-built Chain Drive for a 12½-inch Reflector (*Sky & Tel.*, **42**, No. 6, 378).

10.31 W. H. PICKERING, Reflectors versus Refractors (*Pop. Astr.*, **38**, No. 3, 134).

10.32 W. H. PICKERING, The Practical Use of Small Reflectors (*Pop. Astr.*, **34**, No. 9, 570).

10.33 E. W PIINI, A Well-equipped 8-inch Newtonian Reflector (*Sky & Tel.*, **52**, No. 6, 471).

10.34 T. C. PLATT, Two Economical Transistor Invertor Circuits for Speed Control of Synchronous Motor Clock Drives (*J.B.A.A.*, **80**, No. 4, 270).

10.35 A. PONCET, An Equatorial Table for Astronomical Equipment (*Sky & Tel.*, **53**, No. 1, 64).

10.36 H. L. PUGH, Simple Linear Drive for Polar and Declination Axes (*J.B.A.A.*, **78**, No. 4, 268).

10.37 A. SAULIETIS & P. MALEY, A Pipe Mount for Comet Seeking (*Sky & Tel.*, **56**, No. 1, 73).

10.38 V. P. SAXON, An Electronic Speed Control for Clock Drives (*Sky & Tel.*, **49**, No. 1, 50).

10.39 A. SCHUG, A Pillow-block Mounting for a 12½-inch Cassegrain (*Sky & Tel.*, **51**, No. 5, 349).

10.40 L. SECRETAN, A Simple Yoke Mounting for a 12½-inch Reflector (*Sky & Tel.*, **41**, No. 2, 114). [English mounting.]

10.41 F. J. SELLERS, A Simple, Inexpensive and Effective Clock Drive for Small Telescopes (*J.B.A.A.*, **40**, No. 6, 222).

10.42 B. SHUTT, A Split-ring Mount from Aluminium Extrusions (*Sky & Tel.*, **52**, No. 4, 293).

10.43 R. W. SINNOTT, Some Forerunners of the Sellers Mount (*Sky & Tel.*, **56**, No. 2, 157).

10.44 B. L. SOUTHER, The Precise Adjustment of an Equatorial Mounting (*Sky & Tel.*, **51**, No. 1, 78).

10.45 F. H. THORNTON, A Simple, Controlled, Electric Drive (*J.B.A.A.*, **58**, No. 5, 175).

10.46 H. H. TURNER, Some Notes on the Use and Adjustment of the Coelostat (*M.N.*, **56**, No. 8, 408).

10.47 R. TUTHILL, A Novel Drive on a 20-inch Telescope (*Sky & Tel.*, **44**, No. 1, 47).

10.48 F. L. O. WADSWORTH, On a New Form of Mounting for Reflecting Telescopes devised by the late Arthur Cowper Ranyard (*Ap. J.*, **5**, No. 2, 132).

10.49 R. L. WATERFIELD, The Elimination of Atmospheric Refraction in the Adjustment of Equatorials (*J.B.A.A.*, **45**, No. 7, 282).

10.50 J. B. WEST & R. S. BRADFORD, A Crystal-controlled Oscillator for Telescope Drives (*Sky & Tel.*, **50**, No. 2, 125).

10.51 W. REES WRIGHT, An Altazimuth Mounting for Reflecting Telescopes (*J.B.A.A.*, **50**, No. 3, 113).

11. OBSERVATORIES

11.1 A. K. APPLETON, A Home-built Rotating Observatory (*J.B.A.A.*, **68**, No. 6, 217).

11.2 H. E. BECKETT, The Painting of Observatory Domes as a Protection against Solar Heat (*Obs.*, **59**, No. 740, 14).

11.3 H. E. DALL, Thermal Effects of Observatory Paints (*J.B.A.A.*, **48**, No. 3, 116).

11.4 C. A. DEBERITZ, A Norwegian Observatory with Sliding Roof Sections (*Sky & Tel.*, **47**, No. 2, 122).

11.5 A. W. HEATH, A Very Simple Observatory (*J.B.A.A.*, **75**, No. 1, 63).

11.6 E. G. HILL, A Practical Heated Observatory (*J.B.A.A.*, **72**, No. 3, 102).

11.7 F. G. McINNIS, A Well-built Roll-off-roof Observatory in North Carolina (*Sky & Tel.*, **33**, No. 4, 244).

11.8 B. MANNING, A Glass-fibre Dome for a 265 mm f/8 Reflecting Telescope (*J.B.A.A.*, **85**, No. 3, 245).

11.9 R. W. MIDDLETON, A Cheap Observatory (*J.B.A.A.*, **78**, No. 2, 105).

11.10 E. G. MOORE, A Compact Run-off Roof Observatory (*J.B.A.A.*, **84**, No. 4, 275).

11.11 M. REYNOLDS, An Observatory Made from a Prefabricated Toolhouse (*Sky & Tel.*, **47**, No. 5, 338).

11.12 P. SAHULA, Winter Observing from a Heated Dome (*Sky & Tel.*, **56**, No. 1, 67).

11.13 H. SPENCER JONES, [Letter commenting on B. 11.2] (*Obs.*, **59**, No. 740, 17).

11.14 W. D. WILLIAMS, An Inexpensive Slide-off Roof Observatory (*Sky & Tel.*, **38**, No. 1, 49).

11.15 P. D. WROATH, A Roll-off Roof Observatory (*J.B.A.A.*, **85**, No. 2, 129).

12. MICROMETERS

12.1 R. G. AITKEN, *The Binary Stars* (Dover Books, 1963).

12.2 R. W. ARGYLE, Micrometers for Double Star Measurement (*Double Star Observers' Handbook*, The Webb Society, 1975).

12.3 M. P. CANDY, Observation of Double Stars (*Practical Amateur Astronomy*, Lutterworth, 1975).

12.4 R. E. COX, Light-emitting Diodes for Telescope Illumination (*Sky & Tel.*, **45**, No. 5, 319).

12.5 H. E. DALL, Finding and Guiding on Dark Skies (*J.B.A.A.*, **82**, No. 3, 202).

12.6 C. R. DAVIDSON & L. S. T. SYMMS, A Comparison Image Micrometer. (*M.N.*, **98**, No. 3, 176).

12.7 W. S. FINSEN, Double Star Interferometry (*Pop. Astr.*, **59**, No. 8, 399).

12.8 F. J. HARGREAVES, A Comparison-Image Micrometer (*M.N.*, **92**, No. 1, 72).

BIBLIOGRAPHY

12.9   F. J. HARGREAVES, A Modified Comparison-Image Micrometer (*M.N.*, **92**, No. 5, 453).

12.10   F. HOLDEN, Plastic Threads (*Sky & Tel.*, **52**, No. 2, 140).

12.11   P. MULLER, Techniques for Visual Measurements (*Stars & Stellar Systems*, **3**, 440, University of Chicago Press).

12.12   J. POLMAN, A Homemade Filar Micrometer (*Sky & Tel.*, **53**, No. 5, 391).

12.13   L. RICHARDSON, The Interferometer (*J.B.A.A.*, **35**, No. 3, 105).

12.14   L. RICHARDSON, A Home-made Interferometer (*J.B.A.A.*, **35**, No. 5, 155).

12.15   L. RICHARDSON, Results with a Home-made Interferometer (*J.B.A.A.*, **37**, No. 8, 311).

12.16   L. RICHARDSON, Further Notes on the Interferometer for the Measurement of Double Stars (*J.B.A.A.*, **38**, No. 8, 258).

12.17   K. SCHWARTZCHILD, Ueber Messung von Doppelsternen durch Interferenzen (*Astr. Nach.*, **139**, No. 3335, 353).

12.18   E. C. SILVA, An Amateur Micrometer for Double Stars (*A.T.M.*, **2**, 447).

12.19   G. E. TAYLOR, The Use of Crosswires in Obtaining Positions (*J.B.A.A.*, **69**, No. 1, 33).

12.20   A. J. WESSELINK, Astronomical Objective Gratings (*App. Opt.*, **3**, No. 7, 889).

12.21   C. WORLEY, Visual Observing of Double Stars (*Sky & Tel.*, **22**, No. 2, 73).

12.22   C. WORLEY, The Construction of a Filar Micrometer (*Sky & Tel.*, **22**, No. 3, 140).

12.23   C. WORLEY, Measuring Double Stars with a Filar Micrometer (*Sky & Tel.*, **22**, No. 5, 261).

12.24   H. R. WRIGHT, Bright Wire Illumination (*J.B.A.A.*, **31**, No. 6, 229).

13. SPECTROSCOPES

13.1   P. DELVO, A Spectroscope with a Holographic Grating (*Sky & Tel.*, **54**, No. 1, 65).

13.2   J. EVERSHED, A New Arrangement of Prisms for a Solar Prominence Spectroscope (*J.B.A.A.*, **7**, No. 6, 331).

13.3   H. LEINBACH, A Simple Spectroscope for Solar Prominence Observation (*Sky & Tel.*, **12**, No. 3, 78).

13.4   R. A. SAWYER, *Experimental Spectroscopy* (Chapman & Hall, 1945).

13.5   F. J. SELLERS, *A Solar Prominence Spectroscope, with instructions for making and using* (B.A.A. Sectional Notes, No. 1= *J.B.A.A.*, **45**, No. 9, 362).

13.6   W. REES WRIGHT, Two Spectroscopes for Prominence Observations (*J.B.A.A.*, **52**, No. 4, 130).

14. PHOTOGRAPHY

14.1   I. S. BOWEN & L. T. CLARK, Hypersensitization and Reciprocity Failure of Photographic Plates (*J. Opt. Soc. Amer.*, **30**, 508).

14.2   E. H. COLLINSON, An Automatic Meteor Camera (*J.B.A.A.*, **39**, No. 5, 150).

14.3   E. H. COLLINSON, An Improved Automatic Meteor Camera (*J.B.A.A.*, **44**, No. 4, 157).

14.4   L. J. COMRIE, Note on the Reduction of Photographic Plates (*J.B.A.A.*, **39**, No. 6, 203).

14.5   [L. J. COMRIE, Report of B.A.A. Meeting] (*J.B.A.A.* **39**, No. 6, 173).

14.6   S. R. B. COOKE, A 35 mm Camera for Astrophotography (*Sky & Tel.*, **41**, No. 2, 109). [Incorporating a guiding eyepiece.]

14.7   F. COVITZ, An Automatic Guider for Astrophotography (*Sky & Tel.*, **47**, No. 3, 191).

14.8   H. E. DALL, Astronomical Photography for Amateurs (*Practical Amateur Astronomy*, Lutterworth, 1975).

14.9   D. G. DANIELS, A Portable Driven Mount for 35 mm Cameras (*J.B.A.A.*, **84**, No. 3, 189).

14.10 R. L. DICKINSON, A Simple Knife Edge Focusing Attachment (*Sky & Tel.*, **42**, No. 4, 325).

14.11 H. EGGELING, A Swiss Amateur's Cooled-emulsion Camera (*Sky & Tel.*, **34**, No. 6, 400).

14.12 O. ERNEST, Deep-sky Photography with Cooled Emulsions (*Sky & Tel.*, **45**, No. 3, 189).

14.13 R. FOGELQUIST, A Guiding Reticle for Astrophotography (*Sky & Tel.*, **52**, No. 6, 478).

14.14 T. B. FOWLER, A Nomogram for Astrophotographers (*Sky & Tel.*, **51**, No. 5, 353). [Theoretical limiting magnitudes.]

14.15 F. J. HARGREAVES, The Focussing and Squaring-on of Photographic Objectives (*J.B.A.A.*, **37**, No. 3, 85).

14.16 E. A. HARLAN, Increasing the Sensitivity of Commerical Films (*Sky & Tel.*, **27**, No. 3, 187). [Baking.]

14.17 T. H. JAMES & G. C. HIGGINS, *Fundamentals of Photographic Theory* (Lond., 1948).

14.18 E. L. JONES, A Cold Camera That Needs no Vacuum (*Sky & Tel.*, **50**, No. 2, 122).

14.19 G. F. KELLAWAY, The Measurement of Photographic Plates by Projection (*J.B.A.A.*, **47**, 231).

14.20 E. S. KING, Forms of Images in Stellar Photography (*Harv. Ann.*, **41**, No. 6, 153).

14.21 E. SKINNER KING, *A Manual of Celestial Photography* (Eastern Science Supply Co., Boston, 1931). [Invaluable.]

14.22 C. E. K. MEES, *The Theory of the Photographic Process* (Macmillan, N.Y., 1942).

14.23 G. MERTON, An Interpolation Method for Astronomical Photographs (*J.B.A.A.*, **39**, No. 6, 209).

14.24 G. MERTON, Photography and the Amateur Astronomer (*J.B.A.A.*, **63**, No. 1, 7). [B.A.A. Presidential Address, 1952; useful bibliography.]

14.25 P. M. MILLMAN, The Theoretical Frequency Distribution of Photographic Meteors (*Selected Papers from Wash. Nat. Ac. Proc.*, **19**, 34, 1933).

BIBLIOGRAPHY

14.26 P. M. MILLMAN, Amateur Meteor Photography (*Pop. Astr.*, **41**, No. 6, 298).

14.27 P. M. MILLMAN, An Analysis of Meteor Spectra (*Harv. Ann.*, **82**, No. 6, 113).

14.28 P. M. MILLMAN & D. HOFFLEIT, A Study of Meteor Photographs Taken through a Rotating Shutter (*Harv. Ann.*, **105**, No. 31, 601).

14.29 C. B. NEBLETTE, *Photography—Its Principles and Practice* (Chapman & Hall, 1943).

14.30 J. PATON, Aurora Borealis: Photographic measurements of height (*Weather*, **1**, No. 6, 8, 1946).

14.31 J. PATON, Aurora Borealis (*Science News*, **11**, No. 9, 15, 1949).

14.32 H. E. PAUL, [Field Flattening Lens] (*A.T.M.*, **3**, 313).

14.33 H. E. PAUL, *Outer Space Photography* (Amphoto, New York, 1976).

14.34 T. RACKHAM, *Astronomical Photography at the Telescope* (Faber, 1972).

14.35 T. RACKHAM, A Lunar Camera (*J.B.A.A.*, **66**, No. 2, 61).

14.36 J. M. SANTOS, Hypersensitizing Films for Astrophotography (*Sky & Tel.*, **40**, No. 5, 322). [Ammonia treatment.]

14.37 F. SCHLESINGER, A Short Method for Deriving Positions of Asteroids, Comets etc., from Photographs (*A.J.*, **37**, 872).

14.38 F. SCHLESINGER, Some Aspects of Astronomical Photography of Precision (*M.N.*, **87**, No. 7, 506). [George Darwin Lecture, 1927.]

14.39 E. W. H. SELWYN, *Photography in Astronomy* (Eastman Kodak, 1950).

14.40 H. H. TURNER, On the Measurement of a Meteor Trail on a Photographic Plate (*M.N.*, **67**, No. 9, 562).

14.41 B. D. WALLIS & R. W. PROVIN, [Darkroom Techniques] (*Sky & Tel.*, **53**, No. 6, 484).

14.42 F. L. WHIPPLE & P. J. RUBENSTEIN, The Limiting Magnitudes of Photographic Telescopes (*Pop. Astr.*, **50**, 24).

14.43 W. T. WHITNEY, The Determination of Meteor Velocities (*Pop. Astr.*, **45**, No. 3, 162).

14.44 J. D. WISEMAN, Cooled-emulsion Photography Without a Vacuum (*Sky & Tel.*, **37**, No. 2, 118).

14.45 H. E. WOOD, The Measurement of the Position of Objects on Photographic Plates (*J.B.A.A.*, **39**, No. 6, 196).

14.46 H. E. WOOD, Further Note on the Method of Dependences as Applied to the Reduction of Photographic Plates (*J.B.A.A.*, **43**, No. 1, 25).

14.47 H. E. WOOD, *Light Reading No. 8* (Kodak Ltd., 1977). [Evaluation of different films for astrography; see also *Sky & Tel.*, **50**, No. 1, 61.]

14.48 H. E. WOOD, *Kodak Filters for Scientific and Technical Uses* (Publication B-3, Eastman Kodak, New York).

14.49 H. E. WOOD, *Kodak Plates and Films for Scientific Photography* (Publication P-315, Eastman Kodak, New York).

14.50 H. E. WOOD, The Measurement of the Position of Objects on Photographic Plates (*J.B.A.A.*, **39**, No. 6, 196).

15. PHOTOMETRY

15.1   J. AFRICANO & R. QUIGLEY, What is a Fabry Lens? (*Sky & Tel.*, **54**, No. 1, 60).

15.2   D. S. BROWN, A Combined Microdensitometer and Plate Measuring Machine (*J.B.A.A.*, **72**, No. 1, 3)

15.3   D. S. BROWN, Some Experiments in Photographic Photometry using Comparison Prisms (*J.B.A.A.*, **75**, No. 2, 78).

15.4   H. W. COX, Photoelectric Photometry (*J.B.A.A.*, **58**, No. 3, 101).

15.5   M. A. DANJON, Recherches de Photométrie Astronomique (*Ann. Obs. Strasbourg*, **2**, No. 1, 82).

15.6   D. S. EVANS, *Astronomical Photographic Photometry* (Chapman & Hall, 1952).

15.7   P. B. FELLGETT, An Exploration of Infra-red Stellar Magnitudes using the Photo-conductivity of Lead Sulphide (*M.N.*, **111**, No. 6, 537).

15.8   W. S. HOUSTON, A Visual Photometer for Small Telescopes (*Sky & Tel.*, **52**, No. 1, 64).

15.9   J. B. IRWIN, The Observational Problem of Stellar Magnitudes and Colours (*Obs.*, **67**, No. 837, 62). [See also B. 15.28.]

15.10 G. E. KRON, The Construction and Use of a Photomultiplier Type of Photoelectric Photometer (*Bull. of the Panel on the Orbits of Eclipsing Binaries*, 4).

15.11 G. E. KRON *et al.*, Photoelectric Photometry (*Bull. of the Panel on the Orbits of Eclipsing Binaries*, 6).

15.12 G. E. KRON, Application of the Multiplier Phototube to Astronomical Photoelectric Photometry (*Ap. J.*, **103**, No. 3, 326).

15.13 G. E. KRON, Two Direct-coupled Amplifiers for Use with a Stellar Photoelectric Photometer (*A.T.M.*, **3**, 287).

15.14 M. W. OVENDEN, A Visual Photometer for the Comparison of Stars of Different Colours (*J.B.A.A.*, **54**, No. 8/9, 172, 156).

15.15 H. M. PARKHURST, [Deflection Photometer] (*Harv., Ann.*, **18**, No. 3, 1890).

15.16 W. E. PENNELL, The Use of Modern 35 mm Panchromatic Film for Magnitude Determination (*J.B.A.A.*, **80**, No. 5, 371).

5.17   E. PETTIT & S. B. NICHOLSON, Stellar Radiation Measurements

15.18 E. PETTIT & S. B. NICHOLSON, Lunar Radiation and Temperatures (*Ap. J.*, **71**, No. 2, 102=*M.W.C.*, 392).

5.19   E. PETTIT, Radiation Measurements on the Eclipsed Moon (*Ap. J.*, **91**, No. 4, 408=*M.W.C.*, 627).

5.20   E. C. PICKERING, Photometric Observations (*Harv. Ann.*, **11**, No. 1, 1).

5.21   E. C. PICKERING, A New Form of Stellar Photometer (*Ap. J.*, **2**, No. 2, 89).

5.22   E. C. PICKERING & O. C. WENDELL, Discussion of Observations made with the Meridian Photometer (*Harv. Ann.*, **23**, No. 1, 1).

BIBLIOGRAPHY

15.23 V. C. REDDISH, [Photographic Photometry] (*M.N.*, **114**, 583).

15.24 L. SECRETAN, A Simple Magnitude-measuring Experiment (*Sky & Tel.*, **14**, No. 10, 432). [Bright-field photometer.]

15.25 J. STEBBINS, Measuring Starlight by Photocell (*Sci. Amer.*, **186**, No. 3, 56).

15.26 J. STEBBINS, The Photo-Electric Photometer of the Yerkes Observatory (*Ap. J.*, **74**, No. 5, 289).

15.27 J. STEBBINS, The Electrical Photometry of Stars and Nebulae (*M.N.*, **110**, No. 5, 416). [George Darwin Lecture, 1950.]

15.28 R. H. STOY & R. O. REDMAN, The Observational Problem of Stellar Magnitudes and Colours (*Obs.*, **66**, No. 834, 330). [See B. 15.9.]

15.29 V. VAND, Remarks on Making Thermocouples (*J.B.A.A.*, **53**, No. 2, 85).

15.30 H. F. WEAVER, The Development of Astronomical Photometry (*Pop. Astr.*, **54**, 211, 287, 339, 451, 504).

16. ACCESSORY INSTRUMENTS
[Arranged according to instrument]

16.1 H. E. DALL, Filter Type Solar Prominence Telescopes for Amateurs (*J.B.A.A.*, **77**, No. 2, 94).

16.2 A. GABRIEL, A Narrow-band $H\alpha$ Telescope for Visual and Photographic Solar Observations (*J.B.A.A.*, **86**, No. 2, 140).

16.3 M. A. ELLISON, The Formation of the Monochromatic Image in a Spectrohelioscope (*J.B.A.A.*, **50**, No. 2, 68).

16.4 M. A. ELLISON, The Construction of an Auto-collimating Spectrohelioscope (*J.B.A.A.*, **50**, No. 3, 107).

G. E. HALE, The Spectrohelioscope and its Work:

16.5 I. (*Ap. J.*, **70**, No. 5, 265=*M.W.C.*, 388).

16.6 II. (*Ap. J.*, **71**, No. 2, 73=*M.W.C.*, 393).

16.7 III. (*Ap. J.*, **73**, No. 5, 379=*M.W.C.*, 425).

16.8 IV. (*Ap. J.*, **74**, No. 3, 214=*M.W.C.*, 434).

16.9 G. E. HALE, The Brightness of Prominences as shown by the Spectrohelioscope (*M.N.*, **95**, No. 5, 467).

16.10 G. E. HALE, A Simple Solar Telescope and Spectrohelioscope (*A.T.M.*, **1**, 192).

16.11 H. HILL, An Amateur's Spectrohelioscope (*J.B.A.A.*, **78**, No. 5, 342).

16.12 F. J. SELLERS, A Note on the Spectrohelioscope and a Description of a Vibrating Slit Mechanism (*J.B.A.A.*, **48**, No. 6, 243).

16.13 F. J. SELLERS, Seventeenth Report of the Section for the Observation of the Sun, 1928-1950 (*Mem. B.A.A.*, **37**, No. 2, 1952).

16.14 F. N. VEIO, A Miniaturized Spectrohelioscope (*J.B.A.A.*, **85**, No. 3, 242).

16.15 A. F. COLLINS, A Mains Chronograph with provision for G.M.T. Signals (*J.B.A.A.*, **57**, No. 4, 176).

16.16 W. T. HAY, A Simple Chronograph (*J.B.A.A.*, **43**, No. 2, 80).

16.17 W. T. HAY, An Improved Simple Chronograph (*J.B.A.A.*, **44**, No. 8, 299).

16.18 L. POINTON, A Chronograph for Amateur Use (*J.B.A.A.*, **74**, No. 7, 281).

16.19 R. L. WATERFIELD, A Cheap Chronograph (*J.B.A.A.*, **42**, No. 4, 147).

16.20 H. H. WATERS & D. A. CAMPBELL, A Simple Method of Recording Radio Time-Signals on a Chronograph (*J.B.A.A.*, **42**, No. 4, 147).

16.21 L. F. MAHON, Construction of a Blink Comparator (*Sky & Tel.*, **46**, No. 4, 254).

16.22 C. R. MARTYS, A Method of Blink Comparison using Standard 35 mm Equipment (*J.B.A.A.*, **86**, No. 4, 277).

16.23 [E. G. H. MOBSBY] A Simple Comparator for Star Photographs (*Sky & Tel.*, **32**, No. 6, 373).

16.24 H. VEHRENBERG, A Homebuilt Machine for Scanning Plates (*Sky & Tel.*, **38**, No. 3, 186).

16.25 W. A. GRANGER, An Observing Ladder with Easily Adjustable Seats (*J.B.A.A.*, **60**, No. 4, 107).

16.26 F. J. SELLERS, An Efficient and Easily Constructed Observing Chair (*J.B.A.A.*, **43**, No. 7, 283; repr. with added Note *J.B.A.A.*, **63**, No. 1, 36).

17. VISION

17.1 W. DE. W. ABNEY, *Colour Vision* (Sampson Low, 1895).

17.2 E. M. ANTONIADI, On the Optical Character of Gemination (*J.B.A.A.*, **8**, No. 4, 176).

17.3 E. M. ANTONIADI, Further Considerations on Gemination (*J.B.A.A.*, **8**, No. 7, 308).

17.4 E. M. ANTONIADI, On Some Subjective Phenomena (*J.B.A.A.*, **9**, No. 6, 269).

17.5 E. M. ANTONIADI, Considerations on the Double Canals of Mars (*J.B.A.A.*, **10**, No. 7, 305; **11**, No. 1, 26).

17.6 M. A. DANJON, L'Acuité Visuelle et ses Variations (*Rev. d'Opt.*, **7**, 205).

17.7 D. DUTTON, Naked Eye Observations of Jupiter's Moons (*Sky & Tel.*, **52**, No. 6, 482).

17.8 J. E. EVANS & E. W. MAUNDER, Experiments as to the Actuality of the 'Canals' observed on Mars (*M.N.*, **63**, No. 8, 488).

17.9 F. J. HARGREAVES [Report of Presidential Address to the B.A.A., 1944] (*J.B.A.A.*, **55**, No. 1, 1).

17.10 H. SPENCER JONES, The Eye and Astronomical Observation (*Brit. J. Physiol. Optics*, **8**, No. 2, 93).

17.11 P. LOWELL, The Canals of Mars, Optically and Psychologically Considered: A Reply to Professor Newcomb (*Ap. J.*, **26**, No. 3, 131). [See B. 17.12, 15, 16.]

17.12 P. LOWELL, Reply to Professor Newcomb's Note (*Ap. J.*, **26**, No. 3, 142).

17.13 T. L. MACDONALD, The Observation of Planetary Detail—An Experiment (*J.B.A.A.*, **29**, No. 8, 225).

17.14 E. W. & A. S. D. MAUNDER, Some Experiments on the Limits of Vision for Lines and Spots as applicable to the Question of the Actuality of the Canals of Mars (*J.B.A.A.*, **13**, No. 9, 344).

17.15 S. NEWCOMB, The Optical and Psychological Principles involved in the Interpretation of the so-called Canals of Mars (*Ap. J.*, **26**, No. 1, 1).

17.16 S. NEWCOMB, Note on the Preceding Paper (*Ap. J.*, **26**, No. 3, 141).

17.17 F. W. SHARPLEY, Night Vision (*J.B.A.A.*, **57**, No. 3, 148).

17.18 R. W. SINNOTT, Correcting Astigmatism of the Observer's Eye (*Sky & Tel.*, **52**, No. 5, 376).

17.19 W. S. STILES & B. H. CRAWFORD, The Effect of a Glaring Light Source on Extrafoveal Vision (*Proc. Roy. Soc.*, Series B, **122**, No. 827, 255).

17.20 W. J. WRIGHT, *Photometry and the Eye* (Hatton Press, 1949).

18. PERSONAL EQUATION

18.1 R. L. DUNCOMBE, Personal Equation in Astronomy (*Pop. Astr.*, **53**, No. 2, 63, 110).

18.2 E. C. PHILLIPS, Second Note on Personal Equation in Observing Occultations (*Pop. Astr.*, **36**, 403).

18.3 W. H. PICKERING, Artificial Disks (*Harv. Ann.*, **32**, No. 2, 117).

18.4 T. H. SAFFORD, On the Various Forms of Personal Equation in Meridian Transits of Stars (*M.N.*, **57**, No. 7, 504).

18.5 T. H. SAFFORD, Additional Note on Personal Equation (*M.N.*, **58**, No. 2, 38).

19. ATMOSPHERE AND SEEING

19.1 J. A. ANDERSON, Astronomical Seeing (*J. Opt. Soc, Amer.*, **25**, No. 5, 152).

19.2 D. BRUNT, *Physical and Dynamical Meteorology* (Cambridge University Press, 2nd edn., 1939).

19.3 H. E. BUTLER, Observations of Stellar Scintillation (*Contr. Dunsink Obs.*, 4=*Proc. R. Irish Acad.*, **54**A, 321, 1952).

19.4 H. E. BUTLER, Scintillation and Atmospheric Seeing (*Irish Astr. J.*, **1**, No. 8, 225).

19.5 R. E. COX, Neutralising Atmospheric Dispersion (*Sky & Tel.*, **33**, No. 6, 388).

19.6 H. E. DALL, Atmospheric Dispersion (*J.B.A.A.*, **71**, No. 2, 75). [Eyepiece corrector unit.]

19.7 A. E. DOUGLASS, Atmosphere, Telescope and Observer (*Pop. Astr.*, **5**, No. 2, 64).

19.8 A. E. DOUGLASS, Scales of Seeing (*Pop. Astr.*, **6**, No. 4, 193).

19.9 A. E. DOUGLASS, The Study of Atmospheric Currents by the Aid of Large Telescopes, and the effect of such currents on the Quality of the Seeing (*Meteorological Jnl.*, U.S.A., March 1895).

19.10 G. M. B. DOBSON & A. R. MEETHAM, Atmospheric Pollution (*Nat.*, **151**, No. 3829, 324).

19.11 M. A. ELLISON & H. SEDDON, Some Experiments on the Scintillation of Stars and Planets (*M.N.*, **112**, No. 1, 73).

19.12 W. J. HUMPHREYS, *Physics of the Air* (McGraw-Hill, 1940).

19.13 A. E. LEVIN, The Correction of Declination and Hour Angle for Atmospheric Refraction (*Mem. B.A.A.*, **24**, No. 1, 18).

19.14 A. E. LEVIN, Very High-Resolution Photography with Large Reflectors (*Sky & Tel.*, **42**, No. 1, 11).

20. TIME AND CLOCKS

20.1 D. A. BATEMAN, Time by Radio (*Wireless World*, June 1975, 277).

20.2 J. E. HASWELL, *Horology* (Chapman & Hall, 1947).

20.3 F. HOPE–JONES, *Electric Clocks* (N.A.G. Press, Lond., *N.D.* [1931]).

20.4 F. HOPE-JONES, *Electrical Timekeeping* (N.A.G. Press Lond., 1940).

20.5 W. NOBLE, The Transit Instrument (*J.B.A.A.*, **3**, No. 4, 177).

20.6 M. W. OVENDEN, The Fundamental Principles of the Quartz Oscillation Clock (*J.B.A.A.*, **60**, No. 1, 31).

20.7 R. W. SINNOTT [Sidereal time conversion] (*Sky & Tel.*, **52**, No. 1, 62).

20.8 G. A. TOMLINSON, Recent Developments in Precision Timekeeping (*Obs.*, **57**, 189).

20.9 G. A. TOMLINSON, *Admiralty List of Radio Signals*, Vol. V (Hydrographer of the Navy).

20.10 G. A. TOMLINSON, *Time Service Circulars*, Series A (Royal Greenwich Observatory).

21. MATHEMATICAL

21.1 A. C. AITKEN, *Statistical Mathematics* (Oliver & Boyd, 1939). [Useful general introduction.]

21.2 D. BRUNT, *The Combination of Observations* (Cambridge, 1931).

21.3 J. L. COOLIDGE, *An Introduction to Mathematical Probability* (Oxford University Press, 1925).

21.4 H. R. HULME & L. S. T. SYMMS, The Law of Error and the Combination of Observations (*M.N.*, **99**, No. 8, 642).

21.5 C. G. LAMB, *Elements of Statistics* (Longmans, 1952). [Recommended modern treatment.]

21.6 H. C. LEVINSON, *The Science of Chance* (Faber, 1952). [Introductory.]

21.7 M. J. MORONEY, *Facts from Figures* (Pelican Books, 1951).

21.8 H. C. PLUMMER, *Probability and Frequency* (Macmillan, 1939). [Concise treatment of theory of error, with numerous references to astronomical observations.]

BIBLIOGRAPHY

22. STAR ATLASES AND CHARTS
[See also various entries under 'Star Catalogues'.]

22.1 F. W. A. ARGLELANDER, *Atlas des Nördlichen Gestirnten Himmels 1855.0* (Bonn, 1863; ed. KUSTNER, Bonn, 1899). [40 charts covering the sky N of Dec −2° to mag 9 on scale of 2 cm/degree.]

22.2 *Atlas of the Heavens* (Sky Publishing Corp., Cambridge, Mass.). [Previously known as *Atlas Coeli*; covers the whole sky to about mag 7.7 on a scale of 7.5 mm/degree. A standard atlas used by many amatuers.]

22.3 *Atlas Catalogue* (Sky Publishing Corp., Cambridge, Mass.). [Catalogue of interesting objects shown in B. 22.2.]

22.4 *Atlas Borealis, Eclipticalis & Australis* (Sky Publishing Corp., Cambridge, Mass.). [On a uniform scale of 2 cm/degree, these three atlases cover the sky between Decs +90° and +30°, +30° and −30°, and −30° and −90° respectively. Stars to about mag 9·5 shown, spectral classes colour-coded. No nebulae, clusters, etc.]

22.5 E. E. BARNARD (ed. E. B. FROST & M. R. CALVERT), *A Photographic Atlas of Selected Regions of the Milky Way* (Publ. Carneg. Instn. 247, 1927). [Part I, photographs and descriptions; Part II, charts and tables.]

22.6 M. BEYER & K. GRAFF, *Stern-Atlas* (Hamburg, 1925, repr. 1952). [27 charts covering the whole sky N of Dec −23° to mag 9 (plus brighter nebulae and clusters) on scale of 1 cm/degree, epoch 1855. *B.A.A.H.*, 1926 contains a Table of Precession Corrections, 1855 to 1926, for application to the positions of the Beyer-Graff Atlas.]

22.7 G. BISHOP, *Ecliptic Chart* (Lond., 1848, etc). [24 charts, to mag 10, extending 3° on each side of the ecliptic, epoch 1825.]

22.8 K. F. BOTTLINGER, *Galaktischer Atlas* (Julius Springer, Berlin, 1937). [8 charts showing stars and nebulae to about mag 5.5 in galactic coordinates based on Pole at $12^h 40^m$, +28°.]

22.9 E. DELPORTE, *Atlas Céleste* (I.A.U., Cambridge, 1930). [26 maps covering whole sky to mag 6; I.A.U. constellation boundaries; lists of mags, spectroscopic types, and positions at 1875 and 1925 of all stars to mag 4·5, and principal doubles, variables, nebulae, etc.]

22.10 E. DELPORTE, *Délimitation Scientifique des Constellations* (*Tables et Cartes*) (I.A.U., Cambridge, 1930). [Virtually a quotation from B. 22.9.]

22.11 J. FRANKLIN-ADAMS, *Photographic Chart of the Sky* (R.A.S., 1914). [206 sheets, each 16° square, covering the whole sky to mag 15·5 on scale of about 1 in/1°36. Charts 1-67, Dec −90° to −30°; charts 68-139, −15° to +15°; charts 140-206, +30° to +90°. See also *M.N.*, 64, 608, 1904; *ibid.*, 97, 89, 1936.]

22.12 E. HEIS, *Atlas Coelestis Eclipticus: Octo Continens Tabulas ad Delineandum Lumen Zodiacale* (1878). [8 charts of the zodiacal region, especially made for the Zodiacal Light observer.]

543

22.13 A. P. NORTON, *A Star Atlas and Reference Handbook, Epoch 1950* (Gall & Inglis, 17th edn., 1978). [The stand-by of every amateur; some 7000 stars to mag 6.5, plus nebulae, clusters, etc, and a mass of general information and reference material of use to the amateur.]

22.14 J. PALISA & M. WOLF, *Palisa-Wolf Charts* (Vienna, 1908-31). [210 photographic charts in 11 Series; epoch 1875; scale 36 mm/degree.]

22.15 *Palomar Sky Atlas*. [Two-colour photographic survey of the whole sky visible from Mt Palomar, made with the 122-cm Schmidt.]

22.16 F. E. ROSS & M. R. CALVERT, *Atlas of the Northern Milky Way* (University of Chicago Press, 1934). [39 plates, each about 13¼ ins square, exposed in a D=5, F=35-in camera at Mt Wilson and Flagstaff.]

22.17 E. SCHONFELD, *Atlas der Himmelszone zwischen 1° und 23° südlicher Declination, 1855* (*Bonner Sternkarten, Zweite Serie*) (Bonn, 1887). [24 charts continuing B. 22.1 to Dec −23°.]

22.18 P. STUKER, *Sternatlas für Freunde der Astronomie* (Stuttgart, 1925). [Photographic; to mag 7.5; epoch 1900.]

22.19 O. THOMAS, *Atlas der Sternbilder* (Salzburg, 1945). [32 main charts; also useful section on objects of interest.]

22.20 H. VEHRENBERG, *Photographic Star Atlas* (Hamburg; privately published). [In two volumes of loose photographs taken in blue light, to about mag 14; *Northern* covers Decs +90° to −26°; *Southern* from −14° to −90°. Scale, 15 mm/degree, 464 maps altogether.]

22.21 H. B WEBB, *Atlas of the Stars* (privately printed, N.Y., 2nd edn., 1945). [110 charts covering the sky N of Dec −23° to about mag 9.5 on scale 1 cm/degree; epoch 1920.0, with coordinate intersections for 2000.0. Very useful supplement to B. 22.13.]

23. STAR CATALOGUES−POSITIONAL

23.1 A. KOPF, Star Catalogues, especially those of Fundamental Character (*M.N.*, **96**, No. 8, 714). [George Darwin Lecture, 1936; an extremely useful summary of 19th- and 20th-century work.]

[The following selection of the catalogues of the last 120 years, devoted primarily to star positions, is arranged chronologically.]

23.2 F. BAILY, *A Catalogue of Those Stars in the Histoire Céleste Française of J. Lalande* [etc] (British Association, Lond., 1837). [47,390 stars reduced to epoch 1800.0 from Lalande's observations.]

23.3 S. GROOMBRIDGE (ed. G. B. AIRY), *A Catalogue of Circumpolar Stars* (Murray, Lond., 1838). [4243 stars, epoch 1810.0; see also B. 23.20.]

23.4 F. BAILY, *British Association Catalogue* (Lond., 1845). [8377 stars, epoch 1850.]

23.5 M. WEISSE, *Positiones mediae stellarum fixarum* [etc] (Petropoli, 1846). [Weisse's reductions of 31,085 stars within 15° of the equator to epoch 1825.0, using Bessel's observations; see also B. 23.6. Abbrev: *WB*.]

23.6   M. WEISSE, *Positiones mediae stellarum fixarum* [etc] (Petropoli, 1863) [Continuation of B. 23.5; 31,445 stars between Dec +15° and +45°. Abbrev: *WB2*.]

23.7   W. OELTZEN, *Argelanders Zonen-Beobachtungen vom 45° bis 80° nördlicher Declination in mittleren Positionen für 1842.0* (Wien, 1851-52). [Oeltzen's reductions of 26,425 stars from Argelander's observations. Abbrev: *OA(N)*.]

23.8   W. OELTZEN, *Argelanders Zonen-Beobachtungen vom 15° bis 31° südlicher Declination in mittleren Positionen für 1853.0* (Wien, 1857-58). [Continuation of B. 23.7. Abbrev: *OA(S)*.]

23.9   F. W. A. ARGELANDER, *Bonner Durchmusterung des nördlichen Himmels* (Bonn, 1859-62; reprint 1903; 3 vols). [Approximate positions and visual mags to nearest 0·1 mag, to about mag 9·5. Arranged in successive 1°-wide Dec zones from +90° to −2°. Charts, scale 2 cms/degree, have been issued in photostat. Abbrev: *BD*. See B. 23.12, 15.]

23.10 HEIS, *Atlas [and] Catalogus Coelestis Novus* (Cologne, 1872). [Abbrev: *H'*.]

23.11 J. BIRMINGHAM, The Red Stars: Observations and Catalogue (*Trans. Roy. Irish Acad.*, **26**, 1877; ed. T. E. ESPIN, Dublin, 1890). [Abbrevs: *B* and *E-B* respectively.]

23.12 E. SCHONFELD, *Durchmusterung* (1886). [Continuation of B. 23.9 to Dec −23°. Abbrev: *SD*.]

23.13 B. A. GOULD, The Argentine General Catalogue (*Result. Obs. Nac. Argent.*, **14**, Cordoba, 1886). [32,448 southern stars plus additional stars in clusters. Abbrevs: *CGA, AGC*.]

23.14 *Astronomische Gesellschaft Katalog* (Leipzig, 1890 etc). [The most complete catalogue of precision; epoch 1875. Abbrevs: *AG, AGC, CAG*. See B. 23.38.]

23.15 J. M. THOME, Cordoba Durchmusterung (*Result. Obs. Nac. Argent.*, **16**, 1892 etc). [Continuation of B. 23.12 to Dec −52° in 10° vols. To mag 10 approximately; with visual mags and charts. Abbrev: *CD*.]

23.16 *The Astrographic Catalogue*. [Initiated at the Paris Congress, 1887; work shared by observatories over the world, began 1892, still in progress. Positions taken from photographic charts, scale 6 cm/degree. Abbrev: *AC*.]

23.17 D. GILL & J. C. KAPTEYN, The Cape Photographic Durchmusterung, 1875.0 (*Ann. Cape Obs.*, **3**, 1896- **5**, 1900). [Mags and approximate positions from Dec −19° to −90°, to mag 9. Abbrev: *CPD*.]

23.18 J. SCHEINER, Photographische Himmelskarte (*Publ. Astrophys. Obs. Potsdam*, 1899-1903). [Mags and positions, epoch 1900.0.]

23.19 S. NEWCOMB, Catalogue of Fundamental Stars for the Epochs 1875 and 1900 (*Astr. Pap., Wash.*, **8**, No. 2, 77, 1905).

23.20 F. W. DYSON & W. G. THACKERAY, *New Reduction of Groombridge's Circumpolar Catalogue, Epoch 1810.0* (H.M.S.O., 1905). [See B 23.3.]

23.21 H. B. HEDRICK, Catalogue of Zodiacal Stars, for the Epochs 1900 and 1920 (*Astr. Pap., Wash.*, 8, No. 3, 405, 1905). [1607 stars to mag 7·5. Abbrev: *WZC*. See also B. 23.37.]

23.22 J. BOSSERT, *Catalogue d'Etoiles Brillantes, 1900.0* (Gauthier-Villars, Paris, 1906). [3799 stars in 1° NPD zones.]

23.23 J. & R. AMBRONN, *Sternverzeichnis enthaltend alle Sterne bis zur 6.5 Grösse* (Julius Springer, Berlin, 1907). [Positions of 7796 stars to mag 6·5 for epoch 1900, and proper motions of 2226 stars.]

23.24 A. AUWERS, Neue Fundamentalkatalog des Berliner Astronomischen Jahrbuchs (*Veröff. König. Astron. Rechen-Instit.*, 33, 1910). [925 stars; one of the best fundamental catalogues; abbrev: *NFK*. See also B. 23.33, 40.]

23.25 L. BOSS, *Preliminary General Catalogue of 6188 Stars for the Epoch 1900* (Publ. Carneg. Instn., 115, 1910). [Accurate positions and proper motions of all naked-eye stars; abbrev: *PGC*.]

23.26 T. W. BACKHOUSE, *Catalogue of 9842 Stars, or all Stars Very Conspicuous to the Naked Eye* (Sunderland, 1911). [Epoch 1900; useful for amateurs.]

23.27 *Geschichte des Fixstern-Himmels* (Karlsruhe, 1922-23). [Collection of pre-1900 observations reduced to 1875.0.]

23.28 E. C. PICKERING & J. C. KAPTEYN, Durchmusterung of Selected Areas between δ=0 and δ=+90° (*Harv. Ann.*, 101, 1918).

23.29 E. C. PICKERING, J. C. KAPTEYN & P. J. VAN RHIJN, Durchmusterung of Selected Areas between δ=−15° and δ=−30° (*Harv. Ann.*, 102, 1923).

23.30 E. C. PICKERING, J. C. KAPTEYN & P. J. VAN RHIJN, Durchmusterung of Selected Areas between δ=−45° and δ=−90° (*Harv. Ann.*, 103, 1924).

23.31 L. BOSS & B. BOSS, *San Luis Catalogue of 15,333 Stars for the Epoch 1920.0* (Publ. Carneg. Instn., 386, 1928). [To mag 7; mostly southern.]

23.32 R. SCHORR & W. KRUSE, *Index der Sternörter 1900-25* (Bergedorf, 1928). [Monumental analysis of over 400 catalogues. Vol. 1, northern stars; vol. 2, southern.]

23.33 *Dritter Fundamentalkatalog des Berliner Astronomischen Jahrbuchs* (Berlin, 1934). [Abbrev: *FK3*. See B. 23.24.]

23.34 F. SCHLESINGER, Catalogue of Bright Stars containing all important data known in June 1930 (*Publ. Yale Obs.*, 1930). [9110 stars to visual mag 6·5, and some fainter. Abbrev: *BS*.]

23.35 B. BOSS, *Albany Catalogue of 20,811 Stars for the Epoch 1910* (Carnegie Institution of Washington, 1931).

23.36 B. BOSS, *General Catalogue of 33,342 Stars for the Epoch 1950* (Carnegie Institution of Washington, 1937, 5 vols.).

23.37 J. ROBERTSON, Catalogue of 3539 Zodiacal Stars for the Equinox 1950.0 (*Astr. Pap., Wash.*, 10, No. 2, 175, 1940). [Revision and enlargement of B. 23.21.]

23.38 R. SCHORR & A. KOHLSCHÜTTER, *Zweiter Katalog der Astronomische Gesellschaft, Aquinoktium 1950* (Hamburg-Bergedorf, 1951). [Vols 1-4, Dec +90° to +50°. See B. 23.14. Abbrev: *AGK2*.]

23.39 *Apparent Places of Fundamental Stars* (H.M.S.O., annually since 1941). [Mean and Apparent places of the 1535 stars of the *FK3* (B 23.33) and its Supplement.]

23.40 W. FRICKE & A. KOPFF, *FK4* (Karlsruhe-Brown, 1963). [The most recent revision of B. 23.24; contains positions and proper motions for 1535 stars, mostly brighter than mag 7.5, over the whole sky.]

23.41 *Star Atlas of Reference Stars and Non-Stellar Objects* (MIT Press, Cambridge, U.S.A., 1969). [Based on the *Smithsonian Astrophysical Observatory Catalogue* (*SAOC*); consists of 152 charts covering the whole sky to mag 9, on a scale of $6 \cdot 95' = 1$ mm. Abbrev: *SAOA*.]

24. STAR CATALOGUES—MOTIONS AND PARALLAXES

24.1 W. S. ADAMS & A. H. JOY, The Radial Velocities of 1013 Stars (*Ap. J.*, **57**, No. 3, 149=*M.W.C.*, 258).

24.2 J. BOSSERT, *Catalogue des mouvements propres des 5671 étoiles* (Paris, 1919).

24.3 W. W. CAMPBELL & J. H. MOORE, Radial Velocities of Stars brighter than Visual Magnitude $5 \cdot 51$ (*Publ. Lick Obs.*, **16**, 1928).

24.4 W. S. EICHELBERGER, Positions and Proper Motions of 1504 Standard Stars, 1925.0 (*Astr. Pap., Wash.*, **10**, Part 1, 1925).

24.5 H. KNOX-SHAW & H. G. SCOTT BARRETT, *The Radcliffe Catalogue of Proper Motions in Selected Areas 1 to 115* (Oxford University Press, 1934). [To mag 14 in the Selected Areas on and N of the equator.]

24.6 J. H. MOORE, General Catalogue of Radial Velocities of Stars, Nebulae and Clusters (*Publ. Lick Obs.*, **18**, 1932).

24.7 J. S. PLASKETT & J. A. PEARCE, A Catalogue of the Radial Velocities of O and B Type Stars (*Publ. Dom. Astrophys. Obs.*, **5**, No. 2, 99).

24.8 J. G. PORTER, E. I. YOWELL & E. S. SMITH, A Catalog of 1474 Stars with proper motion exceeding four-tenths of a second per year (*Publ. Cincinn. Obs.*, **20**).

24.9 F. SCHLESINGER, *General Catalogue of Stellar Parallaxes* (Yale, 1924). [Includes all determinations available in 1924 Jan.]

24.10 F. SCHLESINGER & L. F. JENKINS, General Catalogue of Stellar Parallaxes (*Publ. Yale Obs.*, 1935). [B. 24.9 revised to 1935.]

24.11 R. SCHORR, *Eigenbewegungs-Lexikon* (Hamburg Observatory, Bergedorf, 1 vol, 1923; 2 vols, 1936). [2nd edn. includes all proper motions available at the end of 1935; 94,741 stars, with mags and spectral types. Vol. 1, N stars; vol. 2, S stars. Abbrev: *EBL*.]

25. STAR CATALOGUES—PHOTOMETRIC

25.1 S. I. BAILEY, A Catalogue of 7922 Stars observed with the Meridian Photometer, 1889-91 (*Harv. Ann.*, **34**). [The 'Southern Meridian Photometry', abbrev: *SMP*. Continuation of B. 25.16 to the S Pole.]

25.2 V. M. BLANCO et al, *Photoelectric Catalogue* (Publ. U.S. Naval Observatory, Second Series, vol. XXI, Washington, 1968). [U, B, V photometry of more than 2000 stars, some as faint as mag 15.]

25.3 A. BRUNN, *Atlas photométrique des Consellations de +90° à −30°* (privately printed, France, 1949). [55 sheets showing all *BD* stars to mag 7·5, scale 1 cm/degree, epoch 1900. Against each star is printed its visual mag, to 0·01 mag for stars of mag 6·50 and brighter, to 0·1 mag for those fainter than 6·50. Other data include photometric mags of all extragalactic nebulae brighter than mag 12·0 photographic.]

25.4 S. CHAPMAN & P. J. MELOTTE, Photographic Magnitudes of 262 Stars within 25′ of the North Pole (*M.N.*, **74**, No. 1, 40).

25.5 B. G. FESSENKOFF, *Photometrical Catalogue of 1155 Stars* (Kharkow, 1926).

25.6 W. FLEMING, Spectra and Photographic Magnitudes of Stars in Standard Regions (*Harv. Ann.*, **71**, No. 2, 27).

25.7 B. A. GOULD, *Uranometria Argentina* (Buenos Aires, 1879). [Visually determined mags, and positions, of stars to mag 7 South of Dec +10°. Abbrev: *UA, G.*]

25.8 [Harvard], Stars near the North Pole (*Harv. Ann.*, **48**, No. 1, 1).

25.9 Harvard Standard Regions (*Harv. Ann.*, **71**, No. 4, 233).

25.10 D. HOFFLEIT, *Catalogue of Bright Stars* (Yale University Observatory, Conn., 1964). [Photometric data to stars of about mag 7.]

25.11 B. IRIARTE et al, *Arizona-Tonantzintla Catalogue* (*Sky & Tel.*, **30**, No. 1, 21). [Five-colour photometry (U, B, V, R, I) of 1325 bright stars north of about Dec −50°, carried out between 1963 and 1965.]

25.12 H. S. LEAVITT, The North Polar Sequence (*Harv. Ann.*, **71**, No. 3, 47).

25.13 Magnitudes of Stars of the North Polar Sequence (*B.A.A.H.*, 1926, 32).

25.14 G. MULLER & P. KEMPF, Photometrische Durchmusterung des nördlichen Himmels (*Publ. Astrophys. Obs. Potsdam*, Nos. 31, 43, 44, 51, 52, 1894-1907).

25.15 E. C. PICKERING, Adopted Photographic Magnitudes of 96 Polar Stars (*Harv. Circ.*, 170).

25.16 E. C. PICKERING, Observations with the Meridian Photometer, 1879-82 (*Harv. Ann.*, **14**, 1884). [The Harvard Photometry (*HP*): magnitudes of 4260 stars, including all brighter than mag 6 N of Dec −30°.]

25.17 E. C. PICKERING, Revised Harvard Photometry (*Harv. Ann.*, **50**). [Positions, visual mags, and spectral types of 9110 stars, mostly mag 6·5 and brighter. Abbrevs: *HR, RHP.*]

25.18 E. C. PICKERING, A Catalogue of 36,682 Stars Fainter than Magnitude 6·50 . . . forming a Supplement to the Revised Harvard Photometry (*Harv. Ann.*, **54**).

25.19 C. PRITCHARD, *Uranometria Nova Oxoniensis* (Oxford, 1885). [Wedge-photometer redeterminations of Argelander's *Uranometria Nova* magnitudes; 2784 entries. Abbrev: *UO*.]

25.20 F. H. SEARES, Magnitudes of the North Polar Sequence. (Report of the Commission de photométrie stellaire) (*Trans. I.A.U.*, **1**, 69).

25.21 F. H. SEARES, J. C. KAPTEYN & P. J. VAN RHIJN, *Mt Wilson Catalogue of Photographic Magnitudes in Selected Areas 1-139* (Carnegie Institution, Washington, 1930).

25.22 F. H. SEARES, F. E. ROSS & M. C. JOYNER, *Magnitudes and Colours of Stars North of +80°* (Publ. Carneg. Instn., 532, 1941).

25.23 R. W. SINNOTT [Chart of part of North Polar Sequence] (*Sky & Tel.*, **51**, No. 5, 356).

26. STAR CATALOGUES–SPECTROSCOPIC

[Many of the catalogues mentioned elsewhere in this Bibliography quote spectroscopic types. Specially demanding mention, however, are the following.]

26.1 E. C. PICKERING, The Draper Catalogue of Stellar Spectra (*Harv. Ann.*, **27**, 1890).

26.2 A. C. MAURY, Spectra of Bright Stars (*Harv. Ann.*, **28**, 1897). [Together with B. 26.1 constitutes the 'old' Draper Catalogue of over 10,000 stars.]

26.3 A. J. CANNON & E. C. PICKERING, The Henry Draper Catalogue of Stellar Spectra (*Harv. Ann.*, **91-99**, 1918-24).

26.4 A. J. CANNON, The Henry Draper Extension (*Harv. Ann.*, **100**). [Together with B. 26.3 constitutes the 'new' Draper Catalogue (*HD*): mags, spectral types and positions of 225,000 stars to mag 10 approximately.]

26.5 A. SCHWASSMANN & P. J. VAN RHIJN, *Bergedorfer Spektral-Durchmusterung* (Bergedorf, 1935, 1938). [To mag 13 (photographic) in the northern Kapteyn areas.]

27. STAR CATALOGUES–VARIABLE STARS

27.1 A. J. CANNON,* Second Catalogue of Variable Stars (*Harv. Ann.*, **55**, 1907). ['Second' with reference to B. 27.12; 1957 variables.]

27.2 S. C. CHANDLER,* Catalogue of Variable Stars (*A.J.*, **8**, No. 11/12, 82, 1888). [225 variables.]

27.3 S. C. CHANDLER,* Second Catalogue of Variable Stars (*A.J.*, **13**, No. 12, 89, 1893). [260 variables.]

27.4 S. C. CHANDLER,* Third Catalogue of Variable Stars (*A.J.*, **16**, No. 9, 145, 1896). [393 variables.]

27.5 S. C. CHANDLER,* Revision of Elements of the Third Catalogue of Variable Stars (*A.J.*, **24** No. 1, 1, 1904).

*Primarily of historical value, B. 27.9 now being the standard work.

27.6   J. G. HAGEN, *Atlas* [*et Catalogus*] *Stellarum Variabilium* (9
        Series, 1899-1941).   [Charts and lists of about 24,000 com-
        parison stars for 488 variables. The 5th Series (Berlin, 1906)
        is particularly useful, including all variables wholly observable
        with the naked eye or binoculars (minima brighter than mag 7).
        See also B. 27.13.]

27.7   *Katalog und Ephemeriden Veränderlicher Sterne* (Vierteljahr-
        sschrift der Astronomischen Gesellschaft, annually 1870-1926).

27.8   *Katalog und Ephemeriden Veränderlicher Sterne* (Berlin-Babels-
        berg, annually to 1941).

27.9   B. V. KUKARKIN *et al, General Catalogue of Variable Stars*
        (Moscow, 3rd edition, 1970). [Recognised by the I.A.U. as the
        standard work; data on 20,437 stars, extended by subsequent
        Supplements.]

27.10   R. PRAGER, *Geschichte und Literatur des Lichtwechsels der
        Veränderlichen Sterne* (Ferd. Dümmlers Verlagsbuchhandlung,
        Berlin; vol. 1, 1934; vol. 2, 1936). [See also B. 27.11.]

27.11   R. PRAGER, History and Bibliography of the Light Variations of
        Variable Stars (*Harv. Ann.*, **111**, 1941). [Vol. 3 of B. 27.10.]

27.12   Provisional Catalogue of Variable Stars* (*Harv. Ann.*, **48**, No. 3,
        91, 1903). [1227 variables.]

27.13   J. STEIN & J. JUNKERS, [Index to the 9 Series of B. 27.6.]
        (*Ricerche Astronomiche*, **4**; Specola Vaticana, 1941).

28. STAR CATALOGUES–BINARY STARS

28.1   R. G. AITKEN, *New General Catalogue of Double Stars within
        120° of the North Pole* (Publ. Carneg. Instn., 417, 2 vols, 1932).
        [Includes all measures of 17,181 doubles prior to 1927; epochs
        1900·0 and 1950·0. Abbrev: *ADS*.]

28.2   S. W. BURNHAM, *A New General Catalogue of Double Stars
        within 121° of the North Pole* (Publ. Carneg. Instn., 5, 2 vols,
        1906).   [Measures etc of 13,665 doubles; epochs 1880.0 and
        1900.0. Part I, The Catalogue; Part II, Notes to the Catalogue.
        Abbrev: *BGC*.]

28.3   W. W. CAMPBELL & H. D. CURTIS, First Catalogue of Spectro-
        scopic Binaries (*L.O.B.*, **3**, No. 79, 136). [Complete to 1905;
        140 stars.]

28.4   W. W. CAMPBELL, Second Catalogue of Spectroscopic Binary
        Stars (*L.O.B.*, **6**, No. 181, 17). [To 1910; 306 stars. See also B.
        28.3, 28.22-24.]

28.5   F. W. DYSON, *Catalogue of Double Stars from observations made
        at The Royal Observatory Greenwich with the 28-inch Refractor,
        1893-1919* (H.M.S.O., 1921).

28.6   J. F. W. HERSCHEL, Descriptions and approximate Places of 321
        new Double and Triple Stars (*Mem. R.A.S.*, **2**, No. 29, 459).

28.7   J. F. W. HERSCHEL, Approximate Places and Descriptions of 295
        new Double and Triple Stars (*Mem. R.A.S.*, **3**, No. 3, 47).

*Primarily of historical value, B. 27.9 now being the standard work.

BIBLIOGRAPHY

28.8   J. F. W. HERSCHEL, Third Series of Observations . . . Catalogue of 384 new Double and Multiple Stars; completing a first 1000 of Those Objects (*Mem. R.A.S.*, **3**, No. 13, 177).

28.9   J. F. W. HERSCHEL, Fourth Series of Observations . . . containing the Mean Places . . . of 1236 Double Stars [etc] (*Mem. R.A.S.*, **4**, No. 17, 331).

28.10 J. F. W. HERSCHEL, Fifth Catalogue of Double Stars . . . Places, Descriptions, and measured Angles of Position of 2007 of those objects [etc] (*Mem. R.A.S.*, **6**, No. 1, 1).

28.11 J. F. W. HERSCHEL, Sixth Catalogue of Double Stars . . . 286 of those objects (*Mem. R.A.S.*, **9**, No. 7, 193).

28.12 J. F. W. HERSCHEL, Seventh Catalogue of Double Stars (*Mem. R.A.S.*, **38**, 1870).

28.13 J. F. W. HERSCHEL (ed. R. MAIN & C. PRITCHARD), Catalogue of 10,300 Multiple and Double Stars (*Mem. R.A.S.*, **40**, 1874).

28.14 W. HERSCHEL, Catalogue of Double Stars (*Phil. Trans.*, **72**, 112, 1782).

28.15 W. HERSCHEL, Catalogue of Double Stars (*Phil. Trans.*, **75**, 40, 1785).

28.16 W. HERSCHEL, On the places of 145 new Double Stars (*Mem. R.A.S.*, **1**, 166, 1821).

28.17 H. M. JEFFERS, W. H. VAN DEN BOS & F. M. GREEBY, *Index Catalogue of Visual Double Stars, 1961.0* (Publ. Lick Obs., XXI, 1963). [Tabulates basic data for 64,247 double stars, but does not give measures of rapid binaries. Abbrev: *IDS*.]

28.18 R. JONCKHEERE, Catalogue and Measures of Double Stars discovered visually from 1905-1916 within 105° of the North Pole and under 5″ Separation (*Mem. R.A.S.*, **17**, 1917). [Virtually a Supplement to B. 28.2.]

28.19 T. LEWIS, Measures of the Double Stars contained in the Mensurae Micrometricae of F. G. W. Struve (*Mem. R.A.S.*, **56**). [See B. 28.26.]

28.20 J. MEEUS, Some Bright Visual Binary Stars, I (*Sky & Tel.*, **41**, No. 1, 21).

28.21 J. MEEUS, Some Bright Visual Binary Stars, II (*Sky & Tel.*, **41**, No. 2, 88).

28.22 J. H. MOORE, Third Catalogue of Spectroscopic Binary Stars (*L.O.B.*, **11**, No. 355, 141). [To 1024; 1954 stars. See also B. 28.3, 4, 23, 24.]

28.23 J. H. MOORE, Fourth Catalogue of Spectroscopic Binary Stars (*L.O.B.*, **18**, No. 483, 1). [375 stars.]

28.24 J. H. MOORE & F. J. NEUBAUER, Fifth Catalogue of the Orbital Elements of Spectroscopic Binary Stars (*L.O.B.*, No. 521).

28.25 P. MULLER & C. MEYER, *Troisième Catalogue d'Ephémérides d'Etoiles Doubles* (Paris Observatory, 1969). [A standard work.]

28.26 J. SOUTH & J. F. W. HERSCHEL, *Observations of the Apparent Distances and Positions of 380 Double and Triple Stars, made in the Years 1821, 1822 and 1823* (Lond., 1825).

28.27 F. G. W. STRUVE, *Catalogus Novus Stellarum Duplicium et Multiplicium* (Dorpat, 1827). [The Dorpat Catalogue (Σ).]

28.28 F. G. W. STRUVE, *Stellarum Duplicium et Multiplicium Mensurae Micrometricae* (Petropoli, 1837). [See B. 28.19.]

28.29 F. G. W. STRUVE, *Stellarum Fixarum imprimis Duplicium et Multiplicium Positiones Mediae pro Epocha 1830.0* (Petropoli, 1852).

28.30 O. STRUVE, *Revised Pulkova Catalogue* (Pulkova, 1850). [Abbrev: $O\Sigma$; Part II denoted by $O\Sigma\Sigma$.]

29. NEBULAE AND CLUSTERS—CATALOGUES
[See also under 'Star Atlases and Charts'.]

29.1 S. I BAILEY, Globular Clusters—A Provisional Catalogue (*Harv. Ann.*, **76**, No. 4, 43). [113 clusters.]

29.2 E. E. BARNARD, On the Dark Markings of the Sky, with a Catalogue of 182 Such Objects (*Ap. J.*, **49**, No. 1, 1). [See *Ap. J.*, **49**, No. 5, 360 for list of errata.]

29.3 H. D. CURTIS, Descriptions of 762 Nebulae and Clusters Photographed with the Crossley Reflector (*Publ. Lick Obs.*, **13**, No. 1, 9).

29.4 J. L. E. DREYER, New General Catalogue of Nebulae and Clusters of Stars (*Mem. R.A.S.*, **49**, 1888). [NGC. Based on B. 29.5.]

J. L. E. DREYER, Index Catalogue (*Mem. R.A.S.*, **51**, 1895). [IC. Extension of NGC.]

J. L. E. DREYER, Second Index Catalogue (*Mem. R.A.S.*, **59**, 1908). [Extension of NGC.] [See B. 29.11.]

29.5 J. F. W. HERSCHEL, *General Catalogue of Nebulae and Clusters of Stars of the Epoch 1860* (Lond., 1864).

29.6 W. HERSCHEL, Catalogue of 1000 new Nebulae and Clusters of Stars (*Phil. Trans.*, **76**, 457, 1786).

29.7 W. HERSCHEL, Catalogue of a second 1000 Nebulae and Clusters of Stars (*Phil. Trans.*, **79**, 212, 1789).

29.8 W. HERSCHEL, Catalogue of 500 new Nebulae, Nebulous Stars, Planetary Nebulae, and Clusters of Stars (*Phil. Trans.*, 1802, 477).

29.9 J. HOLETSCHEK, Catalogue of Nebular Magnitudes (*Ann. K.K. Univ.-Stern.*, **20**, 114, Vienna, 1907).

29.10 P. J. MELOTTE, A Catalogue of Star Clusters shown on Franklin-Adams Chart Plates (*Mem. R.A.S.*, **60**, No. 5, 175). [245 objects.]

29.11 C. MESSIER, *Catalogue of 103 Nebulae and Clusters* (1771-84). [Abbrev: *M*. See also B. 29.13.]

29.12 J. SULENTIC & W. TIFFT, *Revised New General Catalogue* (University of Arizona, 1975). [Updated version of B. 29.4.]

29.13 H. SHAPLEY & H. DAVIES, Messier's 'Catalogue of Nebulae and Clusters' (*Obs.*, **41**, No. 529, 318). [Reprint of B. 29.11, with corresponding NGC (B. 29.4) numbers.]

29.14 G. DE VAUCOULEURS, *Reference Catalogue of Bright Galaxies* (University of Texas Press, 1964). [Major revision and enlargement of the 1932 Shapley-Ames catalogue; includes 2,599 galaxies, mostly brighter than mag 14.]

29.15 C. WIRTZ, Flächenhelligkeiten von 566 Nebelflecken und Sternhaufen (*Lund. Medd.*, **2**, No. 29, 1923).

APPENDIX

Table of Contents of 'Observational Astronomy for Amateurs'

INDEX